X-RAYS FOR ARCHAEOLOGY

X-rays for Archaeology

Edited by

M. UDA
Waseda University, Tokyo, Japan

G. DEMORTIER
Namur University, Namur, Belgium

and

I. NAKAI
University of Tokyo, Japan

 Springer

A C.I.P. Catalogue record for this book is available from the Library of Congress.

ISBN-10 1-4020-3580-2 (HB)
ISBN-13 978-1-4020-3580-7 (HB)
ISBN-10 1-4020-3581-0 (e-book)
ISBN-13 978-1-4020-3581-4 (e-book)

Published by Springer,
P.O. Box 17, 3300 AA Dordrecht, The Netherlands.

www.springeronline.com

Cover shows an image of the tomb of Amenhotep III in Egypt.
Photo courtesy of Masayuki Uda.

Printed on acid-free paper

Printed in the Netherlands.

Table of contents

Foreword

The First International Symposium on X-ray Archaeometry took place in the conference hall of Waseda University, Tokyo, Japan, on 18–20 July 2002. The participants of the symposium were from Belgium, China, France, Greece, Hungary, Israel, Italy, Japan, Korea, Mexico, Romania, Slovenia, Sri Lanka, Taipei, UK, and USA.

One of the most important aims of the symposium was to combine two scientific fields, i.e. archaeology or art and X-ray science. Finding archaeological sites, dating, analyzing of archaeological objects, and so on needs the help of natural scientists and technicians. Natural scientists have a taste for solving mysteries hidden in archaeology. However, previously, using x-ray techniques was only a small part of the archaeological fieldwork and the x-ray field was largely disinterested in the analysis of archaeological objects. Until this symposium, no attempt has been made on having an international meeting on a worldwide scale to discuss archaeological subjects under equal partnership between the two fields mentioned above.

The symposium provided a broad forum for discussing experimental results of X-ray-based analysis. Of particular interest for the participants of the symposium was the non-destructive analysis of archaeological monuments using several kinds of X-ray techniques, especially under *in situ* and contact-free conditions, as well as the introduction of experimental results using advanced technologies such as ion beam and synchrotron radiation techniques.

This book, named "X-rays for Archaeology", consists of papers selected from presentations in the First International Symposium on X-ray Archaeometry.

Finally, it is an especially great pleasure for me to warmly recommend this book to every reader interested in knowing more about X-ray archaeometry and understanding the importance of joining both scientists in the fields of archaeology or art and X-ray analysis.

Tokyo, Japan

Professor M. Uda
Chairman
Organizing Committee of the First International
Symposium on X-ray Archaeometry

Part I: *In situ* Measurements

Chapter I-1

Characterization of Pigments Used in Ancient Egypt

M. UDA
Waseda University, 3-4-1 Ohkubo, Shinjuku-ku, Tokyo, Japan
muda@waseda.jp

Keywords: pigment, X-rays, XRF, XDR, ED-XRDF, Egypt, Amenhotep III, tomb
wall, huntite, calcite, anhydrite, hematite, goethite, realgar, orpiment,
Egyptian blue, Amarna blue

Abstract

X-ray fluorescence (XRF) and particle induced X-ray emission (PIXE), and
X-ray diffraction (XRD) methods have been used extensively to study the chemical
compositions and crystal structures of monuments, respectively, especially to
characterize pigments used in ancient Egypt. Recent investigations of ancient
Egyptian pigments by these methods are reviewed here. These investigations were
performed by the author and his collaborators on several tens of fragments in the
laboratory, and in the field on monuments exhibited in the Egyptian Museum and on
tomb walls of Amenhotep III. Some fragments were excavated from ruins, and
others were found within ancient tombs in Egypt. For field experiments, portable
X-ray instruments were used: XRF, XRD, and energy dispersive X-ray diffraction
and fluorescence. The latter two instruments were specially designed and produced
for these investigations by the author and his collaborators. Experimental results
obtained in the laboratory were very important in creating optimum experimental
conditions in the field and were used very effectively for prompt analysis of
observed data in the field.

1. Introduction

Analytical investigations of archaeological objects bring us much information
on the development and propagation of the cultures and technologies of the human
race. To conserve and restore these valuable and scarce objects, it is indispensable to
understand their chemical constituents and crystal structures. Among archaeological
objects, pigments are most attractive targets for scientific study because their colors
are yardsticks of a sense of beauty, and they provide a means for estimating ancient
technologies' ability to prepare pigments artificially.

There are many ancient remains all over the world. In most cases, however,
nothing is left of buildings painted with beautiful pigments but ruins. Fortunately,

M. Uda et al. (eds.), X-rays for Archaeology, 3–26.
© 2005 *Springer. Printed in the Netherlands.*

ancient grave contents decorated with pigments and tomb walls painted with pigments are still in almost perfect condition today in Egypt. This is because humidity is very low in the areas where most of the tombs were built in ancient Egypt and because the tombs, located deep underground, effectively cut off harmful sunlight.

Ancient Egyptian pigments have been analyzed extensively in laboratory studies (Barbieri *et al*. 1974; Riederer 1974; Noll 1981; Jaksch *et al*. 1983; Goresy *et al*. 1986; Saleh 1987; Riederer 1988; Lucas and Harris 1989; Uda *et al*. 1993; Green 1995; Weatherhead 1995: Nagashima *et al*. 1996; Goresy 1997; Uda 1998, Uda *et al*. 2000a; Uda *et al*. 2002; Yoshimura *et al*. 2002), and in the field (Uda *et al*. 1999; Uda *et al*. 2000b). It is self-evident that non-destructive and non-contact analysis in the field is indispensable for investigating ancient remains.

Here, experiences accumulated from X-ray analyses performed on ancient Egyptian pigments by the author and his collaborators are reviewed. X-ray fluorescence (XRF) and particle induced X-ray emission (PIXE) methods were used to determine the chemical compositions of pigments. An X-ray diffraction (XRD) technique was used to characterize their structural features. These experiments were performed in the laboratory on fragments excavated from remains. In addition, *in situ* experiments were also performed in the field at the Egyptian Museum and the tomb of Amenhotep III, king or Pharaoh ca. 1400 B.C. For the field experiments, portable XRF, XRD, and energy dispersive X-ray diffraction and fluorescence (ED-XRDF) systems were used together with an optical microscope. The XRF and ED-XRDF systems were specially prepared for these experiments by the author and his coworkers.

2.   Objects

Pigments painted on two types of wall fragments were investigated by PIXE and XRD in the laboratory. The first type of fragments studied, which were made of sun-dried bricks, was excavated from the site at Kom al-Samak, the so-called Hill of the Fish, at the Malqata-South site, 18th Dynasty (ca. 1400 B.C.), New Kingdom. This ruin is located on the west bank of the Nile at Luxor. The second fragment type, also made of sun-dried bricks, was unearthed from the ruins of Malqata palace located at Malqata-South. The palace was built by Amenhotep III, 18th Dynasty (ca. 1400 B.C.), a king of the New Kingdom. Also investigated by XRD were plasters on wall fragments unearthed from the site at Kom al-Samak and from private tombs W6 and 333 at Dra' Abu al-Naga, located on the west bank at Luxor, which also date from the reign of Amenhotep III. In addition, pigments painted on pottery fragments were investigated by PIXE and XRD in the laboratory. These fragments were excavated from the three sites at Kom al-Samak; at Dahshur, located 20 km south of Cairo; and at Abusir, located at Giza near Cairo. The latter two sites are said to be cemeteries of Memphis, the old capital, and to be monuments dedicated to Khaemwaset, a son of Ramesses II, ca. 1300 B.C. These fragments are preserved in the storehouses of the Institute of Egyptology at Waseda University in Tokyo and of the Egyptian government in Luxor.

Three objects exhibited at the Egyptian Museum in Cairo were also studied. The first one, shown in Fig. 1 (a), is the block of Ramesses II holding captives (JF 46189), and is of painted limestone (H: 99.5 cm, L: 89 cm, W: 50 cm). It is exhibited in gallery #10 on the ground floor and was taken from Mit Rahina, 19th Dynasty (ca. 1270 B.C.), New Kingdom. The second is the Wooden Coffin and Lid of Neb-Seny (CG 61016), which is of painted wood (H: 64 cm max., L: 194 cm, W: 50 cm) (Fig. 1(b)). It is exhibited in gallery #48, case E, on the upper floor and was taken from Thebes, 18th Dynasty (ca. 1400 B.C.), New Kingdom. The third, shown in Fig. 1 (c), is the Funeral Stele of Amenemhat (JE 45626), which is of painted limestone (H: 30 cm, L: 50 cm). It is exhibited in gallery #21, case A, on the ground floor and was taken from Asasif (Tomb R.4), 11th Dynasty (ca. 2000 B.C.), Middle Kingdom. These wall fragments were investigated with a portable XRF system and are kept in the storehouse of the Egyptian government in Luxor, Egypt.

Painted tomb walls of Amenhotep III, 18th Dynasty (1402-1364 B.C.), New Kingdom, were also investigated in the field. The tomb is located in the Valley of the Kings, on the west bank of the Nile at Luxor. Pigments and plasters analyzed were selected from the south wall of Room E (Fig. 2 (a)), the north and west walls of Room I, (Figs. 2 (b) and (c), respectively) and the north wall of Room J-2.

3. Experiments

All measurements were performed in open air and were carried out with PIXE, XRF, XRD, ED-XRDF, and optical microscopy methods, even though most PIXE and XRF experiments are done in a vacuum. The reasons the measurements were performed in open air may easily be understood from the facts that 1) the tomb of Amenhotep III is very large, 40 m in length, and its walls can not be moved from their original sites, 2) monuments exhibited in the Egyptian Museum in Cairo are not allowed to be transferred from open air to a vacuum, and 3) most of the wall fragments on which pigments are painted are made of sun-dried bricks, which are extremely fragile and can not be examined in a vacuum.

For PIXE experiments, a 2.0-MeV proton beam was generated in the Pelletron accelerator at the Radiation Laboratory of the Tokyo Metropolitan Industrial Technology Research Institute. The beam was collimated to 2×3 mm^2 through two graphite slits. The proton beam was extracted to air through an exit foil made of Al (6 μm thick). The spot size on the specimen was 3×5 mm^2. Proton beam current was 100 nA or less at the Al exit foil. The energy of the proton beam was reduced from 2.0 to 1.6 MeV after passing through the Al foil and the 10 mm of air between the exit foil and the specimen (Uda et al. 1993; Uda et al. 2000a). Emitted X-rays were detected by a Si(Li) semiconductor detector placed at 135° relative to the incident beam direction after passing through 70 mm of air between the specimen and the 8-mm-thick Be window of the detector. Energy resolution of the detector was 180 eV for the Mn Kα(5.89 keV) X-ray energy.

(a) The block of Ramesses II holding captives, 19th Dynasty (ca. 1270 B.C.).

(b) The Wooden Coffin and Lid of Neb-Seny, 18th Dynasty (ca. 1400 B.C.).

(c) The Funeral Stele of Amenemhat; 11th Dynasty (ca. 2000 B.C.).

Fig. I-1-1. Objects studied here, which are exhibited at the Egyptian Museum in Cairo.
(See also Color plate, p. 296)

(a) The south wall of Room E.

(b) The north wall of Room I.

(c) The west wall of Room I.

Fig. I-1-2. The painted tomb walls of Amenhotep III, 18th Dynasty (1404-1364 B.C.) in Luxor, Egypt.
(See also Color plate, p. 297)

XRD in the laboratory was performed with an automated X-ray diffractometer (Phillips) equipped with a Cu Kα source; the accelerating voltage and the electric current at the Cu anode were 40 kV and 30 mA (1200W), respectively. Scanning speed of the goniometer ($2\theta°s^{-1}$) and scanning step ($\Delta 2\theta°$) were 0.005 and 0.025, respectively.

Experimental conditions in the field were set up to ensure that a necessary and sufficient quality of data for analysis was accumulated in a minimum amount of measurement time. A portable XRF spectrometer and X-ray diffractometer were used to investigate monuments exhibited in the Egyptian Museum. XRF measurements in air were carried out under touch-free conditions with a handy and compact Portarix XRF analyzer (H: 16 cm, L: 28 cm, W: 11 cm), commercially available from Rigaku . The analyzer offers high detection sensitivity and low background noise. These advantages are a result of installation of a cylindrically shaped crystal for X-ray wavelength dispersion and of the introduction of a monochromatic X-ray, i.e., Mo Kα operating at 40 kV and 0.125 mA (5 W), for X-ray emissions to be measured. Measurements of only 5 min. were adequate to obtain well resolved spectra, i.e., $\Delta E \leq 300$ eV at 5.9 eV. In the field, intensities of characteristic X-rays emitted from pigments were measured with the energy-dispersive type of Si PIN photodiode and were processed with Windows 95 software, which was also used to control the experimental conditions. A homemade X-ray diffractometer was utilized in this investigation. The diffractometer is 25 cm in width, 12 cm in length, and 20 cm in height (Fig. 3). It is equipped with a Cr X-ray tube that produces Cr Kα and Kβ X-rays and is operated at 50 kV and 1 mA (50 W). The diffraction angle 2θ, ranging from 35° to 90°, i.e., d =3.8 Å to 1.6 Å (lattice spacing), is changed by a pulse motor. The X-rays were measured by a scintillation counter and recorded on a special paper prepared for this experiment. Goniometer scanning speed was 0.11 ($2\theta°s^{-1}$).

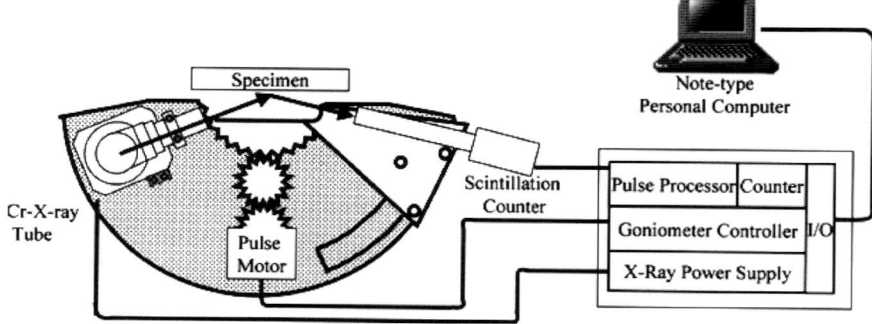

Fig. 1-1-3. The homemade XRD system used in the field.

Another type of X-ray fluorescence spectrometer, shown in Fig. 4, was used to investigate pigments painted on sun-dried bricks excavated from the ruins of Malqata palace. The XRF spectrometer, Model 100F, is commercially available from OURSTEX and has a Pd X-ray tube operating at 40 kV and 0.25 mA (10 W) and a silicon drift detector. This convenient, compact spectrometer is in 25 cm in height, 12 cm in length, and 12 cm in width. It also offers high detection sensitivity, especially in the energy regions of Co, Zn, and Sr Kα X-rays, and offers low background noise in all energy regions. Energy resolution of the spectrometer is 150 eV at 5.9 keV, allowing recording of an adequate amount of intense X-rays within 1 min.

The tomb walls of Amenhotep III were investigated with two portable instruments. The first was an ED-XRDF spectrometer, shown in Fig. 5, which was specially designed and prepared at Waseda University and RIKEN KEIKI for this study. The same areas of pigmentation examined by ED-XRDF were also studied with the second instrument, an optical microscope, Picture Folder for VH-5000, commercially available from Keyence (5x to 40x magnification). The microscope has a long focal length (500 mm) and is characterized by computer-aided operation. Changes in magnification and color can be executed without touching the microscope.

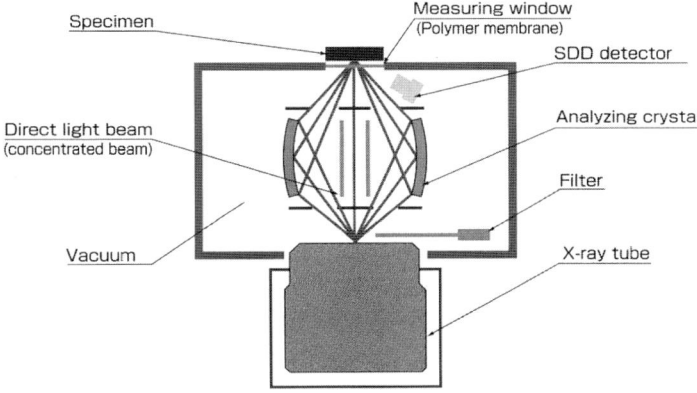

Fig. I-1-4. The Model 100F XRF system used in the field.

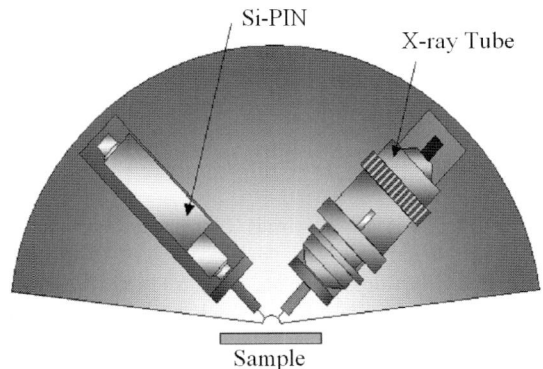

Fig. I-1-5. The portable ED-XRDF spectrometer specially designed and prepared for this investigation.

In the ED-XRDF spectrometer, a Cu (VF-50J-Cu/S, Varian Medical Systems) or W (VF-50J-W/S, Varian Medical Systems) tube was installed as the X-ray emitting source, and a Si-PIN photodiode (XR-100CR, AMPTEK INC.) was used to detect diffracted and fluorescent X-rays. Energy dispersive diffraction (Giessen and Gordon 1968) was performed using white X-rays emitted from the Cu tube, where the diffraction condition can be written as Eq. 1.

$$2d\sin\theta = n\lambda = nc/v = nhc/E \qquad (1)$$

Here, d denotes the lattice spacing of the crystal to be investigated; θ, the diffraction angle between the incident X-ray and the target surface; n, a positive integer; λ, the wavelength of the incident or diffracted X-ray; c, the velocity of light; v, the frequency of the emitted or diffracted X-ray; and E, the energy of the emitted or diffracted X-ray. This formula indicates that the measuring principle of the ED-XRDF spectrometer is completely different from that of a commercially available diffractometer, which is operated under an angle dispersive condition. The Cu tube and the X-ray detector were both originally set at fixed angles of 10° and, if necessary, can be set at angles of 12.5°, 15°, 17.5°, and 20°. The W tube, installed at an angle of 60°, was used only when it was necessary to record pure X-ray fluorescence spectra free from diffracted X-rays and elastically scattered Cu K X-rays. Here, we present only the diffraction data taken at the angle of 10° to avoid complexity. The ED-XRDF spectrometer was used in open air in the tomb of Amenhotep III; thus, only X-rays with energies of 2.0 keV or more, emitted from elements with atomic numbers higher than P, can be detected. This is because X-rays with energies of 2.0 keV or less are absorbed by the air during passage between the sample and the detector, i.e., the Si-PIN photodiode. Parallel and coherent X-ray beams were prepared using two capillaries with 0.2- and 0.15-mm diameters installed in front of the object and the X-ray detector, respectively, and were made of stainless steel or glass.

The ED-XRDF spectrometer used here is characterized by 1) short measuring time, typically 100 secs even for X-ray diffraction, 2) non-contact and non-destructive measurements, 3) no restriction on sample size and shape, 4) small size (300 x 250 x 60 mm^3), 5) light weight (2 kg), 6) no need for special coolant for the X-ray tube, which is equipped with an air-cooling system, and 7) operation assisted by a personal computer. Such noteworthy features suggest that ED-XRDF will be a very strong weapon for investigating monuments in the field and especially for investigating wall paintings.

4.   Results

Concentrations of the elements and the crystal structures of the pigments painted on several sun-dried mud bricks had been accurately determined in the laboratory prior to the field experiments by PIXE and X-ray diffraction methods. These results were very useful for judging whether experiments in the field were working well and for comparing X-ray data taken by the portable XRF and XRD instruments. XRF and XRD data were used to characterize pigments in the Egyptian Museum, the Dahsur excavation sites and the tomb of Amenhotep III.

4-1. PLASTER

Most pigments in ancient Egypt were painted on plaster, which was applied in a thin layer or layers. Thus, plaster must be characterized before determining the chemical constituents and crystal structures of pigments. Before field experiments, the layered structure was confirmed with XRD in the laboratory (Uda *et al.* 1998, Yoshimura *et al.* 2002). Plaster on a wall fragment (601612) excavated from Kom al-Samak was composed of anhydrite ($CaSO_4$) when it was applied in one layer on a brick. A two-layered structure of plaster was also observed on three fragments unearthed from the ruins of Kom al-Samak and two private tombs. The surface layer was composed of anhydrite, whereas the subsurface layer was composed of calcite ($CaCO_3$) and quartz (SiO_2) together with small amounts of anhydrite. A three-layered structure of plaster was also detected on a wall fragment excavated from Kom al-Samak and was composed of fine clay (surface layer), white calcite (middle layer), and coarse clay (base layer).

Based on experimental results obtained in the laboratory, plaster was additionally investigated in the field with the portable ED-XRDF instrument. The walls of the tomb of Amenhotep III were coated with plaster before painting with the pigments and have plaster of a two-layered or, in a few cases, three-layered structure. The outer surface layer appears to be whitish yellow, which was named "point 36" in Room J-2, and is composed of fine grains. An XRF spectrum taken from "point 36" is shown in Fig. 6. S, Ar, Ca and Fe K X-rays were found, together with L X-rays of W. Here, Ar originated from the air between the X-ray tube and the detector, and W originated from the W X-ray tube because one series of XRF experiments with the ED-XRDF system was performed in open air with an X-ray emitter made of W.

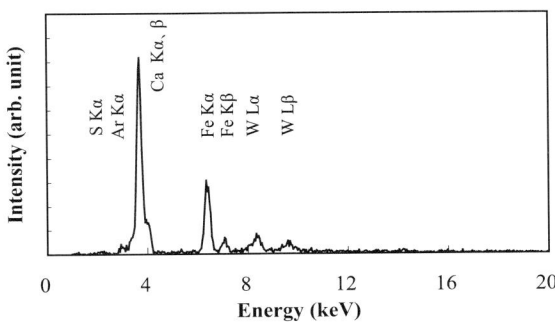

Fig. I-1-6. XRF spectrum taken from "point 36" with a W X-ray tube.

ED-XRD patterns accompanied by XRF spectra taken from reference materials (pure chemicals) of anhydrite ($CaSO_4$), calcite ($CaCO_3$), and quartz (SiO_2) are shown in Figs. 7 (a), (b), and (c), respectively. ED-XRD and XRF spectra taken from "point 36" are shown in Fig. 7 (d). Comparing Fig. 7 (d) with Figs. 7 (a), (b), and (c), we concluded that plaster at "point 36" is composed of anhydrite ($CaSO_4$) and quartz (SiO_2). The main component of the plaster was anhydrite but not gypsum ($CaSO_4 \cdot 2H_2O$). This suggests that the original form of $CaSO_4$ was not $CaSO_4 \cdot 1/2H_2O$ but $CaSO_4$ itself because the volume change from $CaSO_4 \cdot 2H_2O$ to $CaSO_4$ during aging is very large, and it would be very difficult to maintain a wall's form without breakdown. From repeated analysis of plaster components around the area of "point 36", we detected that the mixing ratios of anhydrite ($CaSO_4$) to quartz (SiO_2) varied from place to place. This suggests that surface compositions are not homogeneous even if plaster colors are similar.

The wall of Room J-2 appears incomplete because some parts of the wall were coated twice ("point 36") and some were coated only once ("point 37"). An XRF spectrum and ED-XRD and -XRF spectra shown in Figs. 8 (a) and (b), respectively, were taken from "point 37", which appears to be yellowish white and to be composed of coarse grains, indicating that the inner plaster consists mainly of calcite ($CaCO_3$) together with small amounts of anhydrite ($CaSO_4$). These results, deduced from the experimental data taken in the field, are consistent with those obtained in the laboratory (Yoshimura 2002).

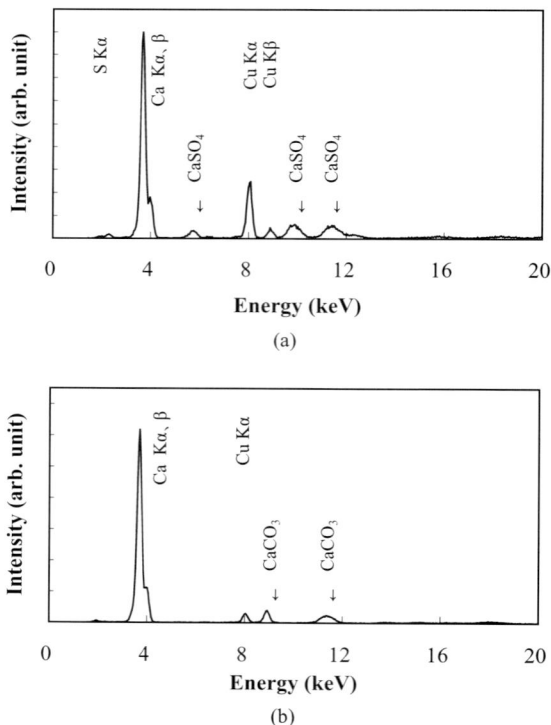

Figs. 1-1-7. (a) and (b). ED-XRDF spectra taken from pure chemicals $CaSO_4$ (a) and $CaCO_3$ (b) with a Cu X-ray tube.

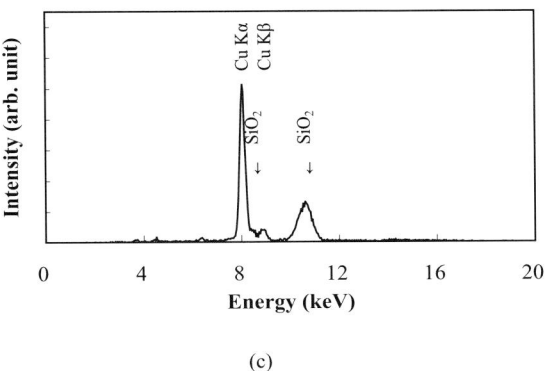

(c)

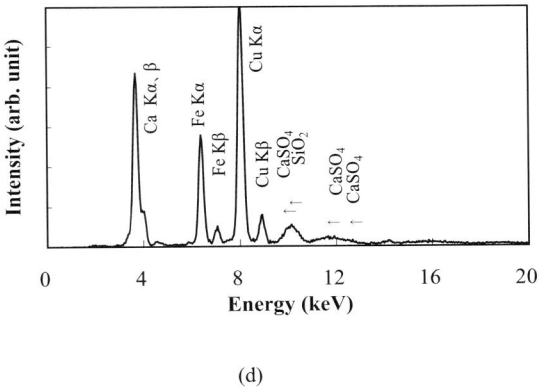

(d)

Figs. I-1-7. (c) and (d). ED-XRDF spectra taken from a pure chemical, SiO$_2$ (c) and from "point 36" (d) with a Cu X-ray tube.

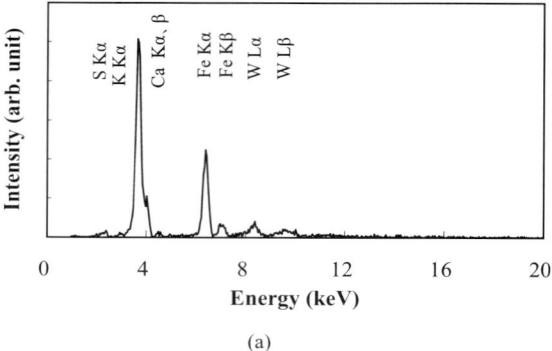

(a)

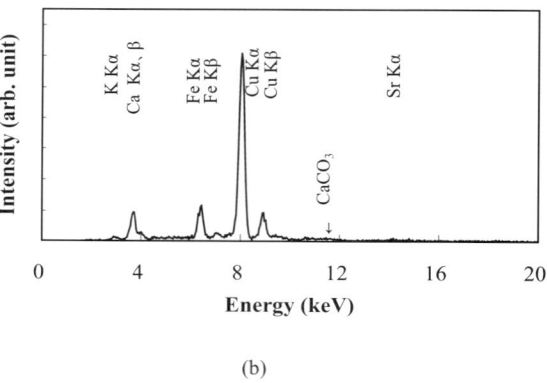

(b)

Fig. 1-1-8. XRF (a) and ED-XRDF (b) spectra taken from "point 37" with W and Cu X-ray tubes, respectively.

4-2. WHITE PIGMENT

PIXE and XRD experiments were performed in the laboratory prior to investigating white pigments in the field. Huntite ($Mg_3Ca(CO_3)_4$) was found in white pigment on a wall fragment excavated from Kom al-Samak and produced well crystallized X-ray diffraction patterns on a very weak background as shown in Fig. 9 (Uda *et al.* 1993; Nagashima *et al.* 1996), indicating that an amorphous phase was not included in the white pigment. Contaminants in the white pigment are composed of small amounts of mud components, i.e., quartz (SiO_2), gismondine ($Ca(Al_2Si_2O_8)$ • $4H_2O$), and dickite ($Al_2O_3 2SiO_2$ • $2H_2O$). Similar results were observed from a white pigment spotted in the Malqata palace (Uda *et al.* 2000a).

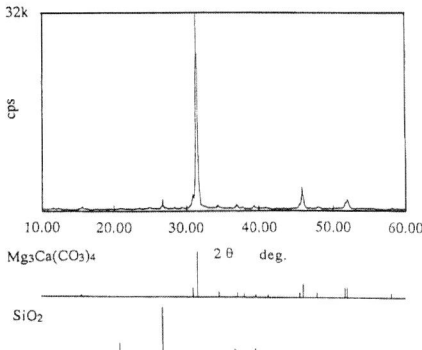

Fig. I-1-9. XRD pattern, taken from a white pigment in the laboratory, compared with expected diffraction patterns from the reference materials, huntite and quartz.

Huntite was detected with the portable homemade diffractometer from the white part of a lady's robe, shown on the right hand side of the Funeral Stele in Fig. 1 (c). The Portarix XRF analyzer showed Ca, Fe and Sr to be the main components of this pigment. Under experimental conditions in the field, however, quantities of Mg, O, and C could not be determined from the observed X-ray fluorescence spectra because the experiment was performed in open air, which absorbs most X-rays with energies lower than 2.0 keV, i.e., Kα X-rays emitted from elements with atomic numbers less than 15 (P). This same phenomenon also applies to the portable XRF spectrometer.

Huntite was also detected by ED-XRDF from the painted tomb walls of Amenhotep III from the areas of 1) the kilt ("point 3") of the second male from the right on the south wall in Room E (Figs. 2 (a) and 10 (a)), 2) the white fan ("point 43") on the right-hand side of the north wall in Room I (Fig. 2 (b)), and 3) the kilt ("point 14" and "point 15") of the fifth male from the left on the west wall of Room I (Fig. 2 (c)). Examples of ED-XRD and -XRF spectra are shown in Figs. 10 (b) and (c), respectively. Huntite, shown in Fig. 10 (c) as $CaCO_3$ • $3Mg CO_3$, is contaminated with small amounts of anhydrite and quartz.

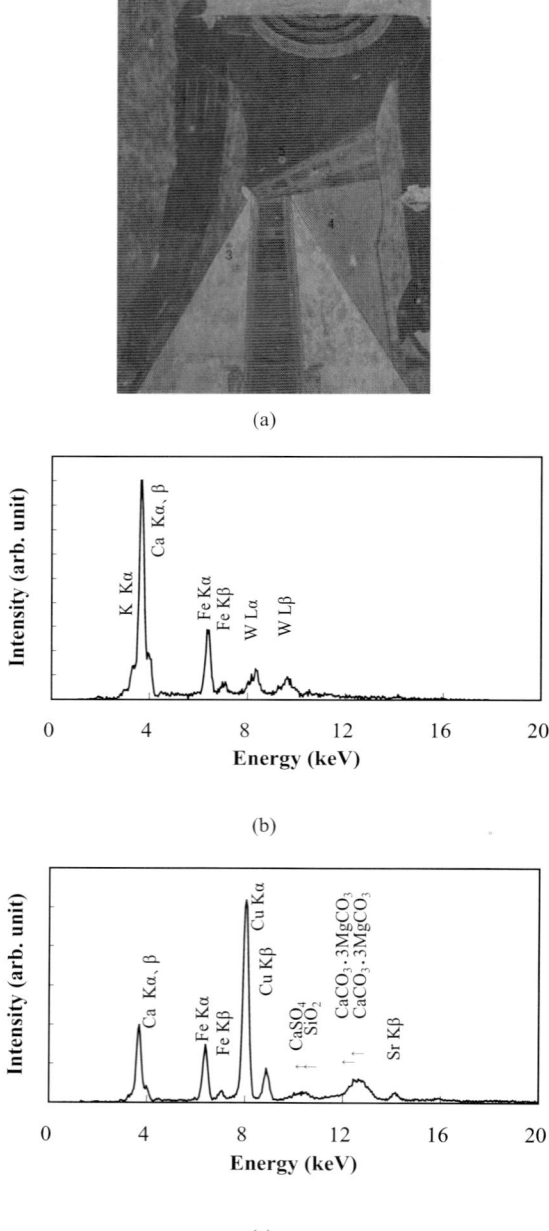

(a)

(b)

(c)

Fig. I-1-10. The kilt ("point 3" (a)) of the second male from the right on the south wall in Room E, and XRF (b) and ED-XRDF (c) spectra taken from "point 3" with W and Cu X-ray tubes, respectively.

The Wooden Coffin and Lid of Neb-Seny (Fig. 1 (b)) are covered with a white pigment over the entire surface. The white layer was confirmed by the homemade XRD system to consist mainly of calcite ($CaCO_3$) together with small amounts of quartz. This result is also supported by an XRF spectrum on which intense Ca K, and weak Fe K, As K, and Sr K peaks were identified. Here, Fe and Sr are thought to be impurities included in the calcite. An As-bearing compound should not, however, be an impurity in calcite; it probably originates from the spreading of adjacent pigments due to wiping during cleaning. This assumption is supported by detection of As over the entire surface, for example, from skin parts painted reddish brown and hair painted green, and also by detection of different amounts of As from parts with the same colors.

4-3. BROWN OR RED AND PINK PIGMENTS

Hematite (α Fe_2O_3) and goethite (α FeO • OH) together with mud components were distinguished on an X-ray diffraction pattern taken from the red part of a wall fragment excavated from Kom al-Samak (Uda *et al.* 1993; Nagashima *et al.* 1996). The main metallic component was Fe, as determined by PIXE. This same result was obtained from a wall fragment excavated from Malqata palace (Uda *et al.* 2000a). A PIXE spectrum taken from a reddish brown part is shown in Fig. 11 (a) together with relative atomic concentration ratios shown in Fig. 11 (b). The high concentration of Fe is consistent with detection of α Fe_2O_3 in an XRD pattern taken from the same area. Small amounts of S and As were detected on a PIXE spectrum taken from the red part of the wall fragment excavated from Kom al-Samak (Uda *et al.* 1993), suggesting that realgar (AsS) is also an ingredient of red pigments.

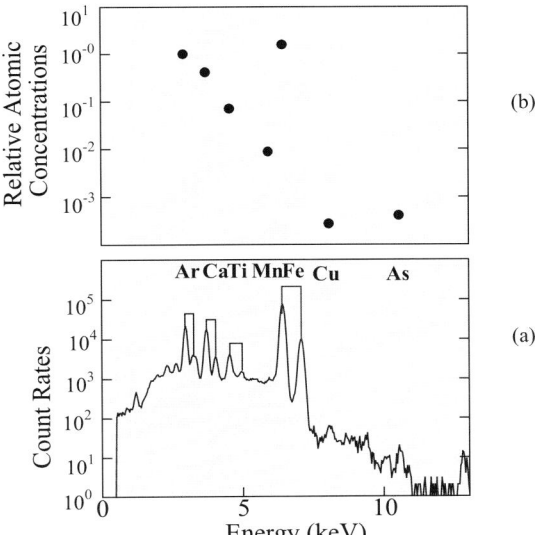

Fig. I-1-11. PIXE spectrum (a) taken from a brown pigment and its atomic ratios (b) calculated from the spectrum.

XRF spectra taken from the painted limestone of the Ramesses II block (Fig. 1 (a)) are shown in Figs. 12 (a) and (b). The spectrum shown in Fig. 12 (a) was taken from a part with no pigment and is characterized by intense Ca Kα, medium Sr Kα, and weak Fe Kα peaks. Here, broad and sharp peaks at 17 and 18 keV, respectively, represent inelastic and elastic peaks of Mo Kα used for excitation. An XRD pattern, not shown here, taken from the same part was characterized by calcite ($CaCO_3$), i.e., limestone. The XRF spectrum shown in Fig. 12 (b) was taken from the reddish brown part (the leg of the captive drawn on the left-hand side) and is characterized by an intense Fe Kα peak. An XRD pattern, also not shown here, indicates that the reddish brown pigment is composed mainly of hematite. The same is also true for the parts painted red or brown representing the male's skin on the Wooden Coffin and Lid of Neb-Seny, shown in Fig. 2 (Uda et al. 1999), and on the Funeral Stele of Amenemhat (Uda et al. 2000b). Enrichment by As was found in the light red parts or the tomb walls of Amenhotep III, although the parts were too thin to give an adequately intense XRD pattern. However, this suggests use of realgar (AsS) for painting the light red color. A mixture of white and red pigments, i.e., huntite ($Mg_3Ca(CO_3)_4$) and hematite (α Fe_2O_3), was detected by X-ray diffraction and PIXE analyses from pink-colored parts on the fragment excavated from Kom al-Samak (Uda et. al. 1993).

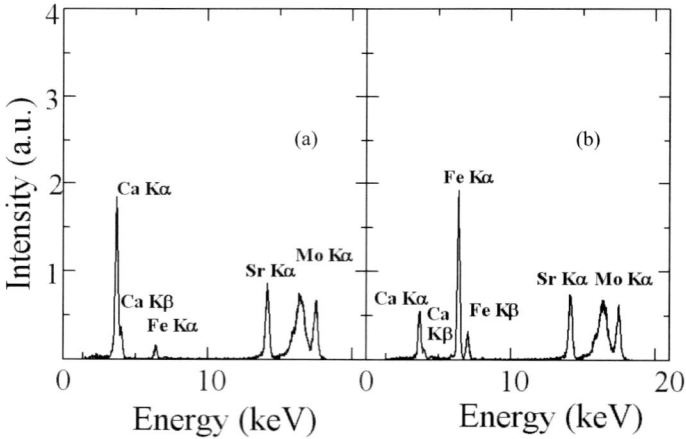

Fig. 1-1-12. XRF spectra taken from a part with no pigment (a) and a reddish brown part (b) on the Ramesses II block with a Mo X-ray tube.

4-4. BLUE AND GREEN PIGMENTS

Egyptian blue (CaO • CuO • 4SiO$_2$) was identified on an X-ray diffraction pattern together with small amounts of mud components, quartz and gismondine. The X-ray pattern was taken from a blue part on a fragment excavated from Kom al-Samak. PIXE spectra taken from the blue part, shown in Figs. 13 (a) and (b), supported the chemical compositions mentioned above. In many cases, chemical compositions are similar for blue and green parts, as seen in Fig. 14 (a). Intense Cu K peaks were found from greenish blue parts ("point 27" and "point 28", as shown in Figs. 14 (b) and (c), respectively) on the robe of the 6th person from the left on the west wall of the tomb of Amenhotep III (Fig.2 (c)). Cu was also detected, as shown in Fig. 15, from hair painted dark green on the Wooden Coffin and Lid shown in Fig. 1 (b).

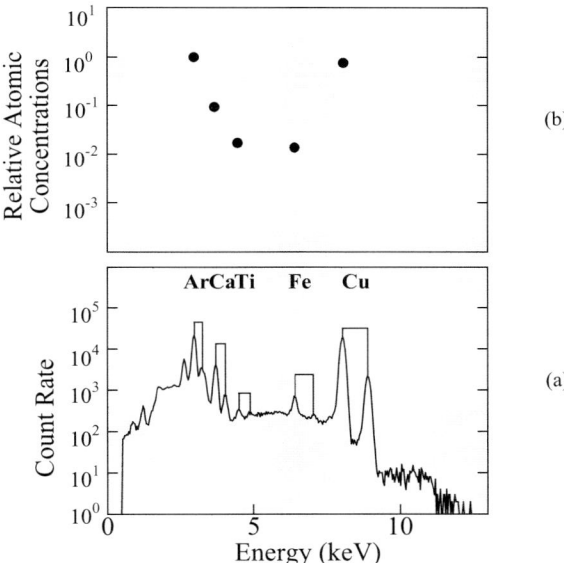

Fig. I-1-13. PIXE spectrum (a) taken from a blue part on a wall fragment excavated from Kom al-Samak and its atomic ratios (b) calculated from the spectrum.

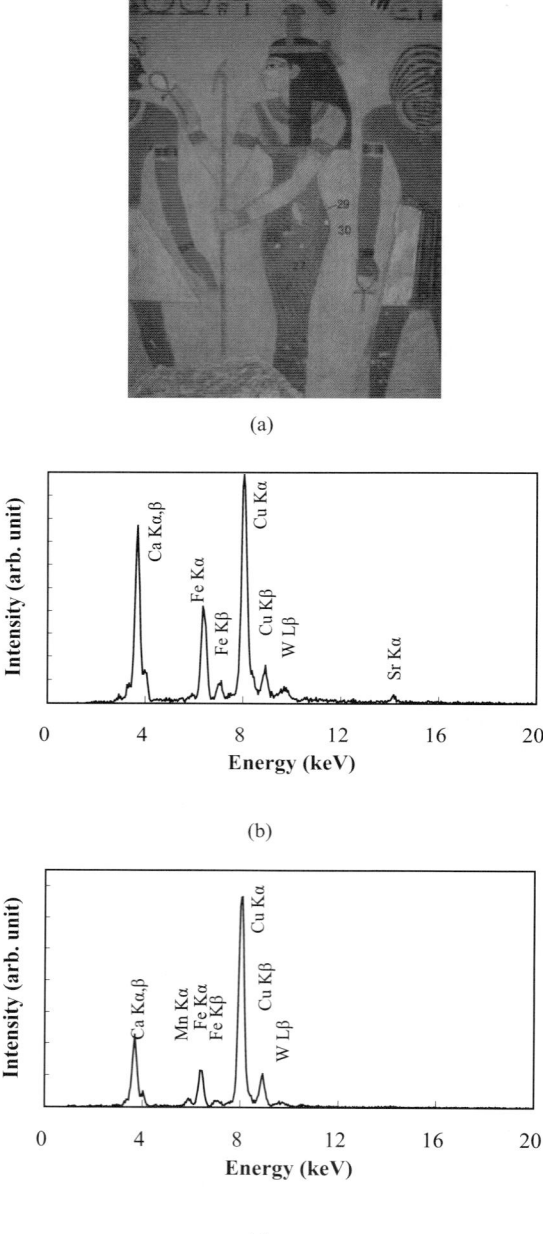

(a)

(b)

(c)

Fig. I-1-14. The greenish blue parts, as shown in (a), of the 6th person from the left on the west wall in Room I, and their XRF spectra taken from "point 27" (b) and "point 28" (c) with a W X-ray tube.

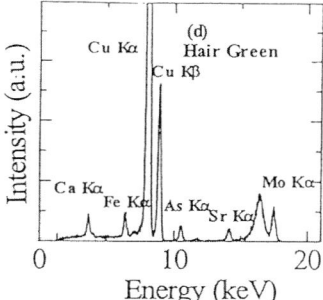

Fig. 1-1-15. XRF spectrum taken with a Mo X-ray tube from hair painted dark green on the Wooden Coffin and Lid.

We also investigated another pigment, Amarna blue, which was painted on pottery fragments unearthed from the ruins of Kom al-Samak, Dahshur, and Abusir. The PIXE spectrum (Uda *et al.* 2002) shown in Fig. 16 was taken from a light blue part on a fragment unearthed from the ruin of Kom al-Samak. Enrichment by S, Cl, Ca, Mn, Co, Ni and Zn was remarkable compared with the PIXE spectrum taken from an unglazed part, not shown here. Sharp diffraction patterns, indicating the large grain sizes of powder specimens, from $CaSO_4$ and $NaCl$ were found from the light blue part, although diffracted intensities from $CaSO_4$ and $NaCl$ were different from place to place. This indicates that $NaCl$ was not added on purpose but is an unavoidable impurity. Broad diffraction peaks at d = 2.44, 1.43, and 1.56 Å from the light blue part were also found, which were characteristic of a spinel type of oxide with a lattice constant of 8.10 Å. The broad peaks suggest the grain sizes of the powder specimens to be small. The constituents of the light blue pigment therefore must be, at least in part, $Co(M)Al_2O_4$ with small grain sizes and $CaSO_4$ with large grain sizes, where M denotes Mn, Fe, Ni and/or Zn. The chemical compositions and crystal structures of the Amarna blue pigment were determined for the first time in this series of experiments.

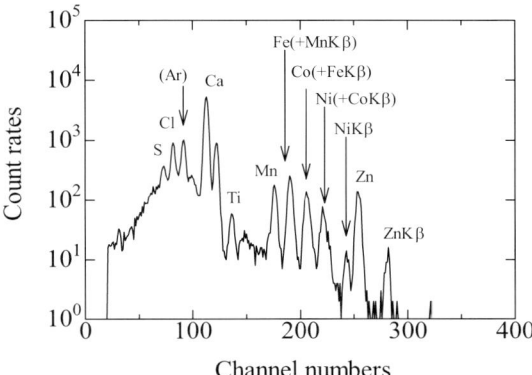

Fig. 1-1-16. PIXE spectrum taken from a light blue part on a fragment excavated from Kom al-Samak.

4-5. YELLOW PIGMENT

Goethite (α-FeO • OH) and orpiment (As_2S_3) were located on X-ray diffraction patterns taken from yellow parts painted on fragments excavated from Kom al-Samak (Uda *et al.* 1993), from which intense Fe K and moderate As K X-ray peaks were revealed by PIXE. A PIXE spectrum taken in the laboratory is shown in Fig. 17. Fundamentally similar XRD pattern and PIXE spectrum were also taken from a yellow-colored part on a wall fragment unearthed from the ruins of Malqata palace (Nagashima *et al.* 1996).

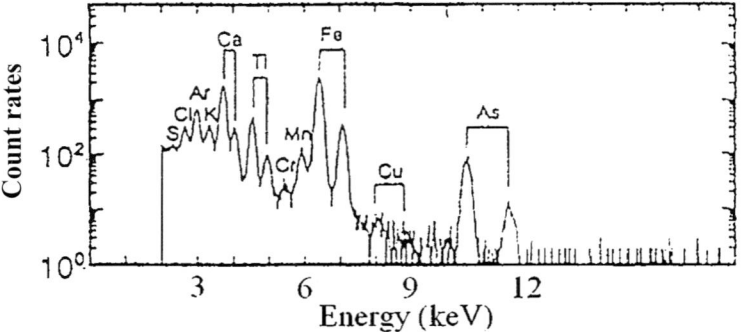

Fig. I-1-17.   PIXE spectrum taken from a yellow part on a fragment excavated from Kom al-Samak.

On the XRF spectrum shown in Fig. 18, Ca, Fe, and Sr K X-rays were detected from the light-yellow part depicting female skin on the Funeral Stele, first person from the left (Fig. 1 (c)), suggesting that the yellow part was painted with a mixture of calcite ($CaCO_3$) and goethite, where Sr may be an inclusion in both pigments.

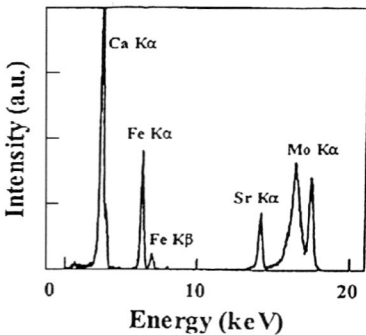

Fig. I-1-18. XRF spectrum taken with a Mo X-ray tube from a light-yellow part on the Funeral Stele.

Orpiment (As_2S_3) was found from brilliant yellow parts, i.e., the left hand of the second person ("point 13") and from the kilt of the fourth person ("point 17", shown in Fig. 19 (a)) from the left on the north wall of Room I (Fig. 2 (b)). ED-XRF and -XRDF spectra taken from "point 17" are shown in Figs. 19 (b) and (c), respectively.

(a)

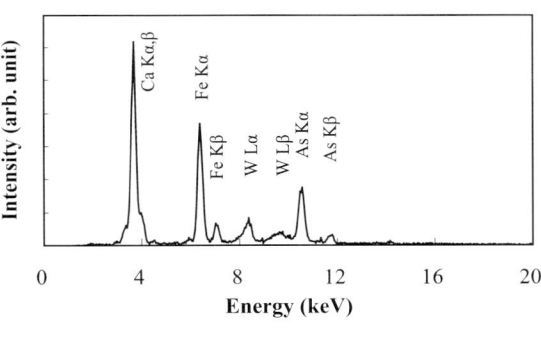

(b)

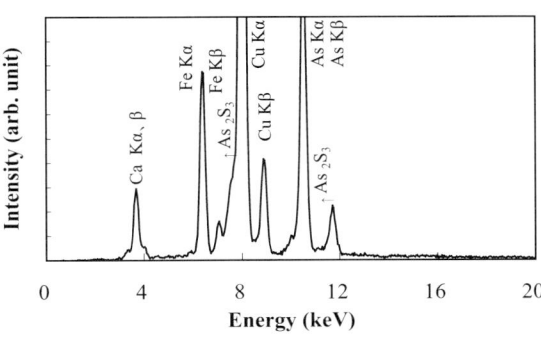

(c)

Fig. I-1-19. The yellowish kilt of the 4th person from the left on the north wall of Room I (a) and XRF (b) and ED-XRDF (c) spectra taken from "point 17" with W and Cu X-ray tubes, respectively.

5.   Conclusion

Characterization of ancient Egyptian pigments painted on fragments excavated from ruins and ancient tombs in Egypt was done with XRF, PIXE and XRD methods. This is the first report of the successful determination of the chemical composition and crystal structure of Amarna blue pigment. Amarna blue is light blue in color and was confirmed to be a mixture of $Co(M)Al_2O_4$ (dark blue) with anhydrite (white) or calcite (white). Non-destructive and non-contact experiments for analyzing pigments were executed for the first time with portable XRF, XRD and ED-XRDF systems in the field.

Plaster coating the outermost layer of the tomb walls of Amenhotep III was composed of anhydrite ($CaSO_4$) and quartz (SiO_2). The inner layer of plaster was composed mainly of calcite ($CaCO_3$) together with small amounts of anhydrite ($CaSO_4$). White pigments containing calcite ($CaCO_3$) and huntite ($CaCO_3 \bullet 3Mg CO_3$) were used on the Wooden Coffin and Lid and on the tomb walls of Amenhotep III, respectively. Hematite (αFe_2O_3) was frequently found from parts red and brown in color. Light red parts found on the tomb walls of Amenhotep III were enriched with As, although these parts were too thin to allow an adequately intense X-ray diffraction pattern. This suggests use of realgar (AsS) for painting this light red color. Pink was made by mixing hematite (αFe_2O_3) with huntite ($CaCO_3 \bullet 3Mg CO_3$). Most blue and green parts were painted with Egyptian blue ($CaO \bullet CuO \bullet 4SiO_2$) and several kinds of amorphous Cu-bearing compounds, respectively. The light-yellow part on the Funeral Stele was painted with a mixture of calcite ($CaCO_3$) and goethite ($FeO \bullet OH$). Brilliant yellow parts on the tomb walls of Amenhotep III were painted with orpiment (As_2S_3).

We confirmed that portable XRF, XRD and ED-XRDF systems can be used very effectively to investigate pigments on monuments under non-destructive and non-contact conditions in the field. It is also highly probable that these portable systems can be used to study surfaces of other monuments in the field without difficulty. These methods may supply important information necessary for the conservation and restoration of the unique monuments of the world.

Acknowledgements

I would like to thank my collaborators, S. Yoshimura, M. Tamada, S. Sassa, S. Nagashima, Y. Sasa, M. Nakamura, K. Taniguchi, J. Kondo, S. Hasegawa, N. Kawai, T. Tsunokami, R. Murai, K. Saito, and M. Saito.

References

Barbieri M, Calderoni G, Cortesi C, Fornaseri M (1974) Huntite, A mineral used in antiquity. *Archaeometry* **16**: 211-220

Faust GT (1953) Huntite, $Mg_3Ca(CO_3)_4$, a new mineral. *Amer. Mineralog.* **38**: 4-24

Giessen BC, Gordon GE (1968) X-ray Diffraction: New High-speed Technique Based on X-ray Spectrography. *Science* **159**: 973-975

Goresy A El, Jaksch H, Razek MA, Weiner KL (1986) Ancient pigments in wall paintings of egyptian tombs and temples. *Rep. M.P.I. H V* **12**: 1-45

Goresy A El (1997) Polychromatic wall painting decorations in monuments of pharaonic Egypt: compositions, chronology, and painting techniques. *1st International Symposium on the Wall Paintings of Thera, August 30- September 4, Santorini, Greece*

Green LR (1955) "Recent Analysis of Pigments from Ancient Egyptian Artifacts", In *Conservation in Ancient Egyptian Collections*, Brown CE, Macalister F, Wright MM. eds. Archetype Publication: 85-91

Heywood A (1996) Color and panting in ancient Egypt. *British Museum Colloqium*, **July**: 11-12

Jaksch H, Seipel W, Weiner KL, Goresy A El (1983) Egyptian blue-cuprorivaite, a window to ancient egyptian technology. *Naturwiss.* **70**: 525-535

Lucas A, Harris JR (1989) *Ancient Egyptian Materials and Industries*, London: Histories & Mysteries of Man, LTD., 338-366

Nagashima S, Kato M, Kotani T, Morito K, Miyazawa M, Kondo J, Yoshimura S, Sasa Y, Uda M (1996) Application of the external PIXE analysis to ancient Egyptian objects.. *Nucl. Instr. Meth. in Phys. Res. B* **109/110**: 658-661

Noll W (1981) Zur Kenntnis altaegyptischen Pigmente und Bindemittel. *N. Jb. Miner. Mh. H.* **9**: 416-432

Riederer J (1974) Recently identified egyptian pigments. *Archaeometry* **16**: 102-109

Riederer J (1988) Pigmente in der Antike. *Pd Naturwiss. Chem.* **37**: 3-10

Saleh SA (1987) "Pigments, plaster and salt analyses" In *Wall Paintings of the Tomb of Nefertari: Scientific Studies for Their Conservation: First Progress Report, July 1987*, Conzo MA, ed. California: Cairo and Century, pp 94-105

Sasa Y, Uda M (1995) Analysis of ancient egyptian pigments by external PIXE and XRD. *J. Egyptian Studies*, **3**: 4-29

Schiegl S, Weiner KL, Goresy A El (1989) Discovery of copper chloride cancer in ancient egyptian polychromatic wall paintings and faience: a developing archaeological disaster. *Naturwiss.* **76**: 393-398

Uda M, Tsunokami T, Murai R, Maeda K, Harigai I, Nakayama Y, Yoshimura S, Kikuchi T, Sakurai K, Sasa Y (1993) Quantitative analysis of ancient Egyptian pigments by external PIXE. *Nucl. Instr. Meth. in Phys. Res. B* **75**: 476-479

Uda M (1998) PIXE and archaeology. *J. Egyptian Studies*, **3**: 37-54

Uda M, Sassa S, Yoshioka Y, Taniguchi K, Nomura S, Yoshimura S, Kondo J, Nakamura M, Iskandar N, Zaghloul B (1999) X-ray analysis of pigments on ancient Egyptian monuments. *Intern. J. PIXE* **9**: 441-451

Uda M, Sassa S, Yoshioka Y, Taniguchi K, Nomura S, Yoshimura S, Kondo J, Nakamura M, Iskandar N, Zaghloul B (2000a) Touch-free in situ investigation of ancient Egyptian pigments. *Naturwiss.* **87**: 260-263

Uda M, Sassa S, Yoshioka Y, Taniguchi K, Nomura S, Yoshimura S, Kondo J, Nakamura M, Ban Y, Adachi H (2000b) Yellow, red and blue pigments from ancient Egyptian palace painted walls. *Nucl. Instr. Meth. in Phys. Res.* **161-163**:758-761

Uda M, Nakamura M, Yoshimura S, Kondo J, Saito M, Shirai Y, Hasegawa S, Oshio H, Yamashita D, Nakajima Y, Utaka T (2002) "Amarna blue" painted on ancient Egyptian pottery. *Nucl. Instr. Meth. in Phys. Res. B* **189**: 382-386

Weatherhead F (1995) "Two studies on amarna pigments". In *Amarna Reports VI. (EES Occasional Papers 10)* Kemp BJ, ed. London: Egyptian Exploration Society, pp 384-398

Yoshimura S, Uda M, Saito K (2002) A study of the use of calcite plaster for plastering walls in ancient Egypt. *Arch. Natural Sci.* **44**:1-15

Chapter I-2

Importance of *in-situ* EDXRF Measurements in the Preservation and Conservation of Material Culture

A.G. Karydas[a], X. Brecoulaki[b], Th. Pantazis[c], E. Aloupi[d], V. Argyropoulos[e], D. Kotzamani[f], R. Bernard[g], Ch. Zarkadas[a] and Th. Paradellis[a†]

[a] *Laboratory for Material Analysis, Institute of Nuclear Physics, NCSR "Demokritos",* Athens 153 10, Greece.

[b] *The Wiener Laboratory, The American School of Classical Studies at Athens,* 54 Souidias Street, Athens 106 76, Greece.

[c] *Amptek, Inc., 6 De Angelo Drive,* Bedford, Mass. 01730, U.S.A

[d] *Thetis Authentics - Science and Techniques for Art History Conservation Ltd,* 41, M. Moussourou str., Athens 116 36, Greece.

[e] *Department of Conservation of Antiquities & Works of Art, Technological Educational Institution of Athens,* Ag. Spyridonas, Aigaleo 12210, Greece.

[f] *Benaki Museum,* 1 Koumbari Street, Athens 106 74, Greece.

[g] *Ecole Polytechnique de l'universite de Nantes,* Rue Christien Pauc, BP 50609, 44306 Nantes Cedex 3, France.

karydas@inp.demokritos.gr

Keywords: EDXRF, *in-situ* measurement, archaeological artifacts, pigments, wall painting, Greece, Cyprus, preservation, conservation

Abstract

The paper presents a series of *in-situ* Energy Dispersive X-ray Fluorescence (EDXRF) measurements performed in Greece and Cyprus for the characterization and quantitative analysis of valuable archaeological and historical artifacts, such as pigments on wall paintings and ceramics as well as gold and bronze alloys on sculpture and jewelry. The various types of instrumentation and analytical procedures used for the quantification of *in-situ* XRF data will be discussed. The experience acquired from the above investigations will be critically reviewed in order to assess the importance, as well as the limitations of *in-situ* EDXRF analyses in the preservation and conservation of our cultural heritage.

---

† This paper is dedicated to the memory of Dr. Themis Paradellis, whose pioneering work on the application of EDXRF techniques in the study of Cultural Heritage in Greece inspired all co-authors in this article.

M. Uda et al. (eds.), X-rays for Archaeology, 27–53.
© 2005 *Springer. Printed in the Netherlands.*

1.   Introduction

X-ray instrumentation for energy dispersive X-ray spectrometry has progressed significantly in the past few years. Miniature X-ray detectors working at room temperatures and low power X-ray tubes with compact design have offered the potential of assembling integrated systems for *in-situ* EDXRF measurements. This capability is very applicable to the field of cultural heritage, where analyses very often must be carried out on-site, since museums may not permit artifacts to leave their premises or the monument is immobile for analysis in the laboratory. Moreover, the technique offers non-destructive and multi-elemental analysis with practically no limitation in the number of samples (positions) analyzed *in-situ*. Moreover, it is a fast method so that a large collection of similar objects can be analyzed in a short time, or a specific artifact can be examined in many different locations. The importance of EDXRF in cultural heritage was stressed in a special volume of X-Ray Spectrometry journal [1], which was dedicated to the use of EDXRF in solving problems related to art and archaeometry. Towards this direction, EDXRF spectroscopy faces, very often, demanding analytical problems arising either from the widespread nature of archaeological materials (pigments, alloys, ceramics, glasses) or due to their physical state. It is well known that accurate quantification procedures in EDXRF spectroscopy can be applied only in homogeneous materials. In most cases the ancient artifacts are not very well preserved, exhibit significant surface inhomogeneities (wall-paintings) or corrosion effects (the copper, silver and iron based alloys). Therefore, even with the application of μ-EDXRF techniques a question may arise as to what extent the results from a limited analyzed volume are representative of the whole artifact.

The analytical detection range of EDXRF with the most sophisticated portable instruments covers elements from sodium (Z=11) up to uranium (Z=92). Since the technique exhibits high sensitivity for certain elements, provenance studies can be done not only by means of major elements but also using fingerprint trace elements. One important disadvantage of the EDXRF method is its weakness to identify the oxidation state of the analyzed elements and to provide information about contained minerals or compound forms of the constituents. Therefore, the application of complementary techniques (i.e. XRD, FTIR, μ-Raman, SEM/EPMA) is required for many analytical problems that appear in the field of art conservation and archaeometry. In the case of characterization of pigments from various types of paintings, a problem may also arise from the limited analytical depth resolution of the EDXRF method. This topic is discussed in detail by M. Mantler and M. Schreiner [2]. In the case of a layered structure of different pigments, it is rather difficult, not only to identify their arrangement, but also to discriminate this case from the presence of a single admixture layer. C. Neelmejer *et al* [3] reports some special cases where the X-ray techniques (PIXE and EDXRF) may overcome the above ambiguity.

Various applications and possibilities of portable EDXRF instrumentation in art and archaeometry have been presented and discussed in a number of publications mostly since 1996. R. Cesareo *et al* [4] employed portable instrumentation in numerous studies, such as analysis of bronzes from the Nuragic (Sardinia, one millennium B.C.) and prehistoric periods, analysis of Etruscan gold artifacts from the VII century B.C. [5] and characterization of pigments in frescos. A small-beam portable EDXRF instrument combined by a simple micro-polishing technique was found effective by G. Vittiglio *et al* [6] in the determination of the bulk composition of a series of Egyptian corroded bronze objects dating from *ca.* 1090 BC to the Roman era (30 BC to 640 AD). S. Sciuti [7] examined portable EDXRF instruments in the characterization of the colored pigments, medium and plaster of a recently discovered Roman wall–painting in Rome. Using an EDXRF and XRD spectrometer Uda *et al* [8] analyzed *in-situ* under touch-free conditions the white painted parts of two Egyptian ancient monuments and reported for the first time the occurrence of huntite mineral, as well as an As-bearing phase in yellow painted parts of one of the monuments. Recently, Uda and co-workers [9] identified the elemental and mineral composition of "Amarna blue" pigments painted on pottery fragments. The analysis was performed *in-situ* at the excavation site using a portable compact and handy EDXRF spectrometer, as well as using laboratory X-Ray methods (PIXE, XRD).

2.   Current status of portable XRF instrumentation for Art and Archaeometry

Novel instrumentation, developed mainly in the past few years, offers many of the well-known analytical features of laboratory EDXRF techniques for *in-situ* analyses [6, 10-12]. The current status of portable XRF spectrometers cannot be considered as conventional. Large-beam, sub-millimeter or even μ-beam XRF instruments have been developed in a commercial or research basis for application in the field of the non-destructive analysis and testing of archaeological artifacts and/or art objects. K. Janssens *et al* [10] overviewed the XRF instrumentation used in this field and discussed future potential of the technique, especially towards the μ-XRF investigations. Commercial and experimental XRF set-ups used for *in-situ* measurements may include components (X-ray detectors, X-ray sources) with significant differences in their characteristics and capabilities. Of course, the breakthrough in X-Ray technology was the development of new generation solid-state detectors operating at room temperatures and exhibiting energy resolution competitive to the conventional liquid nitrogen cooled Si(Li) detectors [13-15]. Without the need of liquid nitrogen and a bulky Dewar vessel for operating X-ray detectors, the portability of XRF spectrometers became feasible. Nowadays, two types of X-ray detectors have been developed and are used in portable EDXRF spectrometers, the PIN diodes [14] and the silicon drift detectors (SDD) [15] Concerning their analytical performance, they exhibit competitive energy resolution (158 eV for the PIN and about 140 eV for the SDD detectors respectively at MnK_α), and for the new generation PIN detectors a peak - to - valley ratio of about 6000:1 has been measured with a Fe-55 source and various monochromatic X-ray beams [16]. This characteristic produces a significant improvement of the peak - to - background ratio in the recorded XRF spectra and subsequently better accuracy in the determination of trace elements or of minor amounts of light elements. It should be also mentioned that the SDD energy resolution remains almost constant even for high input count rates up to 30 kHz, which however, is not a typical operating condition for archaeometric applications in the field. Another basic component for a portable XRF spectrometer is the type X-ray source. Miniaturized low power, air-cooled (<1 Watt – 100 Watt) X-ray tubes with different anode materials and different excitation geometries (side window or end-window design) are commercially available. It has been reported [17] that the available flux from these tubes can produce μ-beams at the sample position with satisfactory intensity if appropriate optical elements like polycapillary lenses are used for the focusing. Their compact design allows a more efficient usage of the available X-ray flux at the anode position. Nowadays, X-ray tubes of new technology increase the capabilities of portable XRF instrumentation for archaeometrical studies. For example, the potential to generate the electron emission by a laser beam that activates the cathode lowers the total operating power and allows for battery operation.

Various approaches are used for the modification of the primary tube excitation spectrum. It is well known that the production of a monochromatic and/or polarized excitation X-ray beam instead of a polychromatic one is preferred, not only for achieving low detection limits in EDXRF analysis, but also for extending the detection range to light elements like Na and Mg. These methods can also be applied in portable XRF spectrometers, but the necessary elements (crystals, scatterers, multilayers) should be carefully selected and optimized, since their incorporation in the XRF set-up produces a reduction of the available incident photon flux at the sample position. Another important reason is that for most archaeological artifacts, except for metal alloys, there is always a strong light matrix content (glasses, ceramics, pigments) which produces inherent difficulties for achieving good peak - to - background ratio in the whole spectrum. However, a simple technique to improve the quality of the exciting beam, especially for portable instruments, is the insertion of an appropriate filter between the source and the analyzed sample. The filter can significantly reduce the continuous tube radiation especially in the energy region up to 15 keV. In other portable spectrometers, a crystal or a multilayer is used between the anode and sample, having the property not only to reflect with high efficiency the main K or L emission characteristic X-ray lines of the anode material, but also to focus the exciting beam at the sample position.

For the *in-situ* XRF applications, it is essential for the analyst to have a visual inspection of the analyzed area in order to confirm certain details of the artifacts to be analyzed. Very often, two laser beams and in some cases a CCD camera are used for the inspection of the analysis spot. Beam dimensions of about few millimeters at the sample position are rather satisfactory for many XRF applications in the field of archaeometry. However, for the analysis of smaller details in ancient artifacts and objects of art a μ-beam or sub-millimeter beam is more suitable. For example, the possibility to analyze the bulk composition of corroded bronze objects was feasible by applying *in-situ* a micro-polishing technique and an X-ray beam diameter of 0.5 mm [6].

3. Experimental

3.1 APPARATUS

Three different portable XRF instruments were used in various non-destructive analyses of artifacts/materials with archaeological or historical value. All the set-ups included a peltier cooled PIN detector working at room temperature with a crystal thickness of 300 μm. Low power and air-cooled X-ray tubes, as well as radioisotopes were used for the production of the excitation X-ray beam. More specifically, the first system used (hereafter called as PXRF-A), consisted of a PIN detector with 240 eV resolution at MnK_α (Model XR-100T, Amptek Inc.), an Rh-anode side-window X-ray tube of 50 Watt power with 125 μm Be window (Model XTF5011, Oxford) and a MCA for data acquisition (Model MCA8000A, Amptek Inc.). The tube emission spectrum was modified by inserting a composite set of different filters. Two laser pointers were mounted in the PXRF-A set-up in such a way that the intersection point of their beams coincides with the cross-point of the incident X-ray beam and detector axis. The second system (PXRF-B) was composed by a transmission Au anode X-ray tube with power supply (Model PRS400 Photoelectron, Corp.), a PIN detector with 170 eV resolution at MnK_α (Model XR-100CR, Amptek Inc) and the same MCA8000A acquisition card. The X-ray detector and tube were mounted on a bracket to create a fixed and reproducible geometry. The third XRF set-up (PXRF-C1) included two point radioactive sources, a Cd-109 and an Am-241, a PIN X-ray detector exhibiting a resolution of 240 eV at 5.9 keV, (XR-100T, Amptek Inc) and the MCA8000, acquisition card by Amptek Inc. The X-ray source and the detector were mounted following a specially designed compact geometry that optimizes both excitation and detection conditions. The irradiated area was restricted by proper collimation to about 0.5 cm^2. Another version of the same instrumentation (PXRF-C2) was based on a different design and positioning of the various components. In this case, the irradiated area was about 7 mm^2.

3.2 ELEMENTAL SENSITIVITIES OF THE PORTABLE XRF INSTRUMENTS

Two combinations of filters were inserted to modify the Rh-anode emission spectrum in the PXRF-A system. The configuration F1 extends the exciting beam in the range between 14 keV up to almost the operational voltage, whereas the configuration F2 restricts the exciting spectrum in the energy range 15-24 keV (fig. 1). Therefore, filter configuration F2 is more suitable for the analysis of matrices with significant light elemental content, but exhibits very low sensitivity in the excitation of medium Z elements (i.e. silver, tin). On the other hand, configuration F1 offers a more extended analytical range and is more suitable for the analysis of ancient metal alloys. In PXRF-B spectrometer the tube radiation was used in the unfiltered mode, whereas the radioactive sources Cd-109 and Am-241 (PXRF-C1 and C2) are known to emit mono-energetic radiation. In the case of the Am-241 source the Np L-X-rays are filtered inherently by the source metal enclosure.

The use of X-ray detectors with small crystal thickness (nominal value 300 μm), a common characteristic of all the XRF portable instruments used in this work, plays a key role for their elemental response. There are some basic characteristic consequences. First of all, since their intrinsic efficiency decreases significantly as the energy of the detected photon increases over 15 keV, medium atomic number elements such as silver, tin, antimony have low detection efficiency. If the objectives of the analytical investigation depend crucially on the detection of trace amounts of these elements, then silicon detectors with larger crystals should be used or detectors with crystals composed of different elements (CdTe crystals). On the other hand, the same "negative" characteristic offers two compensating analytical advantages: the sensitivities of heavier elements, excited favorably by the X-ray source (absorption edge close to the energy of the exciting beam) are relatively reduced compared to those of elements with lower atomic numbers, due to their former mentioned lower detection efficiency. Therefore, on a qualitative basis, the energy X-ray spectrum and the measured X-ray intensities reflect more closely the sample elemental composition. Another significant advantage of small crystals is that they are more or less transparent for the scattered exciting radiation, thus reducing this background component that usually restricts the sensitivity and peak to background ratio in the EDXRF technique. Of course, most of the available detectors in the market are LN2 cooled and are superior concerning their analytical performance compared to the small peltier cooled detectors, but the above discussion shows that peltier cooled detectors exhibit remarkable properties and the progress in their design will improve their overall performance in the future.

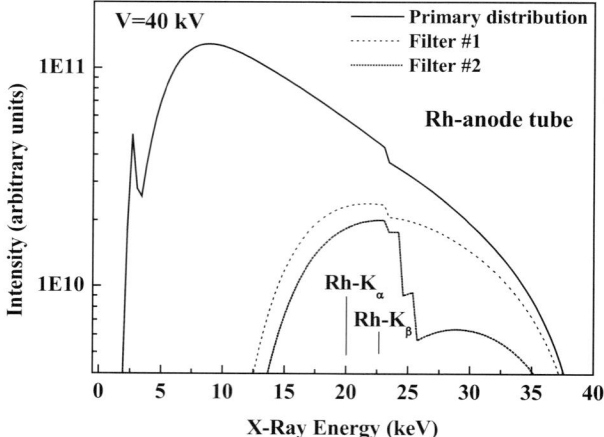

Fig I-2-1. Theoretical calculations of the Rh anode tube emission spectrum (bremsstrahlung), before and after the insertion of two different composite sets of materials in the beam path. (filter configurations #1 and #2).

In Fig. 2 the theoretical and experimental elemental sensitivities of PXRF-A spectrometer (filter configuration F1) are presented as a function of the atomic number of the excited element. The theoretical elemental sensitivities were determined by using a reliable semi-empirical model reported in the literature for the description of the tube continuous and characteristic X-ray radiation [18]. The elemental sensitivities express the intensity (in arbitrary units) of characteristic K_α or L_α radiation emitted per unit area density ($\mu g/cm^2$) of the analyzed pure element. The theoretical elemental sensitivities (solid line in Fig. 2), have been calculated using the tube emission model proposed by Ebel [18] and X-Ray fundamental parameters, whereas experimental sensitivities have been measured using thin or thick pure targets of the corresponding elements. Both the theoretical and experimental elemental sensitivities of Fig. 2 have been normalized with respect to the corresponding sensitivity of Fe. In Fig. 3 the theoretical elemental sensitivities of the PXRF-A set-up (filter configuration F2) are compared for the two operational voltages of 30 kV and 40 kV. It is evident that the operational voltage of 40 kV improves by an order of magnitude the excitation capability of medium atomic number elements. The calculated and measured elemental sensitivities of the PXRF-C portable spectrometers are presented in Fig. 4 for both geometric configurations C1 and C2, normalized with respect to the corresponding sensitivity of Cu. The analytical performance of both systems is suitable for the detection of elements with atomic numbers between Z=19-42 using the characteristic K_α-radiation and Z=73-92 with the L_α-radiation

It should be mentioned that the good agreement observed in Figs. 2 and 4 between theoretical calculations and experimental values is very promising and shows the reliability of the fundamental parameter approach in quantification procedures of XRF data obtained *in-situ*.

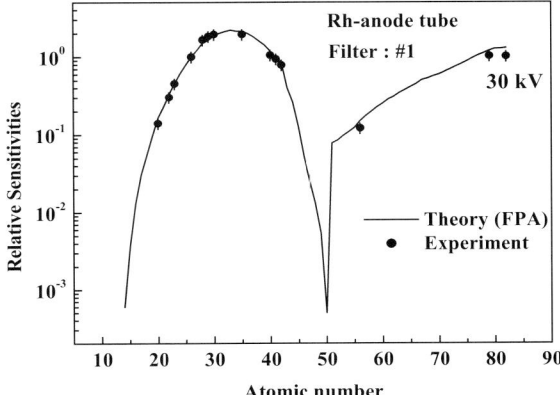

Fig. I-2-2. Theoretically calculated (line) and experimentally determined (symbols) elemental sensitivities for the PXRF-A spectrometer (Rh anode tube, PIN detector). The presented elemental sensitivities have been normalized with respect to the iron (Z=26) sensitivity.

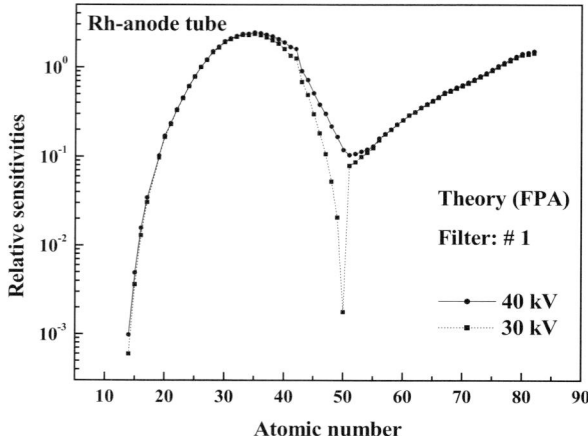

Fig. 1-2-3. Theoretically calculated elemental sensitivities for two operational voltages.

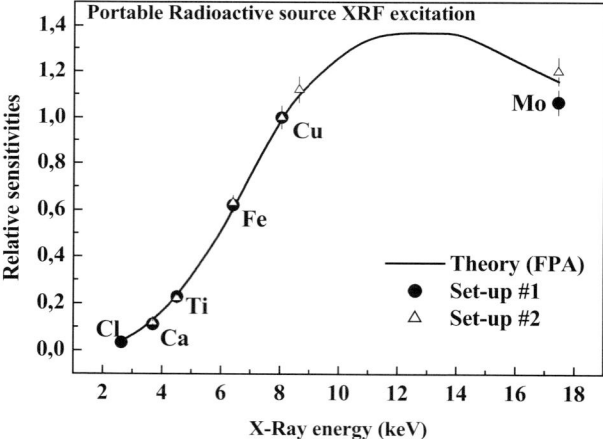

Fig. 1-2-4. Theoretically calculated (line) and experimentally determined (symbols) elemental sensitivities for the PXRF-C spectrometer at two geometrical configurations C1 and C2 (Cd-109 point radioactive, P_N detector). The presented elemental sensitivities have been normalized with respect to the copper (Z=29) sensitivity.

3.3   *IN-SITU* XRF INVESTIGATIONS

The PXRF-A system has been used so far for the *in-situ* characterization of pigments on Hellenistic funerary monuments (mostly wall-paintings) from Macedonia and for the measurement of the chemical composition of the gold jewelry collection (7th - 1st BC) located in Benaki museum, Athens. The PXRF-B was used by a group from Amptek Inc. and Photoelectron Corp. at the request and invitation of the Thera Foundation, in the analysis of various artifacts found at the archaeological site of Akrotiri in Thera, Greece. Among the analyzed artifacts, a very important archaeological find was included, a unique gold ibex, in addition to a bronze dagger and blade, pigments from the wall paintings of room 3 in Xeste 3 and other artifacts. The radioactive source based portable XRF set-up (PXRF-C1) was used for the patina examination and measurement of the alloy composition of an outdoor bronze monument located in the center of municipality of Athens. The PXRF-C2 was used in an *in-situ* survey of 75 ceramics of the Nicosia Museum Collection, Cyprus.

3.3.1 *IN-SITU EDXRF CHARACTERIZATION OF WALL-PAINTING PIGMENTS*

Wall-painting pigments represent an archaeological material that requires *in-situ* non-destructive characterization. Due to the uniqueness and integrity of the representations executed on wall paintings, the sampling is restricted. Therefore, in order to examine the exact gamut of all the pigments used, many locations have to be analyzed fast, and under touch-free conditions. The portable EDXRF technique can contribute significantly towards a complete characterization that needs, however, the use of complementary analytical techniques. A question may arise as to which extent the portable EDXRF method can contribute to the characterization of wall-painting pigments. First of all, it should be stressed that EDXRF is an elemental technique and thus cannot determine directly, neither qualitatively nor in a quantitative basis, the mineral or compound composition of a pigment. On the other hand, some elements or combinations of specific elements can be considered as fingerprints of the presence of a specific mineral-pigment. For example, the detection of mercury indicates the presence of cinnabar red pigment, copper and silicon the presence of the Egyptian blue pigment, barium and sulphur the presence of barite, iron of natural ochre etc. Regarding the quantification procedures, EDXRF faces some analytical problems. Experimental parameters, such as the pigment density, the mixing ratio of the pigment with the binder, and the pigment layer thickness are unknown. The analytical problem is illustrated clearly in Figs. 5 and 6. The analytical yield (intensity of PbL$_\alpha$ radiation for lead white pigment, CuK$_\alpha$ for Egyptian blue and FeK$_\alpha$ for a yellow ochre) exhibits a strong dependence on the pigment thickness and the mixing ratio with the binder. Having only one measured parameter for each case, the quantification problem looks very difficult. However, there are still some possibilities, for example the estimation in a semi-quantitative basis of the weight ratios that two or more pigments are mixed together to produce a composite color.

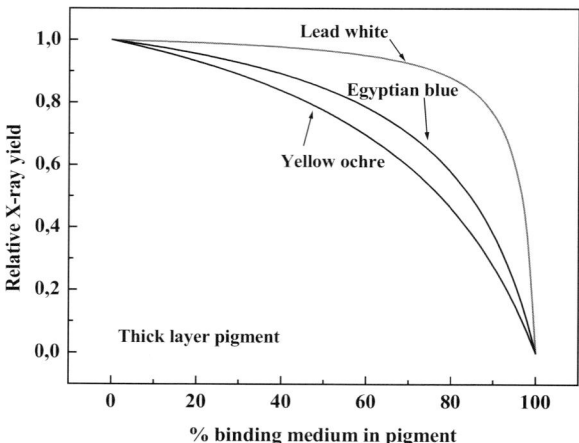

F g. I-2-5. Calculations of various pigments X-ray yield (intensity of PbL$_\alpha$ radiation for lead white pigment, CuK$_\alpha$ for Egyptian blue and FeK$_\alpha$ for a yellow ochre) as a function of the binding medium (cellulose) mixing ratio with the pigment. The pigment layer has been assumed to have infinite thickness and the X-Ray yield has been normalized with respect to the pure pigment yield.

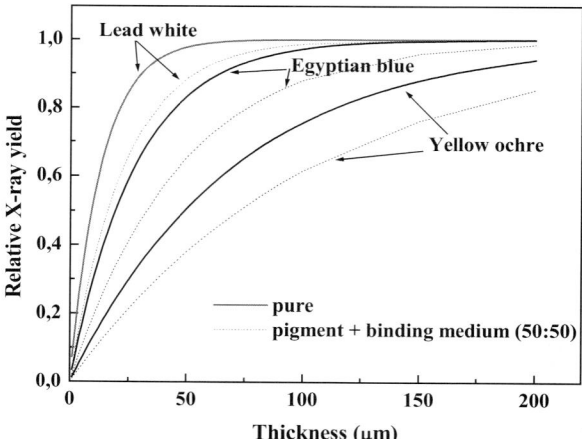

Fig. I-2-6. Calculations of various pigments X-ray yield (intensity of PbL$_\alpha$ radiation for lead white p gment, CuK$_\alpha$ for Egyptian blue and FeK$_\alpha$ for a yellow ochre) as a function of the pigment layer thickness for pure pigments (solid lines) and 50% mixing ratio of the binding medium (cellulose) with the pigments (dot lines). The X-Ray yield has been normalized with respect to the pigment yield of an infinite layer.

Examples of the application of the portable PXRF-A and PXRF-B spectrometers in the characterization of wall-painting pigments from Macedonian Funerary monuments and the Akrotiri archaeological site in Thera demonstrate some of these issues and solutions.

The funerary paintings discovered in Macedonia during the last thirty years have considerably enriched our knowledge on ancient Greek painting during the late Classical and the early Hellenistic periods. These paintings are executed on the walls of monumental tombs and often decorate their sumptuous furniture (couches and thrones), the interior of smaller graves and the surfaces of less impressive gravestones. Scientific analysis carried out on a representative number of samples taken from Macedonian funerary monuments by means of optical microscopy, XRD and SEM-EDS [19, 20] has revealed the use of a range of pigments, either deriving from naturally occurring mineral sources, or artificially made. Organic pigments such as blacks, pinks and violet lakes were also used. More specifically, the gamut of pigments used more frequently is mainly composed of calcium carbonate white, carbon black, natural ochre's and Egyptian blue, while the range of less frequent pigments includes lead white, kaolinite, bone white, cinnabar, realgar, malachite, serpentinite, conichalcite and gold leaf.

The main objective of the portable EDXRF analysis on specific Macedonian monuments (Macedonian tomb at Aghios Athanassios, funerary painted couch from the tomb of Potidaia and "Tomb of Persephone" at Vergina-Aigai) was the creation of a record of the materials applied on the whole surface of the paintings, taking advantage of the possibilities for a non destructive analysis of an unlimited number of areas. Some XRF spectra are shown in Figs. 7 and 8, measured from the Aghios Athanassios tomb after a preliminary analytical survey. In Fig. 7 the EDXRF spectrum from an Egyptian blue pigment is shown. The measurement was performed at two operational conditions. Using 30 kV, calcium and copper are identified as major elements, whereas iron and strontium are contained in trace amounts. With the low high-voltage condition (4.2 kV), silicon can be measured together with calcium proving evidence for the presence of the artificial pigment Egyptian blue ($CuCaSi_4O_{10}$). A broad use of this pigment has been attested in Macedonian paintings, in a variety of hues ranging from the darker to the lighter ones obtained either from grinding, or from the application of Egyptian blue on an undercoat of carbon black [20]. In Figs. (8a) and (8b) the EDXRF spectra from a green and a pale-green pigment are presented. Operating the tube at 30 kV (fig. 8b), we can observe that iron and copper are the major elements in both pigments with the difference that in the pale-green pigment mercury was also detected as a minor element. At the low high voltage condition (4.2 kV) silicon was detected in both spectra. Therefore, it can be assumed that for the production of the green color, a yellow ochre and Egyptian blue pigments were mixed. This is in accordance with previous investigations [19] according to which green colors with restricted use on Macedonian figural paintings, occurred mostly either from mixtures of yellow ochre and carbon black for a warmer hue, or from mixtures of yellow ochre and Egyptian blue for colder hues.

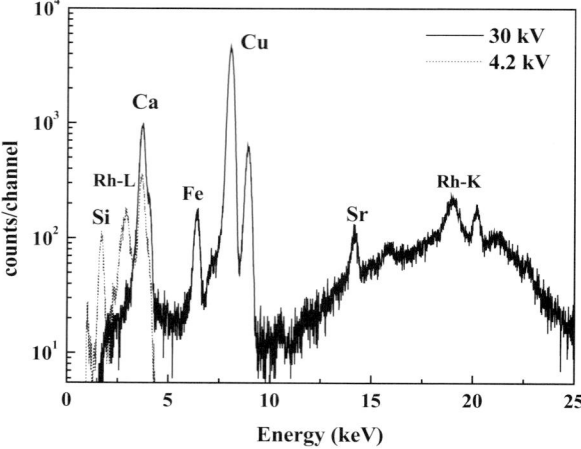

Fig I-2-7. Energy dispersive XRF spectra obtained from the analysis of a blue wall-painting pigment of Agh_os Athanassions funeral monument. The spectra were recorded using PXRF-A portable spectrometer at 30 kV (solid line) and 4.2 kV (dot line) operational voltages.

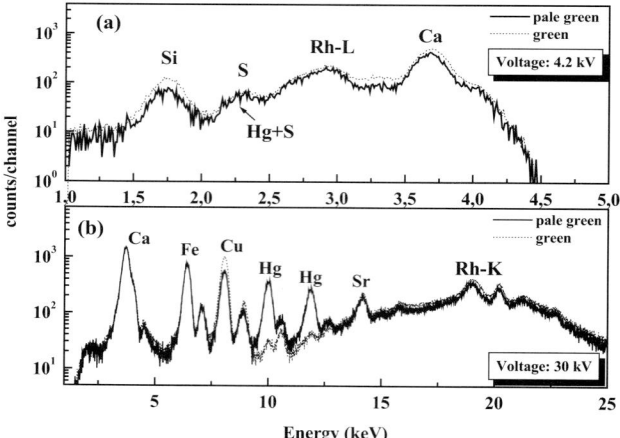

Fig. I-2-8. Energy dispersive XRF spectra obtained from the analysis of green and pale-green wall-painting pigments of Aghios Athanassions funeral monument. The spectra were recorded using PXRF-A portable spectrometer at 30 kV (solid line) and 4.2 kV (dot line) operational voltages.

The wall paintings of Thera due to their unique preservation by Aegean standards are the most impressive finds from the site, and because of their thematic range the most loquacious in terms of the information they yield [21]. The blue Theran pigments were classified first by Phillipakis *et al.* [22], and the red, pink, yellow, orange and brown pigments by Profi *et al.* [23]. Perdikatsis [24] and more recently Perdikatsis *et al.* [25] conducted a more complete approach based on XRD, SEM-EDAX, and optical microscopy examination. The aim of the *in-situ* XRF measurements of Thera wall paintings pigments [26] was to identify fingerprint elements for specific mineral pigments and to assess the applicability of the portable PXRF-B spectrometer.

The XRF analyses of the white background at various locations of the Xeste 3 Room 3 showed high Ca concentration (calcite), with minor amount of Fe. Furthermore, the XRF analyses of a yellow and red pigment preserved inside a pot were found calcite free. This fact supports the argument that since secondary calcite was always detected in the pigments, the ochre's was ground before use to a fine powder and then mixed with lime in variable proportions according to the intensity of the required color. Fig. 9 shows the analysis of a yellow (mustard) pigment and the corresponding XRF spectrum in Fig. 10 proves the presence of a Fe based yellow pigment (goethite) mixed with Ca (calcite), whereas the presence of minor amounts of potassium may indicate the presence of the mineral illite.

An analytical table of results from the *in-situ* XRF analysis of Thera wall paintings is presented in table 1 [26]. For comparison, representative analytical data from XRD/SEM microprobe analyses are shown in the last column, obtained at the locations indicated by the acronym AKR in [25]. The comparison shows clearly the strengths and weaknesses of the technique concerning the identification of pigments. However, it should be stressed that in many cases the *in-situ* XRF data may support interesting archaeological arguments. For example, the fact that Fe can be used as a fingerprint element for the presence of amphiboles in a blue pigment and Cu for Egyptian blue was used initially [22] for grouping the different types of blue pigments found in Thera, Knossos, and mainland Greece. From these data, it has been suggested that the Theran artists introduced the amphibole group minerals to the color palette of Minoan wall paintings [24, 27].

Fig. 1-2-9. In-situ EDXRF analysis at the Akrotiri archaeological site in Thera of a yellow (mustard) wall-painting pigment using PXRF-B portable spectrometer.
(See also Color plate, p. 298)

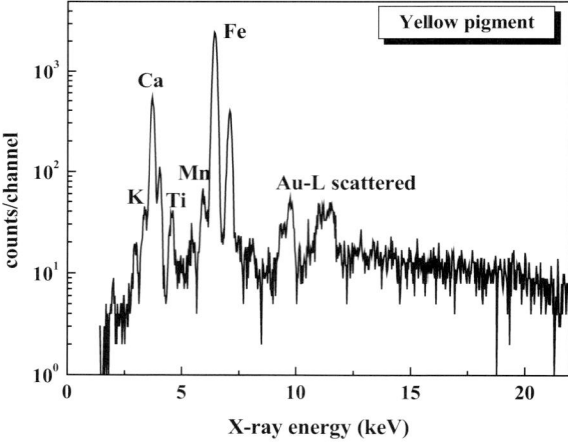

Fig. 1-2-10. Energy dispersive XRF spectrum obtained from the analysis of the yellow (mustard) wall-painting pigment using PXRF-B portable spectrometer.

Table. I-2-1. Results from various Theran wall-painting pigments obtained by performing in-situ XRF analysis (T. Pantazis *et al*). For comparison, representative analytical data from XRD/SEM microprobe analyses are shown in the last column, obtained at the locations indicated by the acronym AKR in V. Perdikatsis et al., 2000.

Pigment	Positions analysed *in-situ*	Analytical data	
		In-situ analysis [26]	XRD, SEM/microprobe analysis [25]
		Elements	Minerals
White Background	Xeste 3, Room 3: Reedbed, Lady with the lilies, Lady with the belt	Ca (Calcite) with minor amounts of Fe (Amphibole)	Calcite, Kaolinite, Chlorite, Amphibole (AKR4)
Blue	Xeste 3, Room 3: Blue pigment on the Lady's belt, Shaven head of the youngest boy	Ca, Cu (Egyptian blue) and Fe (Amphibole)	Egyptian blue, Amphibole (mostly of riebeckite type), Talc, Chlorite, Calcite (AKR47)
Red	From inside a pot	Fe (Hematite) with traces of Mn and Ti	Hematite (AKR31)
Yellow	From inside a pot	Fe (Goethite) with traces of Mn and Ti	-
Yellow (mustard)	Xeste 3: Boy with the yellow base	Fe (Goethite) and Ca (Calcite) as major elements, minor amounts of K (Illite?) and trace amounts of Ti and Mn	Goethite, Calcite, Quartz, Chlorite, Illite, Kaolinite, Amphibole and Talc (AKR16, AKR17)
Brown	Xeste 3: Brown pigment from the arm of the boy	Fe (Hematite) and Ca (Calcite) and trace amounts of Ti and Mn	Hematite, Calcite, Kaolinite, Quartz (AKR27)
Green	Lady with bunch of lilies	Fe (Goethite), minor amounts of K (Illite?) and trace amounts of Ti, Mn, Ca and Cu (Egyptian blue)	Amphiboles (hornblende) as colorant, possibly painted on top of yellow ochre (AKR50)

3.3.2 *IN-SITU EDXRF CHARACTERIZATION OF ANCIENT GOLD ALLOYS*

In this section, results of the compositional analysis of gold alloys using portable EDXRF spectrometers will be presented. The PXRF-B portable spectrometer was used for the analysis of the gold ibex artifact found at the Akrotiri archaeological site in Thera, whereas PXRF-A portable spectrometer was used for the analysis of the gold jewels of the Benaki museum in Athens.

For the quantification of XRF analytical data, the fundamental parameter approach (FPA) is a useful tool for obtaining the composition of homogeneous materials. When a tube is used as an X-ray source in an unfiltered or filtered direct excitation mode, the knowledge of the tube spectrum distribution, not in absolute terms, but at least in relative intensity units is essential. For the PXRF-B spectrometer composed by a transmission Au anode X-ray tube with power supply (Model PRS400 Photoelectron, Corp.) a new semi-empirical model was developed for the description of the exciting beam. The parameters of this model were determined by fitting experimental tube spectra measured at 30 and 40 kV by a CdTe detector (Amptek, Inc) placed inside a collimator kit (Amptek Inc.) and using two diaphragms having diameters of 400 and 200 microns respectively. The tube spectrum was also measured at 30 kV with a Si-PIN detector having a 300 μm crystal thickness and a conventional entrance collimator.

Having the knowledge of the initial tube spectrum distribution, a FPA code was developed to predict measured X-ray intensities or inversely, to determine the composition of the analyzed sample from measured X-ray intensities. The reliability of the FPA code was checked by comparing the theoretically calculated and experimentally determined X-ray intensities from standard samples (gold alloys or pure elements). The agreement was very satisfactory (within 5-10%).

Among the various artifacts that were analyzed at the request and invitation of the Thera Foundation by a group from Amptek Inc. and Photoelectron Corp. was a unique gold ibex [26]. According to Chr. Doumas the gold ibex was perhaps the most intriguing object analyzed in that it is the only artifact made of precious metals found at Akrotiri. The inhabitants, before the eruption that devastated the island and buried the settlement beneath meters of volcanic ash, removed all other precious objects from the site. Found during the construction of the pillars for the new shelter in December of 1999, the ibex was in perfect condition. The fact that the ibex was found *in-situ* makes its importance even greater, since this object is stylistically unique in the Aegean.

The XRF analysis of the ibex was performed by the PXRF-B spectrometer (fig. 11). From the XRF spectra obtained from a clean metal and welding area (fig. 12) it can be observed that at the welding area the Cu signal has been increased significantly relatively to the Au signal.

Fig. I-2-11. In-situ EDXRF analysis at the Akrotiri archaeological site in Thera of the gold ibex figurine using PXRF-B portable spectrometer.

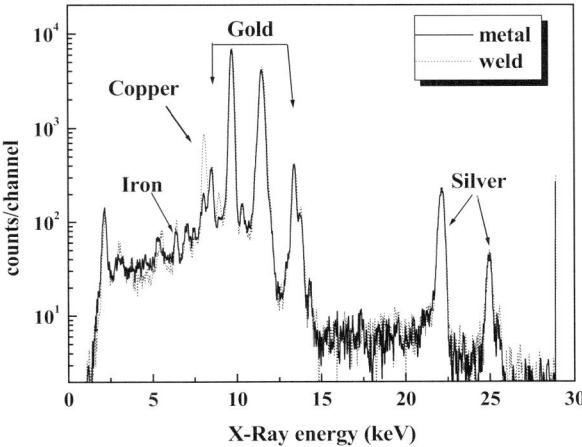

Fig. I-2-12. Energy dispersive XRF spectra obtained from the analysis of the gold ibex at a metal area (solid line) and a welding one (dot line).

The average compositional results from pure metal locations and welding areas are shown in table 2. It is important to note that Cu can generally be detected in native gold up to levels of 2.5%. Typically, it is present in quantities less than 1% [28]. Data obtained from gold grains/nuggets, as well as from objects made possibly by native ore, indicate that Cu values rarely exceed 1%. All of the ibex spectra from the non-weld locations average 0.55% and this support the argument that the artifact was made from native gold.

Table. I-2-2. Average chemical composition of the ibex at the metal area and at the weldings.

Location	u (%)	Ag (%)	Cu (%)	Fe (%)
metal	84.2	15.0	0.55	0.34
welding	82.5	14.9	2.53	0.34

In table 3, data from literature concerning the composition of ancient gold alloys are presented. For an appropriate comparison with the ibex data, the criterion that a Cu content less than 1.5% implies native provenance of the gold was adopted. Therefore, in Table 2 only those data obtained from Maniatis *et al* [29] and A. Hartmann [30] that satisfy this criterion were included and averaged. These data correspond in the first case to 23 of a total 27 alloys, and 40 of 61 for the second. The average Ag concentration of 14.9% of the ibex figurine from Akrotiri is a value that brings it chemically close to the gold artifacts from the Aegean and Balkan regions in general. This is of particular importance since typologically the gold ibex does not seem to have parallels in the Aegean.

Table. I-2-3. Summarizing the Ag and Cu content of ancient gold artifacts artifacts from various sites.

Site	Period	Source	Number	Ag (%)	Cu (%)
Varna	Neolithic	A. Hartman [25]	125	11.0 ± 2.6	0.50 ± 0.37
Confiscated Hoard	Late Neolithic	Y. Maniatis *et al.,* [29]	23	6.61 ± 2.66	0.36 ± 0.32
Mycenae	Mycenaean	A. Hartman [25]	40	16.0 ± 7.5	0.51 ± 0.34
Ibex	Thera	--	--	14.9	0.55

Many welding processes have been reported in literature [31]. Since, in our case, an increased Cu concentration was measured at the welding areas and no statistically significant difference in the Ag/Au concentration ratio was observed, the welding was most likely done using a copper mineral (chrysocolla or malachite) and the solid-state diffusion bonding process. This welding procedure begins by heating, but not melting, the high purity gold alloy in the presence of a copper powder. This forms an Au-Cu alloy with a melting point considerably lower than the gold itself, which fills the gap between the two components [32].

Within the framework of the project "jewelmed" - ICA3-1999-10020 and of identifying the techniques and composition for jewelry from the 7th to the 1st century BC, the metals conservation department of the Benaki Museum, Athens selected 40 items from the museum's collection, in accordance with criteria regarding the metal from which they were manufactured, their typological characteristics, their use, the technique and naturally the period in which they were made.

The analysis of the gold jewels was performed using the spectrometer PXRF-A *in-situ* at the Benaki Museum during a two days survey. The quantitative jewel's composition was based on a semi-empirical model, predicted and formulated by using the results of the fundamental parameter approach, when it is applied on the problem of the bulk analysis of gold alloys. In order to improve the accuracy in the determination of the gold alloys composition, the parameters of the semi-empirical model were calculated after fitting of the measured spectra from a series of gold standard samples. The compositional results of the gold-jewels are presented in a ternary diagram (fig. 13). Although the evaluation of the data is still in progress, we may estimate that for the majority of the museum gold artifacts two manufacturing techniques were employed: in the first, unrefined gold and most likely native gold taken from placer deposits was used, while in the second technique a refining process was undertaken. There is a great possibility that this second technique was applied for most of the jewels, in order to separate, at first the precious metals from the base metal and in a following stage the gold from the silver.

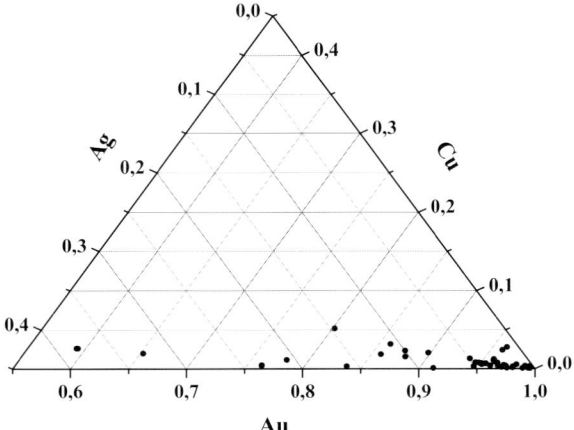

Fig. I-2-13. Ternary diagram presenting in absolute concentration units the composition of 40 gold jewels from the collection of the Benaki museum in Athens (7th – 1st BC).

3.2.3 *IN-SITU EDXRF INVESTIGATION OF CERAMICS DECORATION TECHNIQUES AT NICOSIA MUSEUM, CYPRUS*

Ceramic artifacts provide the ideal material against which cultural interactions can be studied since they contain multidimensional information with respect to the shape, the style of decoration (incised, painted, plastic), the fabric, the raw materials used, the manufacturing techniques etc [33]. The investigation of ancient ceramic technology, which was usually based on the analysis of the ceramic body in the past, can be complemented through the analysis of the pigments used for surface decoration [34]. As it has been demonstrated by E. Aloupi *et al* [33], emphasis must be given to the technique applied for the production of the dark colour (black or brown) whether in slip or painted decoration, i.e. the 'iron reduction', the 'manganese black' techniques each of which presents its special requirements in terms of raw materials and firing conditions (i.e. combination of temperature and atmosphere): a) As is well known, iron rich clays with low calcium oxide and relatively high potassium oxide content, produce dark coloured pigments when fired in a reducing atmosphere (the iron reduction technique). Although the outcome of the application of the technique is uncertain since the final colour depends on several factors, the result is both attractive and functional since in such cases it shows high contrast with the body, fine details and a gloss finish which seals the ceramic surface. b) The manganese black technique is much more reliable, in that it invariably produces a dull matt black result independently of the firing process. It also allows the production of a bichrome effect easily by Mn-ores (umbrae) with varying Mn_3O_4 (2.5-15%) and Fe_2O_3 (20-65%) contents for the black or brown and Fe-oxides for the red. It should be clear from the above that the choice of the technique reflects a rich interplay between cultural, utilitarian and functional considerations. This is particularly true because in both cases the desired result is the same i.e. the production of black painted decoration or surface coating. As has been stressed by E. Aloupi *et al* [33] it is rather surprising that up to date the potential wealth of information that can emerge from this very simple, almost elementary, dichotomy has escaped the attention of the archaeological community. The identification of the nature of the black pigment relies on a very simple non-destructive analytical procedure by means of portable X-ray fluorescence systems. We would even assert that in most cases a simple microscopic examination could provide a decisive test to the experienced eye.

The ceramics in the Cyprus Museum in Nicosia provided a complete and comprehensive archaeological collection for the study, which spans more than 40 centuries from Neolithic to Hellenistic times (5000-325 BC). Due to the nature and wealth of the material, PXRF-C1 portable spectrometer was used for an *in-situ* survey [27, 35]. A measuring time of about 500 sec was enough for identifying the major and minor elements (K, Ca, Ti, Mn and Fe) in ceramics black or dark decoration layer, as well as some traces elements (i.e. Cu, Zn, Rb, Sr, Zr and Pb) mostly common in earthen materials. Typical XRF spectra of a Fe-rich and Mn-rich black pigments on Cypriot ceramics from the Nicosia Museum are shown in Fig. 14. The EDXRF analysis of 75 ceramic artifacts revealed a very clear chronological pattern with respect to the nature of the ubiquitous black or dark colour. Essentially, all dark decorations in Cypriot pottery from the Neolithic to the Middle Bronze Age

(5000-1625 BC) are based on the use of iron rich materials. From the end of the Late Bronze Age onwards (1050-325 BC), the dark colours were achieved using of Mn-ores (umbrae). The results showed that the transition between the two dark colour techniques occurred during the Late Bronze Age on the so-called White Slip (WSI, WSII) and Bichrome wheel made wares. Recent detailed analyses [35, 36] on well-documented sequences of the above characteristic Cypriot pottery revealed that the change from Fe-black, in WSI, to Mn-black, in WSII monochrome ware, was introduced through the bichrome WSI and the contemporary bichrome wheel made [36] wares in order to facilitate the simultaneous production of red and black on the same object. Although the ancient craftsmen were able to produce black and red, separately, using iron based pigments, when called upon to produce a bichrome effect they found it more convenient to use Mn for the black. This is understandable if we consider the difficulties of the fine-tuning between firing atmosphere and temperature required to produce a bichrome effect based on Fe- based materials only [37].

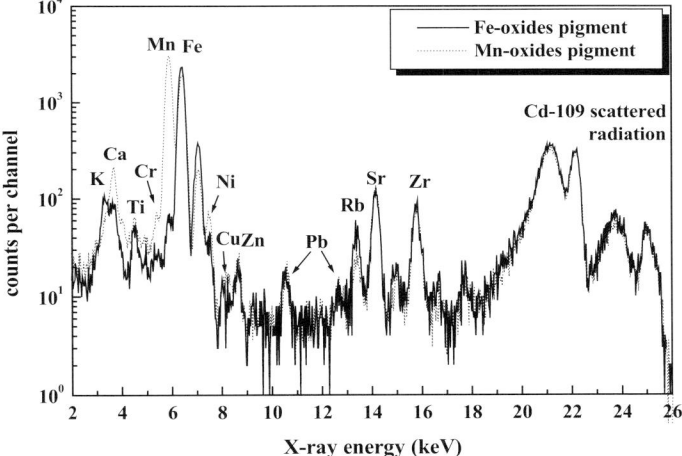

Fig. I-2-14. Typical XRF spectra of Fe-rich and Mn-rich black pigments on Cypriot ceramics from the Nicosia Museum. The spectra were obtained *in situ* at the Nicosia Museum with the PXRF-C1 portable spectrometer, consisting of a Peltier cooled X-ray detector and a point Cd-109 radioactive source.

3.3.4 *IN-SITU EDXRF CHARACTERIZATION OF AN OUTDOOR BRONZE MONUMENT*

The outdoor bronze monument of Theodoros Kolokotronis (fig. 15) created by the famous Greek sculptor Lazaros Sochos, was placed in Athens in 1904. It represents the war hero on his horse pointing the way to victory during the Greek war of independence in 1821. This monument is one of the most popular in Greece, and another copy exists in the city of Nauplion. The aim and objectives of this project was to document and record the condition and the construction of the monument, so that the best strategy could be applied for its conservation. The resulting scientific data could then be grafted into a monitoring program for outdoor monuments in Athens to help protect them long-term.

Fig. 1-2-15. Photo of Th. Kolokotronis outdoors bronze monument located in the center of Athens.

A team of conservators, scientists, historical researcher, photographer and cameramen worked on the documentation and treatment of Kolokotronis monument. *In-situ* visual examination, schematic diagrams, photographs and video were used to record the condition of the monument, before, during and after the treatment. Also scientific analyses using portable X-Ray Fluorescence (XRF), X-Ray diffraction (XRD), scanning electron microscope (SEM) and color chroma measurements were employed [38].

Concerning the role of EDXRF in the surface analysis of corroded objects, it should be stated that such an analysis is inherently unreliable since different elements can be enriched or depleted by the corrosion process and the degree and direction of such compositional changes depend on many variables. EDXRF can be used to identify the nature of the alloy, for example, if it is a tin bronze or leaded tin bronze, as well as corrosion by-products supporting the conservation treatment. Weight ratios between metal elements measured at patina and outer corroded surface layers can be compared with those from a cleaned metal area of the same artifact (the light matrix content of corrosion layers does not affect significantly such weight ratios especially for adjacent metals like Cu, Zn), in order to extract in a semi-quantitative basis information about the mechanisms involved.

The PXRF-C2 portable spectrometer was used to analyze pre-cleaned small areas of the bronze sculpture, as well as areas with patina in place (fig. 16). Eighteen (18) areas on the bronze sculpture were analyzed and typical measuring times were 10-15 minutes. For some areas the measurements were conducted before and after the conservation treatment. Table 4 presents the results of the XRF alloy analysis on pre-cleaned small areas. The value for Pb denotes the corresponding limit of detection of the method for a 10 minute measuring time. In Table 4, the analysis in two separate regions of the cast lower body of the monument (positions #1 and #2) gave similar alloy composition. In addition, the helmet and upper body respectively (positions #3 and # 4), were cast as separate pieces, and are more similar in composition than the lower body. Tin was used to increase the hardness of the metal and the quality of polishing the material, while Zn to improve the pourability.

Except the alloy elements, other metals, such as Fe and Pb, were also identified in many of the patina or green corrosion areas. However, it should be noted that these elements were not detected after the conservation treatment on the uncleaned surfaces. This leads us to believe that these elements most probably are coming from the surrounding environment. The Zn/Cu ratios obtained at positions with corrosion products reveal Zn enrichment in the patinated surface. This is in agreement with the study by Strandberg et al. [39] where a bronze sample high in Zn content was exposed to humid air containing SO_2 and NO_2. After the conservation treatment the Zn/Cu ratios showed a significant decrease indicating that the protective patina layer rich in copper oxides has been revealed above the metal surface. The XRF data showed also that the patinated surface of the metal was richer in tin (most probably in the form of SnO_2) than the base metal, indicating that due to corrosion the metal has become enriched in Sn.

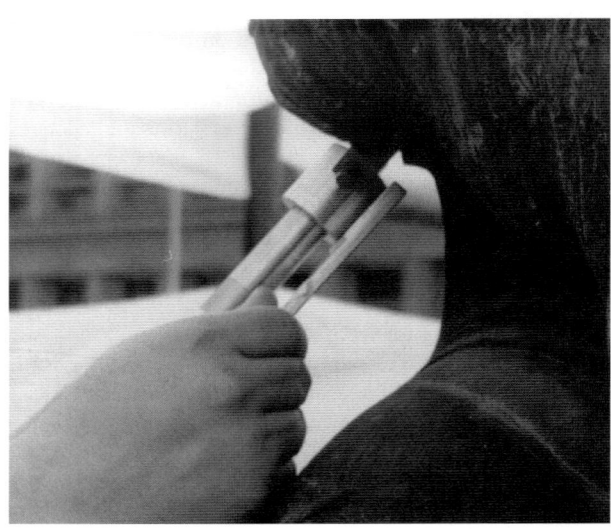

Fig. I-2-16. In-situ examination of the patina surface of the outdoors bronze monument Th Kolokotronis using PXRF-C2 portable spectrometer.

Table. I-2-4. Chemical composition of the outdoors bronze monument of Th. Kolokotronis at various pre-cleaned areas.

A/A	Position	Cu (%)	Zn (%)	Sn (%)	Pb (%)
1	Kolokotronis, left stir-up, heel	89.5	5.0	5.5	< 0.7
2	Horse, right front leg	90.0	4.9	5.1	< 0.7
3	Kolokotronis, helmet next to epigraphy	90.4	3.4	6.2	< 0.7
4	Kolokotronis, upper body	91.2	3.2	5.6	< 0.7

Conclusions

The application of portable EDXRF spectroscopy in the analysis of ancient or historical value artifacts, demonstrates the potential of the method. Even when complementary analytical techniques are required towards a complete characterization, the interpretation of the *in-situ* EDXRF data can provide insight to archaeological posed questions and can motivate the establishment of new approaches in the archaeological research.

Through the examples presented, it is evident that in order to maximize the advantages of portable EDXRF analysis, multidisciplinary collaboration teams are necessary. Natural scientists (i.e. expert analysts, geologists, archaeometrists etc), archaeologists and conservators should collaborate from the initial questioning and data acquisition, until the formulation of the conclusions. Over the few last years, this trend has been strengthened in Europe through E.U. actions such as COST-G8 (http://srs.dl.ac.uk/arch/cost-g8/), which was dedicated to "Non-destructive Analysis and Testing of Museum Objects". It is thus very encouraging, that for cultural heritage issues, a new basis of scientific communication and data management has already been achieved, allowing exchange of knowledge amongst different experts.

Acknowledgements

X. B. and A.G K. would like to thank M. Tsimbidou-Avloniti, A. Kottaridou and K. Sismanidis for their help during the in-situ XRF survey of Macedonian wall-painting pigments. This project was supported by a Short Term Scientific Mission in the frame of the COST-G8 action.

A.G. K. would like to thank Mr. P. Nomikos from Thera Foundation, Prof. Ch. Doumas, Dr. A. Vlachopoulos, Amptek Inc. and Photoelectron Corp. for the analyses that took place in the Archaeological site of Akrotiri, Thera.

The non-destructive analysis of the ceramic collection at the Nicosia Museum was part of a research project directed by THETIS Authentics, funded by the A.G. Leventis Foundation (1996-2000) and a Greek-Cypriot Collaboration project (GSRT, Ministry of Development 1996-1998).

The analyses of the golden jewelry at the Benaki Museum were performed in the course of the EU project- JEWLMED (ICA3-1999-10020).

The analysis of the Kolokotroni's outdoor bronze statue was part of a conservation project, directed by the Technological Educational Institution of Athens, funded by the E.U. Collaboration Programme Culture 2000 and the Municipality of Athens.

References

1) *X-Ray Spectrometry*, Vol. **29** (2000).

2) M. Mantler and M. Schreiner, I **29** (2000) 3

3   C. Neelmejer, I. Brissaud, T. Calligaro, G. Demortier, A. Hautojarvi, M. Mader, L. Martinot, M. Schreiner, T. Tuurnala and G. Weber, *X-Ray Spectrometry* **29** (2000) 101

4   R. Cesareo, G.E. Gigante, A. Castellano, M.A. Rosales, M. Aliphat, F. DeLa Fuente, J.J. Meitin, A. Mendoza, J.S. Iwanczyk and J.A. Pantazis, *J. Trace and Microprobe Techniques*, **14** (1996) 711

5   Caruso, R. Cesareo, C. Giardino, G.E. Gigante, *Supplement de La Revue d'Archeomètrie* (1996) 157

6   G. Vittiglio, K. Janssens, B. Vekemans, F. Adams, A. *Oost, Spectrochimica Acta*, B **54** (1999) 1697

7   S. Sciuti, G. Fronterotta, M. Vendittelli, A. Longoni and C. Fiorini, *Studies in Conservation*, **46** (2001) 132

8   M. Uda, S. Sassa, K. Tanigushi, S. Nomura, S. Yoshimura, J. Kondo, N. Iskander, B. Zaglhloul, *Naturwissenschaften* **87** (2000) 260

9   M. Uda, M. Nakamura, S. Yoshimura, J. Kondo, S. Sassa, M. Saita, Y. Shirai, S. Hasegawa, Y. Baba, K. Ikeda, Y. Ban, A. Matsuo, M. Tamada, H. Sunaga, H. Oshio, D. Yamashita, Y. Nakajima, T. Utaka, *Nucl. Instr. Meth.* B **189** (2002) 382

10) K. Janssens, G. Vittiglio, I. Deraedt, A. Aerts, B. Vekemans, L. Vincze, F. Wei, I. Deryck, O. Schalm, F. Adams, A. Rindby, A. Knochel, A. Simionovici and A. Snigirev, *X-Ray Spectrometry* **29** (2000) 73

1.) R. Cesareo, G. E. Gigante, A. Castellano, *Nucl. Instr. Meth.* A**428** (1999) 171

12) P. Leutenegger, A. Longoni, C. Fiorini, L. Struder, J. Kemmer, P. Lechner, S. Sciuti,   R. Cesareo, *Nucl. Instr. Meth.* A, **439** (2000) 458

13) J. A. Huber, J. A. Pantazis, V. Jordanov, *Nucl. Instr. and Meth.*   B**99** (1995) 665

14) R. Redus, J. Pantazis, A. Huber, T. Pantazis, *IEEE Trans. Nucl. Sci.*, in press

15) L. Struder, N. Meidinger, D. Stotter, J. Kemmer, P. Lechner, P. Leutenegger, H. Soltau, F. Eggert, M. Rohde, and T. Schulein, *Microsc. Microanal.* 4 (1999) 622

16) A. G. Karydas, Ch. Zarkadas, A. Kyriakis, J.Pantazis, A. Huber, R. Redus C. Potiriadis, Th. Paradellis, *X-Ray Spectrometry*, in press

17) M. A. Kumakhov, *X-Ray Spectrometry*, **29** (2000) 343

18) H. Ebel, *X-ray Spectrometry*, **28** (1999) 255

19) H. Brecoulaki, V. Perdikatsis, "Ancient Painting on Macedonian Funerary Monuments, IV-III centuries BC : A comparative study on the use of color", in *Color in Ancient Greece*, International Symposium, The J. Paul Getty Museum and the Aristotelian University of Thessaloniki, 12-16 April 2000, Thessaloniki, in press.

20) H. Brecoulaki , *"Sur la techné de la peinture grecque ancienne d'après les monuments funéraires de Macédoine"*, *Bulletin de Correspondance Hellénique* **124**, 2000, p. 189-216.

2.) Chr. Doumas, *Proceedings of the first International symposium on The Wall Paintings of Thera*, Vol. **I**, Athens 2000, p. 15-20

22) Philippakis, V. Perdikatsis, Th. Paradellis, *Studies in Conservation* **21** (1976) 143.

23) S. Profi, V. Perdikatsis, S. Philippakis, *Studies in Conservation* **22** (1977) 107.

24) V. Perdikatsis, *La couleur Dans Peinture L'Emaillage De L'Egypte Ancienne*, Centro Universitario Europeo, Bari, 1998 p. 103-108.

25) V. Perdikatsis, V. Kilikoglou, S. Sotiropoulou, E. Chryssikopoulou, Proceedings of the first International symposium on The Wall Paintings of Thera, Vol. I, Athens 2000, p. 103-118

26) T. Pantazis, A. G. Karydas, C. Doumas, A. Vlachopoulos, P. Nomikos, E. Thomson, C.Vecoli, M. Dinsmore, Proceedings of the 4th Int. Aegean Conf. , Metron, Aegaeum 25, New Haven, USA, in press

27) E. Aloupi, A.G. Karydas and T. Paradellis, *X-Ray Spectrometry,* **29** (2000) 18

28) P. Craddock, N. Meek's, M. Cowell, A. Middleton, D. Hook, A. Ramage, E. Geckinli, *Art of the Greek* Goldsmith, 1998.

29) Y. Maniatis, A.G. Karydas, E. Mangou, Th. Paradellis, Ion beam study of art and archaeological objects, European Cooperation in the field of Scientific and Technical research", EUR 19218, 2000, p. 110-116

30) A. Hartmann, *Prahistorishe Goldfunde Aus Europa II,* Gebr. Mann Verlag, 1982.

31) G. Demortier, Proceedings of the Materials Research Society Symposium, 1989, p. 193-198

32) R. Tylecote, *The early history of metallurgy in Europe*, 1987.

33) E. Aloupi, A.G. Karydas, P. Kokkinias, T. Paradellis, A. Lekka and V. Karageorgis, Non-destructive analysis and visual recording survey of the pottery collection in the Nicosia Museum, Cyprus in Archaeometry issues in Greek prehistory and antiquity, The Hellenic Society of Archaeometry (Eds. Y. Bassiakos, E. Aloupi, Y. Facorellis), Athens 2001, p. 397- 410.

34) W. Noll, R. Holm, and L. Born, Painting of ancient ceramics, *Angewandte Chemie, (Internat. edn.),* **14** (9), 1975, p. 602-613.

35) E. Aloupi, V. Perdikatsis and A. Lekka, Assessment of the White Slip classification slip based on physicochemical aspects of the technique, in White Slip Ware, ed. V. Karageorghis, The A.G. Leventis Foundation, Nicosia, 1999, p.15-26

36) E. Aloupi, *The nature of pigments in Bichrome wheel-made ware*, KVHAA Konferenser 54:215-219. Stockholm 2001.

37) R.E. Jones, Greek and Cypriot Pottery, *The British School at Athens,* Fitch Laboratory OP1,1986, p. 765, 808.

38) V. Argyropoulos, Z. Antonopoulou, A. G. Karydas, B. Perdikatsis, G. Economou, Proceedings of the 1st International Conference held in Rhodes "Vulnerabiltiy of 20th Century Heritage to Hazards and Prevention Measures", in press

39) H. Strandberg,, L-G. Johansson and O.Lindqvist, The Atmospheric corrosion of statue bronzes exposed to SO_2 and NO_2, Materials and Corrosion, 1997 p. 48

Chapter I-3

Scientific Field Research in Egypt

Results from research undertaken by the Institute of Egyptology, Waseda University

Sakuji Yoshimura
School of Human Sciences, Waseda University, Japan
General Director, Institute of Egyptology, Waseda University, Japan
1-6-1 Nishiwaseda, shinjyuku-ku, Tokyo 169-0051, Japan
maat@tky3.3web.ne.jp

Keywords: Egypt, archaeology, nondestructive methods of exploration, interdisciplinary research, electromagnetic underground radar, satellite images, conductivity measurements, gravity probes, carbon 14, soft X-rays

The First International Symposium on X-ray Archaeometry focused on the question of how radiological techniques can be integrated into archaeology. Harnessing the science of radiology to the requirements of archaeology, a subject that traditionally falls within the arts, promises to be the first step in opening up new and unexplored horizons for colleagues in our field.

Until now it has been common practice for archaeologists who wish to analyze finds or establish a fact to hand over the job to a university or private-sector lab, then, upon delivery of the results, publish their conclusions in a paper or report. In other words, they go about their research without the foggiest idea of the analytical techniques used to obtain the results that support it. In working with Professor Masayuki Uda of Waseda University's School of Science and Engineering, and various other investigators over the past two decades, I have become increasingly convinced of the need to remedy this one-way communication process, whereby one side simply accommodates to the wants of the other. Unless something is done about this state of affairs, archaeology will never be able to fulfil its true potential.

The public regards archaeology as little more than the art of excavation. Yet, the word "archaeology" itself has nothing to do with excavating, as a glance at the etymology shows that it means simply the study of ancient things. It would not be so bad if just the person in the street were under this delusion; but even the majority of archaeologists themselves share the misguided notion of themselves as just glorified diggers. What excavation really entails is the discovery of fresh evidence and data that no one else has.

This does not imply that all an archaeologist has to do is to identify a site using

M. Uda et al. (eds.), X-rays for Archaeology, 55–63.
© 2005 *Springer. Printed in the Netherlands.*

this or that technique or dig up everything he can — and if he finds something big, well, so much the better. That is mere excavating. Japanese archaeology currently devotes much of its energy to salvage digs supervised by the local authorities on sites slated for development. This system of government-run excavations certainly had a role to play back in the days when the Japanese economy was steadily expanding, but surely it is time to bid farewell to the old ways. Individual archaeologists should rather be systematically pursuing specific interests.

The archaeologists of tomorrow will need to take advantage of their own planning and organizational skills to discover new sites. Moreover, they will need to excavate them with painstaking care and analyze the structural remains and objects they uncover from a host of angles. Devastation of the environment is of global concern these days, and non-destructive methods of exploration are indispensable, among other things, to help preserve the planet. That is where techniques like X-ray analysis come into their own. Traditionally the archaeologist's job was completed once he had published the results of the dig or provided a briefing, but now it is of growing importance to post findings on the Web, where they can be accessed from across the globe. It is thus of the essence. The task facing us is to transform archaeology into a complete discipline in its own right.

The first step in achieving that goal is to turn archaeology into a science. Japanese universities stick archaeology in among the arts, which is a source of profound embarrassment. If archaeology is an art, then who can find any fault with the man who fabricated findings of Stone-Age implements in a series of incidents that sent public mistrust of the discipline soaring? After all, art is all about fabrication. That fact alone demonstrates the necessity of making archaeology a science.

Objectivity is among the keys to elevating archaeology to a science. Individual observation is not enough. Everyone must be able to obtain the same results under the same conditions: that is the cornerstone of science. Also important is interdisciplinary research involving the fusion of arts and sciences being promoted by this symposium. I have advocated this goal throughout the forty-odd years I have been involved in Egyptian archaeology, and I believe that it is real progress to have been able to organize this conference with Professor Uda.

Next, I will provide a breakdown of scientific work that has been done in Egypt by my own institution, the Waseda University.

We are currently conducting research in four locations, Giza, Abusir, Dahshur, and Luxor. In Giza we investigated the interior of Khufu pyramid and examined his Solar Barque using electromagnetic radar waves. Once we get things organized, we plan to carry out further investigations on the Solar Barque. We identified a set of archaeological remains at Abusir-South by using electromagnetic underground radar and a second set of remains at Dahshur North by analyzing satellite images. We have also conducted several projects on the West Bank of the Nile at Luxor, a town 700 km south of Cairo. At present we are doing restoration and conservation work on the tomb of Amenhetep III in the west branch of the Valley of the Kings.

The Egypt Archaeological Mission of Waseda University first landed in Egypt 36 years ago, in 1966. Our debut project was a general survey by jeep, performed with the late Kiichi Kawamura, who was then studying occidental archaeology at the

Waseda University's School of Literature.

On January 15, 1974, we made a discovery of global importance at the so-called Hill of the Fish at Malqata-South on the west bank of the Nile at Luxor: a painted staircase belonging to a larger structure. This discovery attracted a great deal of attention, and thanks to donations from university benefactors we were able to build research and accommodation facilities in Luxor. This marked the beginning of serious Japanese excavation work in Egypt. In 1976 research and accommodation facilities were built in Cairo as well.

In the Luxor district in Upper Egypt, the discovery of the remains at Malqata-South and of a sanctuary of Amenhetep III led us to focus on that pharaoh's reign. We implemented a comparative study of wall paintings in some sixty tombs of nobles in the village of Qurna dating to his era. We also explored Amenhetep III's tomb in the Valley of the Kings.

At first we were not quite sure what we had found at Malqata. However, with advice from the members of a German expedition and various other individuals, a team under Yasutada Watanabe, professor of architectural history with the School of Science and Engineering at Waseda University, we succeeded in making a replica of the original structure.

We went on to discover ten tombs belonging to nobles dating from the reign of Amenhetep III in seven locations on the West Bank at Luxor, including Dra' Abu al-Naga', Shaikh 'Abd al-Qurna, and Asassif. Asassif also yielded a mummy, on which we performed a CT scan that enabled us to reconstruct its facial features using computer graphics.

Next, the expedition relocated to the Valley of the Kings. The Valley is divided into an east and a west branch; to date 58 tombs have been found in the East Valley and four in the West Valley. Suspecting that there still might be some tombs in the rock waiting to be discovered, we carried out a survey using scientific equipment like conductivity measurements, gravity probes, and electromagnetic underground radar. Gravity probes and electromagnetic underground radar had already proven their worth by revealing the existence of a hitherto-unknown chamber in the Great Pyramid of Khufu. Currently, working in partnership with UNESCO, we are cleaning the tomb of Amenhetep III, which we earlier explored, as well as implementing a restoration and conservation project on the fragile paintings of its walls in order to preserve them for future generations. Meanwhile Professor Uda is analyzing the colors used in these paintings. He has already nailed down some new facts: the pigments used differ from area to area, even when the color is the same; some of the paints have been imported; and certain of the colors have been mixed.

In Lower Egypt we have conducted survey work on three sites, Giza, Abusir, and Dahshur. At Giza, as already mentioned, we investigated the possible existence of a hidden chamber inside the pyramid of Khufu using electromagnetic underground radar and other technology. We carried out a similar probe on the anterior of the Sphinx. Electromagnetic underground radar consists of an antenna and a display. If reflectance is abnormally high, it shows up in red on the display. The probe revealed previously undiscovered chambers in the Pyramid of Khufu and the forepaws of the Sphinx. We also used electromagnetic underground radar to verify the existence of a second boat pit rumored to be located beside the pyramid.

Later, to ensure impartiality, an American team drilled a hole and photographed the inside. Five years later, the Waseda team inserted a fiberscope into the pit via the same hole, taking photographs and removing three fragments from the boat. One of these was analyzed by the Egyptians, the other two by the Japanese. Kyoto University's Wood Research Institute studied the timber of the latter two fragments, and they were dated using carbon 14 at Nagoya University; one was determined to be 4180±80 years old, the other 4460±45 years old. These results confirmed that the wooden vessel was constructed at the same period as the Pyramid of Khufu itself. The samples and photographs also led to a further conclusion: the boat should be removed and conserved as soon as possible.

The remains at Abusir-South were during a survey of the area by jeep equipped with electromagnetic underground radar. A 3300-year-old mortuary temple erected in honor of Ramses II's fourth son Khaemwaset was discovered at the top of a hill. Later a sundried-brick structure was found on the west side of the hill, which dated to the era of Amenhetep II-Thotmes IV 150 years prior. Unfortunately however, this had been partially destroyed by an anti-aircraft bunker built over one of its sections during one of the Middle East wars. Now that work is almost complete on the summit where the mortuary temple and brick structure are situated, we are cleaning up the west face of the hill. In August 2001 a structure built into a cave was unearthed on that slope; it yielded a statue of the goddess Sekhmet and a carving of a sphinx, both inscribed with the name of Khufu. These are important discoveries, for they are only the second and third sculptures bearing that the Pharaoh's name ever to have been found. We are currently investigating the reason why these objects should have turned up at this particular location.

The third site is that at Dahshur. The remains here were discovered by satellite imaging analysis. First, we homed in on a specific area by analyzing images from an observation satellite with the collaboration of Professor Toshibumi Sakata, director of the Tokai University Research & Information Center, and then we pin-pointed the exact excavation site using electromagnetic underground radar. The remains were buried under a 'blanket-like' layer of a meter-thick sand, which we dug through to find some 5000 highly valuable artifacts since commencing archaeological operations in 1996 — including a ring belonging to Tutankhamen. Stratum II yielded a sarcophagus belonging to one Mes, a senior official under Ramses II, some 120-130 after the reign of Tutankhamen. The body of the sarcophagus is painted with figures of the gods, the pigments used are some of which Professor Uda is analyzing.

Another example of the application of scientific technology was a project that involved photographing two statues of Tutankhamen using various types of soft X-rays. This was carried out to test the assumption of the British scholar Nicholas Reeves that the statues might conceal written texts. His theory was based on the fact that whereas the tombs of other ancient Egyptian Pharaohs invariably have contained papyrus texts such as the Book of the Dead, none were ever found in Tutankhamen's sepulcher. In the end, the results were negative, but the project did add soft X-rays to our extensive arsenal of scientific technology.

There are numerous other technological advances waiting out there to be tapped in addition to the many enumerated here. The application of these technological advances to archaeology will no doubt increase as researchers in different fields come to understand each other better. The prospects for the future look bright.

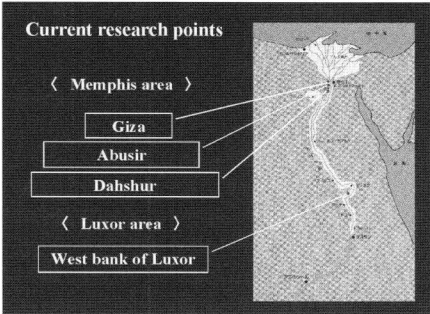

Figure I-3-1(a).
Current research points

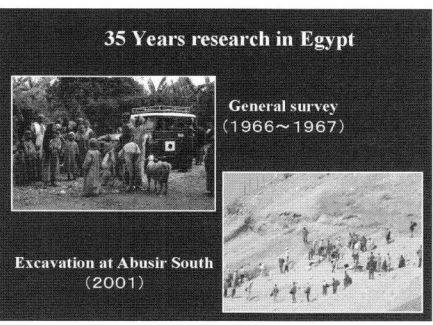

Figure I-3-1(b).
35 Years research in Egypt

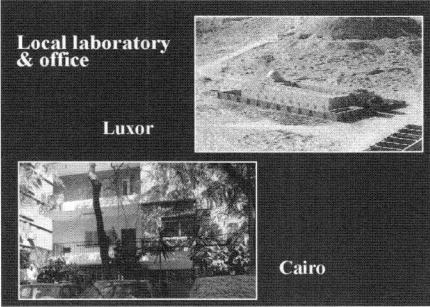

Figure I-3-1(c).
Local laboratory & office

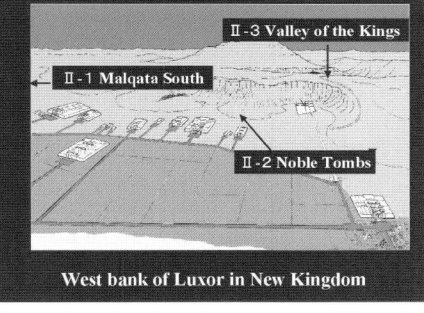

Figure I-3-1(d).
West bank of Luxor in New Kingdom

Figure I-3-1(e).
Malqata South

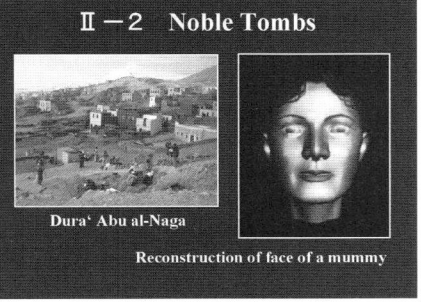

Figure I-3-1(f).
Noble Tombs

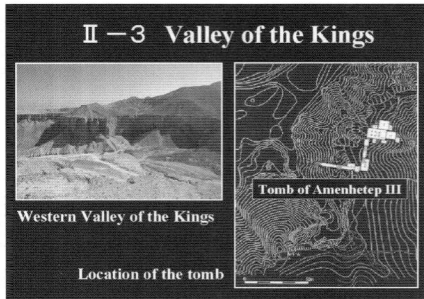

Figure I-3-2(a).
Valley of the Kings

Figure I-3-2(b).
High-Tech' research on site distribution

Figure I-3-2(c).
Conservation works in tomb of Amenhetep III

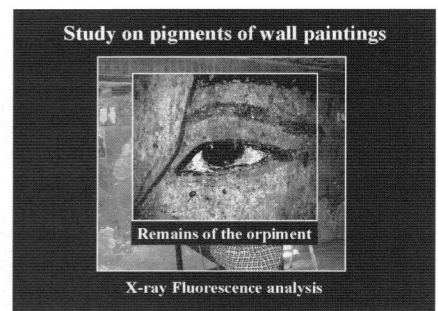

Figure I-3-2(d).
Study on pigments of wall paintings

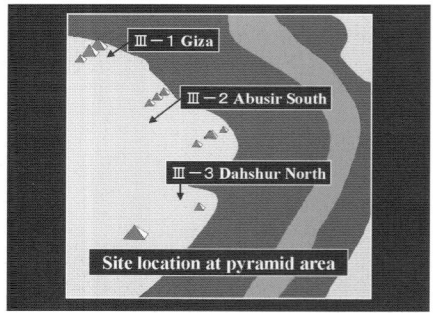

Figure I-3-2(e).
Site location at pyramid area

Figure I-3-2(f).
Giza

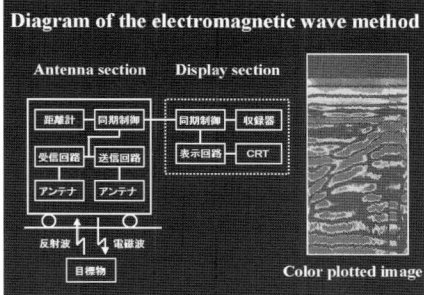

Figure I-3-3(a).
Diagram of the electromagnetic wave method

Figure I-3-3(b).
Survey of Khufu's second boat pit

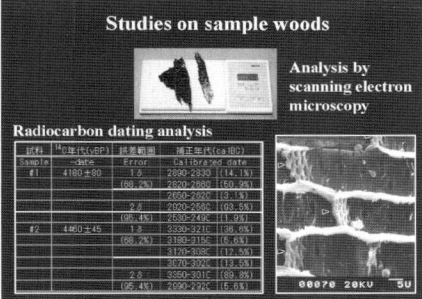

Figure I-3-3(c).
Studies on sample woods

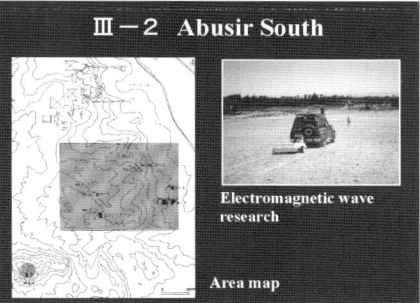

Figure I-3-3(d).
Abusir South

Figure I-3-3(e).
Discovery of the New Kingdom chapel

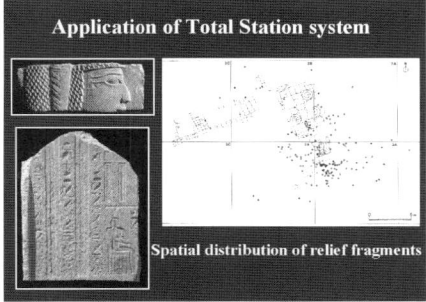

Figure I-3-3(f).
Application of Total Station system

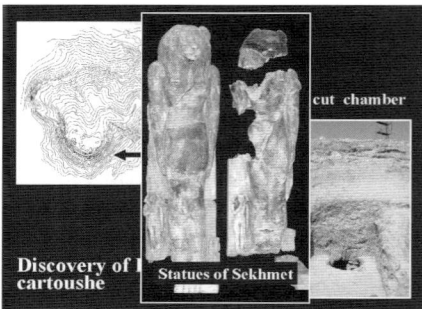

Figure I-3-4(a).
Discovery of Khufu's cartoushe

Figure I-3-4(b).
Dahshur North

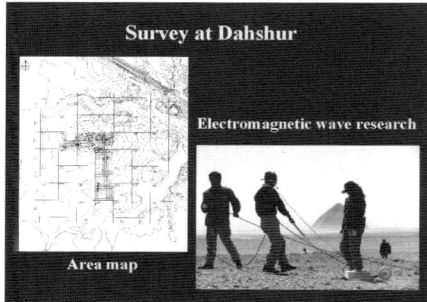

Figure I-3-4(c).
Survey at Dahshur

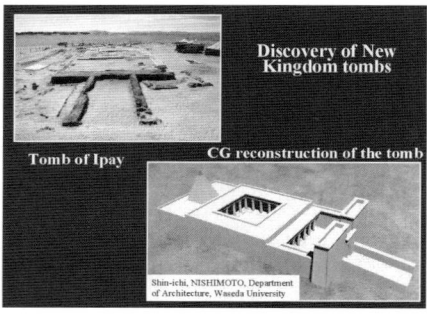

Figure I-3-4(d).
Discovery of New Kingdom tombs

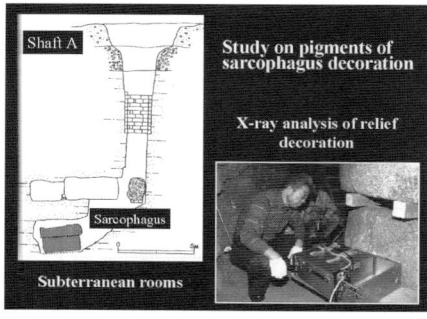

Figure I-3-4(e).
Study on pigments of sarcophagus decoration

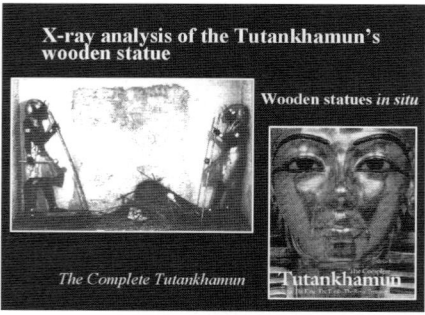

Figure I-3-4(f).
X-ray analysis of the Tutankhamun's wooden statue

(See also Color plate, p. 299)

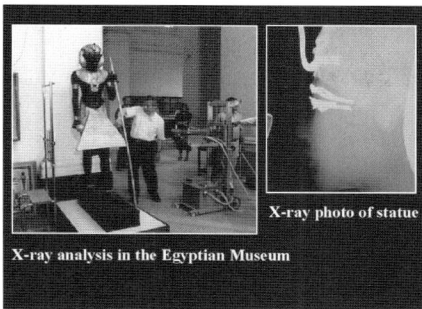

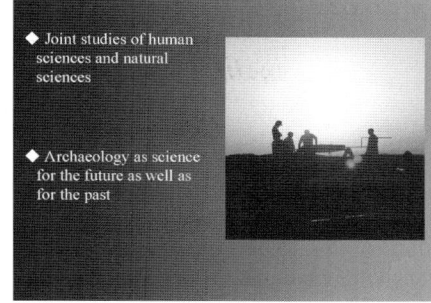

Figure I-3-5(a).
X-ray analysis in the Egyptian Museum

Figure I-3-5(b).
Future Perspective

Part II: Use of Ion Beam

Chapter II-1

Ion Beam Techniques for the Non-destructive Analysis of Archaelogical Materials

Guy Demortier

LARN, Laboratoire d'Analyses par Réactions Nucléaires, Facultés Universitaires Notre-Dame de la Paix, 61 rue de Bruxelles, B-5000 Namur (Belgium)
Guy.Demortier@fundp.ac.be

Keywords: PIXE, RBS, NRA, ion beam technique, non-destructive analysis, archaeological materials

Abstract

Proton-Induced X-ray Emission (PIXE) has been proved to be a very rapid and accurate method for the elemental analysis of ancient artefacts. Using protons whose energy is lower than 3 MeV, it is possible to obtain, in a few minutes, the actual composition of narrow parts of artefacts, without any sampling, even at microscopic level. Any element with an atomic weight higher than 20 may be analysed. Rutherford backscattering spectroscopy (RBS) and Nuclear Reaction Analysis (NRA) are complementary techniques mainly useful to control the homogeneity in depth and then to study superficial layers. RBS is mainly powerful for the profiling of heavy elements, but NRA is only used for light elements. Examples of applications will mainly concern, in this contribution, the study of metallic artefacts.

1. Introduction

Quantitative elemental analyses of archaeological materials are now used to explain the provenance of artefacts and understand the workmanship of ancient craftsmen. When applied to unique archaeological items these analytical methods require to be mainly non-destructive and able to give the chemical composition of different parts of the artefact without any sampling, even at microscopic level. PIXE (Particle-Induced-X-ray-Emission) [1], PIGE (Particle-Induced-Gamma-ray-Emission) [2] and Nuclear Reactions [2] involving the detection of a charged particle (NRA) (three methods sometimes used simultaneously) possess most of these qualities [3]. PIXE and PIGE may be applied to artefacts of large dimensions, kept at atmospheric pressure during the irradiation. If necessary, the investigations may be undertaken in a microprobe assembly (beam diameter down to 3 μm) [4, 5, 6]. RBS may sometimes give complementary information on the distribution of

M. Uda et al. (eds.), X-rays for Archaeology, 67–100.

heavy elements. PIXE and RBS are then suitable to study materials containing medium and heavy atoms, PIGE and NRA to study light atoms [2]. The energy of the incident particles is low enough (1 to 3 MeV) to avoid any residual activity. We intend to present in this paper a general description of PIXE and RBS and only point out elementary information on NRA and PIGE, which are used as IBA technique when PIXE or RBS cannot be applied.

2. Basic Principles of PIXE

The irradiation of all materials with charged particles leads to an ionization of electronic shells and consequently to emission of X-rays whose energy is characteristic of the atom. The technique using this phenomenon induced by protons is called PIXE. In principle, the analysis of all the elements may be performed but, practically, only elements with an atomic weight greater than 20 may be quantitatively determined. Elements with an atomic weight lower than 20 are more accurately analysed by detection of X-rays produced by excitation of the internal structure of the atomic nuclei (PIGE) [2] or detection of charged particles arising from the interaction between the incident particle and the nucleus. The characteristic K X-rays of light elements are indeed so highly absorbed in the sample that corrections cannot be made to obtain a sufficient accuracy with PIXE.

PIXE is very similar in its principles to X-ray analyses with an electron microprobe. The main differences lie in: the trajectories of the incident particles (protons penetrate in any material by gradual decrease of their energy along a straight trajectory while electrons zigzag along a complicated path), the depths of the analysed material layers (5 to 10μm under the surface for PIXE but only 1μm with conventional electron probes) (figure 1), the sensitivity (2 orders of magnitude better for PIXE because proton Bremsstrahlung is nearly absent) (figure 2). Furthermore, PIXE analyses may be performed at atmospheric pressure, preventing very large (like parchments, books, paintings,...) or fragile objects (like paper, intricated jewellery items,...) from being introduced in vacuum [7].

A vast amount of measured data for X-ray production intensities with protons on numerous chemical elements has been published. Several models also very well predict them theoretically. The complete calculation takes into account the energy loss (E), the depth deflection and the velocity change of the projectile in the Coulomb field (C) of the nucleus, the perturbation (P) of the atomic stationary states (SS) by these incident projectiles including relativistic effects (R), the ECPSSR model. Accurate data are now available [8] in nearly exhaustive tables.

Selected values of ionization cross-sections for incident protons up to 3 MeV and corrected for the fluorescence yields to produce Kα or Lα X-ray lines are given in figure 3. We can observe that those X-ray emission cross sections vary by nearly 5 orders of magnitude when the target atomic number varies from 15 to 80. K X-ray emission intensities are then only interesting for quick analyses of elements up to Z = 50. L X-rays are more convenient to achieve a sufficient sensitivity for minor and trace elements of high atomic weight.

In figures 4 to 9, typical X-ray spectra obtained during our study of potteries, bronzes, silver and gold objects are shown. Note that the vertical axis is given in a logarithmic scale in order to appreciate the large variation of intensities.

Each element ionized by the incident proton generated several X-ray signals. When K lines are selected for the analysis, we may choose 2 different signals Kα and Kβ with known intensity ratio. If L lines are selected (for heavy elements) one may select Lα, Lβ or Lγ, depending on the eventuality of coinciding energies from various elements; see below for Zn and Au interferences.

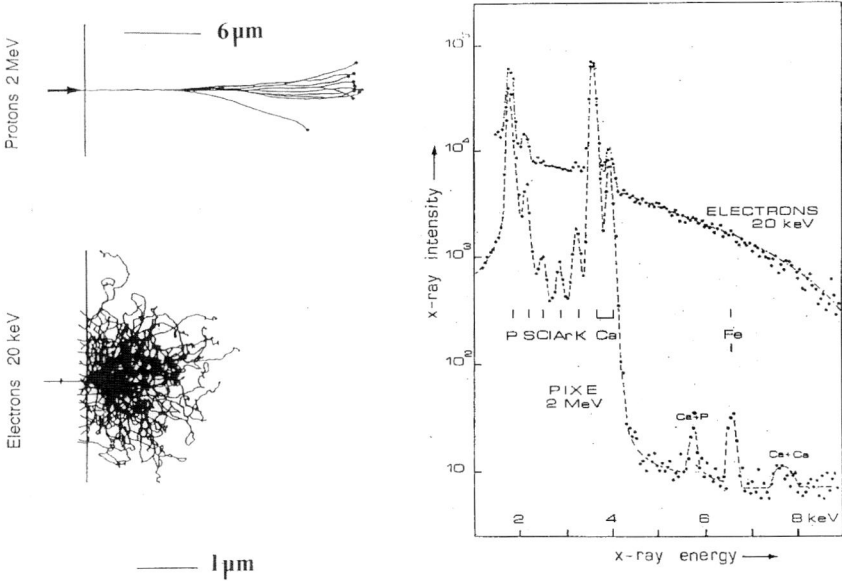

Figure II-1-1. Trajectories of electrons (20 keV) and protons (2 MeV) in a copper matrix.

Figure II-1-2. Comparison of X-rays spectra induced by 20 keV electronsand 2 MeV protons in hydroxiapatite.

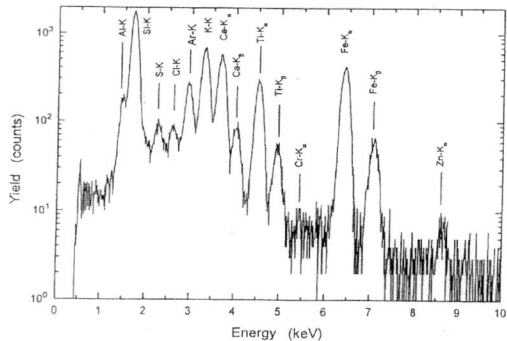

Figure *II-1-4*. PIXE spectrum of ablack shard (Braives-Belgium).

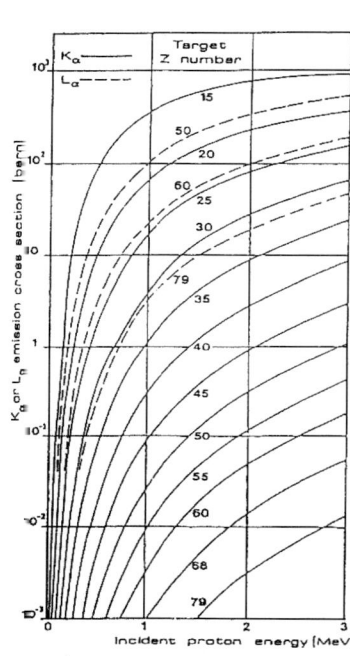

Figure II-1-3. Relative intensities of Kα and Lα X-rays Produced by protons up to 3 MeV on various elements.

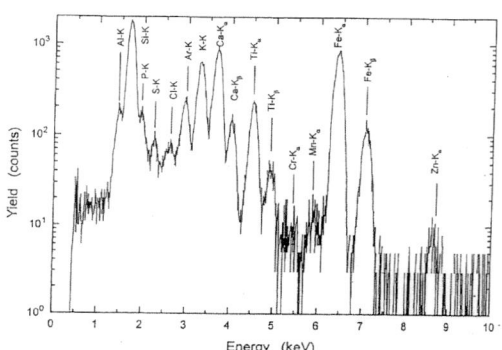

Figure *II-1-5*.   PIXE spectrum of a red shard (Braives-Belgium).

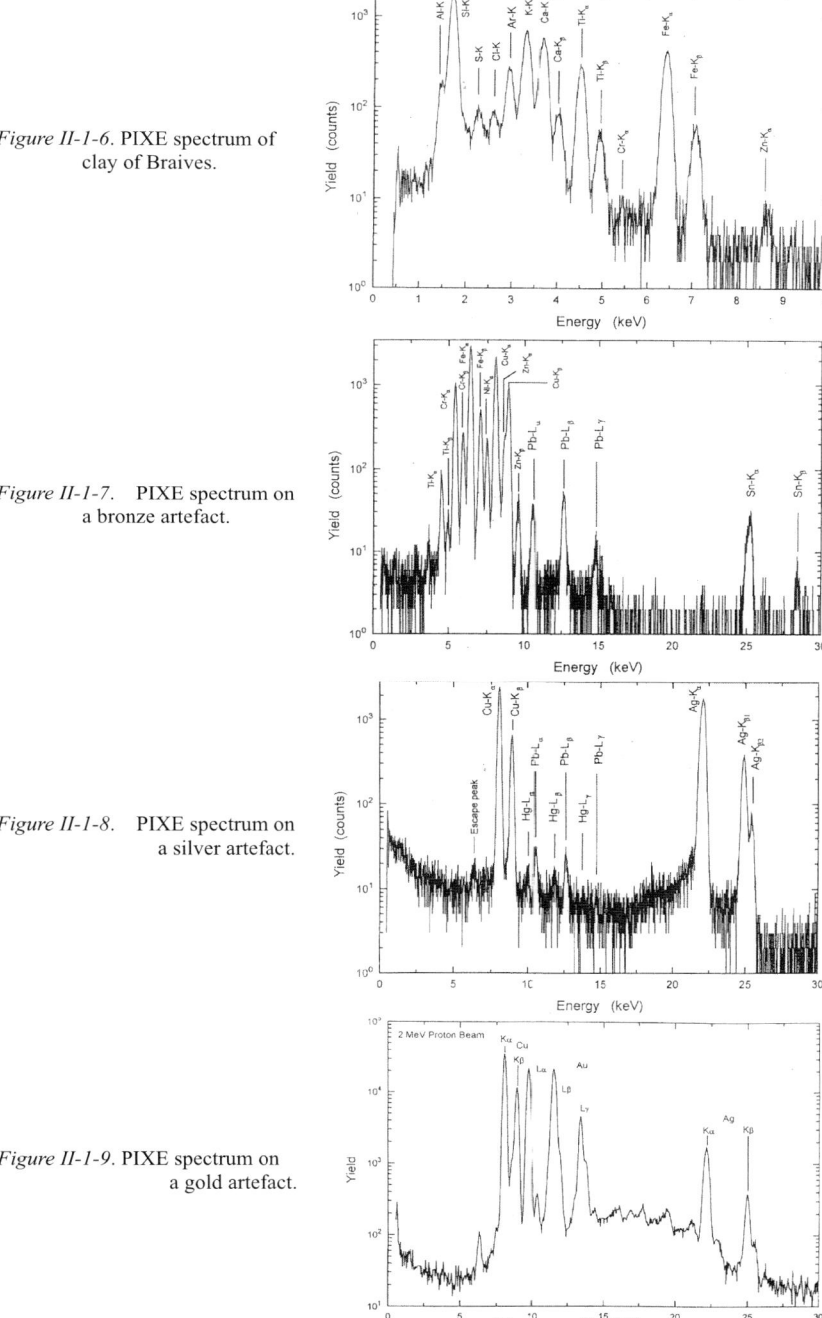

Figure II-1-6. PIXE spectrum of clay of Braives.

Figure II-1-7. PIXE spectrum on a bronze artefact.

Figure II-1-8. PIXE spectrum on a silver artefact.

Figure II-1-9. PIXE spectrum on a gold artefact.

The analysed depth of the irradiated artefact is less than 10μm, even for particles crossing 25 to 30 μm in the material due to the decrease of X-ray production with decreasing energy of the projectiles (see figure 3). Archaeological applications are then restricted to non-corroded materials or to drillings involving partial destruction of the sample only. PIXE is very fast and accurate for the non-destructive characterization of the main components of narrow regions on gold jewellery artefacts in a non-vacuum "milliprobe" arrangement: gold jewellery items are indeed well preserved and not affected by a surface corrosion in a depth comparable with the thickness of the analysed layer.

3. Experimental Procedure for PIXE

Figure 10 illustrates the experimental set up for non-vacuum PIXE at LARN (Namur, Belgium). The proton current from the accelerator is maintained at a low level (0,1 to 10 nA) in order to keep the total counting rate of X-rays compatible with the best energy resolution of the solid state Si(Li) detector (500 to 700 counts per second). Using appropriate filters can solve the problem of different counting rates in characteristic X-ray peaks of main and minor elements here. Foils of selected materials are used to reduce selectively the X-rays of the main component, and therefore enhance the relative intensities of all other useful X-ray lines.

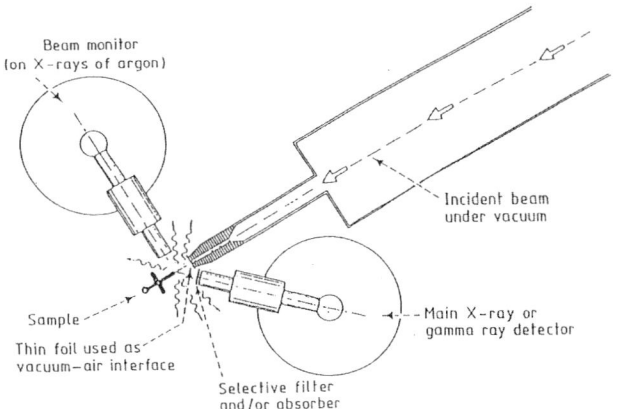

Figure II-1-10. Typical arrangement installed at LARN (Namur) for non-vacuum PIXE. The incident proton beam crosses a thin foil of Al before reaching the target sample situated at 1cm in the air. The X-rays of the sample are collected in the main detector. A lateral monitor collects X-rays of Argon (in air), which gives relative intensities of the incident proton beam.

The presence of the filter between the irradiated target and the detector may give rise to an additional X-ray fluorescence. As the characteristic X-rays of the element(s) of the filter(s) appear in the spectrum, the determination of such element(s) in the target is then impossible. Figure 11 illustrates typical spectra obtained during proton irradiation (E_p=2.8 MeV) of an artefact containing Cu, Au, Ag, in the presence of Zn lines produced in the selective filter.

With the help of a small computer, each X-ray spectrum is examined in order to subtract the actual background (mainly from secondary electron Bremsstrahlung origin) giving rise to a continuum up to 6 keV, to identify all peak regions, and to calculate peak areas. The computation of actual X-ray intensities considers several parameters such as: X-ray cross-sections, X-ray absorption, sum and escape peaks.

All the useful physical data are available to allow quantitative analysis by calculation. In order to avoid calculating all the geometrical corrections (proton beam intensity, solid angle of the detector, its efficiency for different X-ray energies and mainly absorption of X-rays in the materials between the target and the detector), comparison with known reference materials is more straightforward.

Modern reference samples containing most of the elements of interest in archaeological artefacts are commercially available, and may be irradiated in the geometrical conditions used for the studied artefacts.

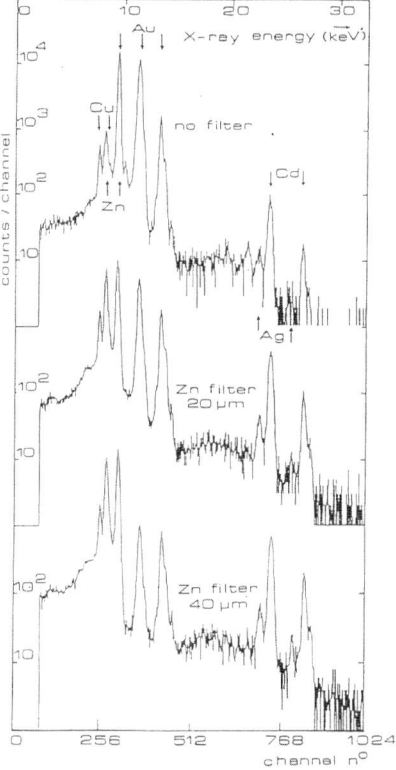

Figure II-1-11. PIXE spectrum of a gold artefact collected in the presence of a zinc filter.

Consequently, each measurement is reduced to the determination of relative peak areas. The actual concentrations have to be calculated from the peak area ratios taking all the target internal parameters into account: projectile range, cross section for X-ray production, differential X-ray absorption and secondary X-ray fluorescence yields. Any variation in the relative concentrations of only two components in an alloy containing more than these two elements modifies the X-ray intensities of all other elements. Several steps in achieving the precision in the actual composition have then to be followed. All the X-ray intensities collected during the analysis of the samples are then determined using several steps of sophistication in the calculation: (1) - concentrations are first calculated using a linear rule with the characteristic peak areas, (2) - these results are then used to calculate the stopping power, and the X-ray absorption in this new material, (3) - this procedure is repeated until a new step gives a modification lower than the statistical accuracy and (4) - finally the secondary fluorescence is calculated. Even with a reference material containing elements whose concentrations may vary from a factor 3 compared with those in the sample, the complete procedure of calculation on the X-ray spectra finally yields concentrations with an accuracy of 5% or better. A control has been made using PIGE. In this latter case, γ-rays produced in the 20-30 first microns under the irradiated surface are not absorbed in the material itself because of their high energy (100 keV and more). As cross sections for γ-ray emissions on medium and heavy elements are around 3 orders of magnitude lower than X-ray ones, this procedure takes a long time and is only used as ascertaining procedure in limited cases.

Nuclear reactions leading to the emission of charged particles may also be used for the quantitative determination of light elements. Characteristic protons spectra induced by deuteron irradiation allow depth profile analysis of Si, S, Al ... in matrices containing heavy elements [2]. All the calculations in PIXE experiments are based on the hypothesis that the material is homogeneous in all the depth of the analysed material (5-10μm). Nevertheless, heterogeneity in those thicknesses may be identified by comparing the relative intensities of all the characteristic X-ray lines of each element. Elements concentrated at the surface (or under a surface layer) show K□/K□ and L□/L□ intensity ratios greater (or lower) than those calculated by the basic program [9].

4. Basic Principles of RBS

Rutherford backscattering spectroscopy is based on the elastic collision of an incident charged projectile with the nucleus of an atom. In this elastic process, the incident and the collided nucleus remain in their fundamental energy state after the interaction: the whole kinetic energy is then conserved. Equations of conservation lead to a simple relation between the energy of the backscattered particle (E_{sc}) (emitted angle close to 180°) and the incident energy (E_0):

$$E_{sc} = E_o \left(\frac{M - m}{M + m} \right)^2$$

where m and M are the masses of the projectile and of the nucleus respectively (in this procedure m < M).

One observes that E_{sc} is proportional to E_O and is closer to E_O when m is very light by comparison with M. If this parameter is to be chosen to distinguish scattered particles of mass m from various nuclei of mass M, m must be the highest as possible but always smaller than M.

Practically, for both acceleration and detection purposes, the incident particle is of a proton (mass 1) or an alpha particle (mass 4). In figures 12 and 13, the scattered energies of protons and α-particles on various nuclei are shown. In this figures we have also taken into account that the probability of elastic scattering is proportional to the square of the atomic number of the target. This true proportionality to Z^2 is expected for high Z values but large discrepancies with that simple rule would arise for low Z nuclei (for example carbon and oxygen). Note the logarithmic scale in the ordinates of those figures.

The relative intensities of scattered particles on Au and O are then in a ratio of $(79/8)^2 = 97.5$. RBS is then 2 orders of magnitude more sensitive for gold than for oxygen. Comparing figure 12 with 13, one clearly observes that the mass separation is better for α-particles than for protons but this separation becomes often insufficient to distinguish neighbouring elements like Cu and Zn, and even Ag and Au.

The study of materials using RBS is not straightforward for bulk samples, due to the decrease in energy of the incident projectiles when they penetrate in the material. The measured energy of a particle scattered in a layer below the surface is not given by the simple formula due to the energy loss of the projectile when penetrating the material and the energy loss of the scattered particle before going out of the target. Typical RBS spectra of α-particles and protons on gold, silver and bronze homogeneous alloys are demonstrated in figures 14 to 19; one observes that the safe identification of one element is only easy for the heaviest of them.

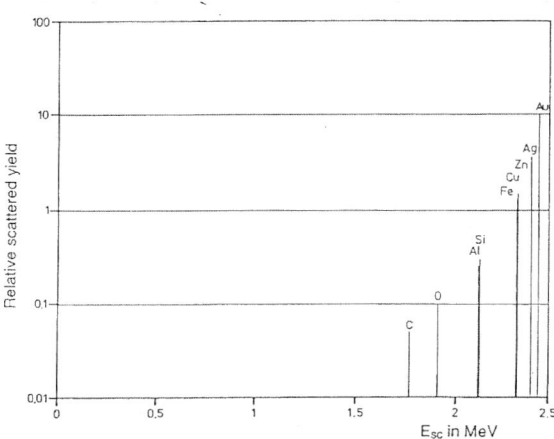

Figure II-1-12. Calculated energies E_{sc} of backscattered protons ($E_i = 2.5$ MeV) on various masses.

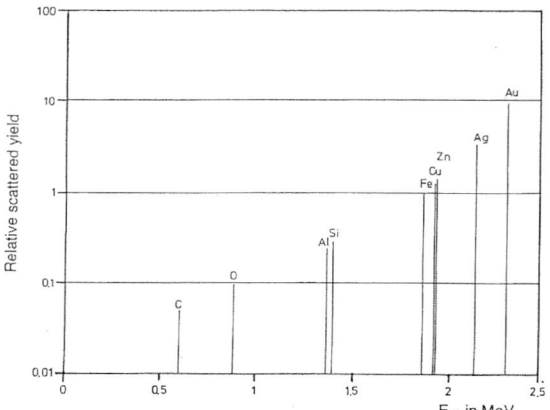

Figure II-1-13. Calculated energies E_{sc} of backscattered α-particles ($E_i = 2.5$ MeV) on various masses.

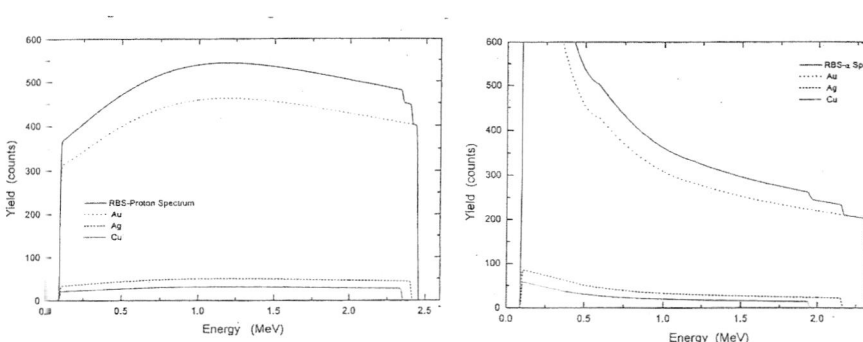

Figure II-1-14. Typical spectrum of backscattered protons (E_i=2.5 MeV) on a homogeneous gold-silver copper (Au 75%; Ag 12.5%; Cu 12.5%).

Figure II-1-15. Typical spectrum of backscattered α-particles (E_i=2.5 MeV) on a homogeneous alloy gold-silver-copper alloy (Au 75%; Ag 12.5%; Cu 12.5%).

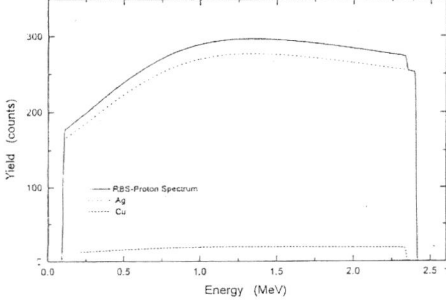

Figure II-1-16. Typical spectrum of backscattered protons (E_i=2.5 MeV) on a homogeneous silver-copper alloy (Ag 90%; Cu 10%).

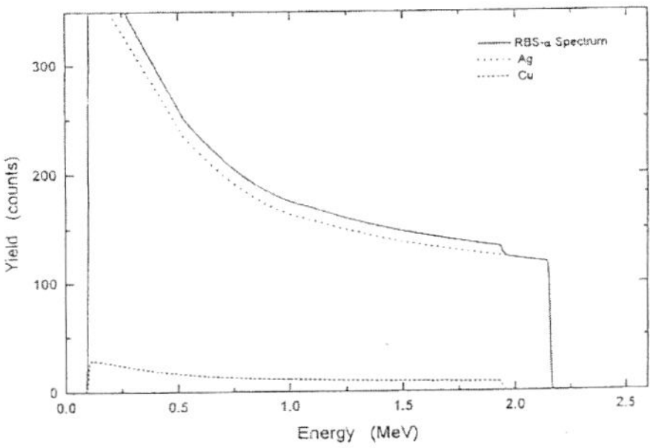

Figure II-1-17. Typical spectrum of backscattered α-particles (E_i=2.5 MeV) on a homogeneous silver-copper alloy (Ag 90%; Cu 10%).

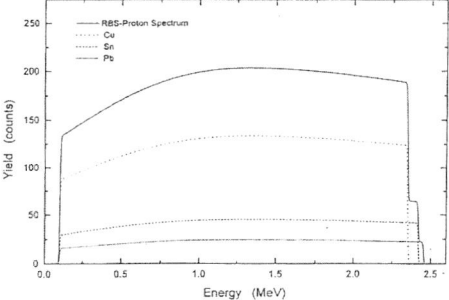
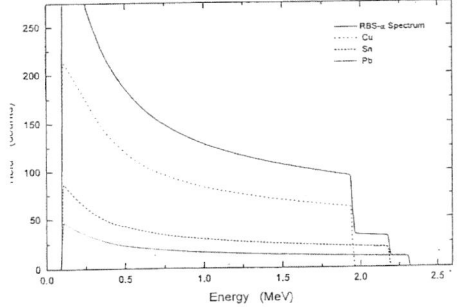

Figure II-1-18. Typical spectrum of backscattered protons (E_i=2.5 M eV) on a bronze artefact (Cu 78%; Sn 18%; Pb 4%).

Figure II-1-19. Typical spectrum of backscattered α-particles (E_i=2.5 MeV) on a bronze artefact (Cu 78%; Sn 18%; Pb 4%).

As the energy loss of α-particles in materials is higher than that of protons, the total analysed depth width is much lower than with protons but the profile at the surface is much easier to determine by α-particles than by protons. The choice of α or protons and the choice of the typical incident energy are then very dependent on the layer we intend to study. RBS is very simple in use for the study of thin layers of heavy elements on bulk matrix containing light or medium elements, but RBS becomes very complicate for materials containing elements with neighbouring atomic masses and is of no use for light elements in heavy matrices [10].

RBS is then of very limited application for archaeological purposes but is ideally used as a complementary method of PIXE as it will be discussed in the study of gold plating or gold depletion gilding in the chapter on metallic samples [11].

5. Typical Illustrations of Non-Destructive and Non-Radioactive Analytical Methods of Archaeological Metals

5.1 COINS

Coins are very valuable testimonies of the past. They keep accurate information on chronology and often on habits and customs of the period they have been struck. In addition to studies of decorations and inscriptions, the compositional analysis can be useful to understand the metallurgical skill of their area.

The circulation of monetary values and the wearing away by time have led to important surface alterations from their original composition. Although activation analyses (with neutrons or high energy charged particles) were often used for the determination of the bulk composition, they do not always give the true bulk values, mainly when the corrosion is important.

5...1 *Gold coins.*

Activation analysis using high-energy protons and/or neutrons has been proven to be a very powerful method of analysis of all the elements in coins, up to trace level [12]. Using PIXE we have demonstrated that the same goal may be reached for the study of the main elements in gold coins. Table 1 illustrates the comparison of the results for similar specimens. A general description of the potentialities of ion beam techniques in numismatics may be found in the work of J.-N.Barrandon and his co-workers of Orléans.

TABLE II-1-1.   Analysis of gold Byzantine coins by PIXE and proton activation.

SAMPLE	PIXE				Proton activation [12]		
	Cu	Ag	Au	N.I.	Cu	Ag	Au
ARCADIUS (semissis)	0.32	1.16	98.5	(5)	0.20	1.00	98.8
LEO I (tremissis)	0.27	1.14	98.6	(4)			
ZENO (solidus)	0.37	1.01	98.6	(5)			
ZENO (tremissis)	0.40	1.32	98.3	(4)			
ANASTASIUS (tremissis)	0.39	1.24	98.4	(5)	0.5	1.40	98.3
JUSTINIEN (1/2 solidus)	0.43	1.49	98.1	(5)	0.35	1.35	98.3
MAURICE TIBERE	0.25	1.00	98.8	(4)	0.35	1.30	98.35
PHOCAS (solidus)	0.43	1.54	98.0	(4)	0.4	1.60	97.8
HERACLIUS (tremissis)	0.30	1.10	98.6	(5)	0.30	1.50	98.2
CONSTANTIN IV (solidus)	0.33	1.40	98.3	(2)	0.17	1.25	98.6
CONSTANTIN VII	0.31	2.80	96.9	(7)	0.41	3.3	96.3
MICHEL IV (solidus)	1.75	8.10	90.1	(12)			

5.1 2 *Silver coins*

The usefulness of non-destructive PIXE on unaltered surfaces of the coins has been demonstrated to classify silver coins struck between 875 and 1456 by 22 Bishops Princes of the Principality of Liège (Belgium) [13].

They are distributed in three sets. The first one consists of 17 coins of Hughes de

Pierrepont (1203-1229), the second of 27 coins of Englebert de la Marck (1334-1364), the third of 20 coins belonging to 3 sovereigns and to 17 other Princes, from Charles le Chauve (869-870) to Jean de Heinsberg (1419-1456) (Table 2).

Each coin was maintained by a frame in order to keep the geometrical conditions constant during their analyses or the irradiation of reference materials. These references were homogeneous Ag-Cu, Ag-Au, Ag-Pb alloys or powder mixtures. The surface of the irradiated materials is parallel to the window of the Si(Li) detector in order to reduce the path of the outgoing X-rays to its minimum value. A 10 μm Co filter is inserted between the sample and the detector in order to selectively reduce the intensity of the Kα and Kβ lines of Cu, one of the most X-ray emitting elements in all coins. At least four different impact regions were selected in flat and bright regions of each coin.

All observed characteristic X-ray lines of the elements were taken into account in order to determine the elemental composition: Kα and Kβ lines of Fe, Cu, Zn, As and Lα, Lβ and Lγ lines of Au, Hg, Pb, Bi. The region of characteristic X-rays of Au, Hg, Pb, Bi collected in each measurement was reconstructed by the addition of typical spectra collected on individual elements, which were irradiated in the experimental conditions in which the coins were analysed. This procedure allows us to control if the material is homogeneous on the whole depth of the analysed region (5-10 μm) under the surface. As the X-ray absorption coefficients of these L lines of Au, Hg, Bi, Pb in their own pure matrix and in the silver matrix are very close, the relative intensities of Lα, Lβ and Lγ lines are nearly identical in the samples and the standards. Any increase of the Lα/Lβ ratio with respect to the value obtained on a homogeneous standard sample would indicate that the studied element is only in the first microns under the surface of the coin. A typical spectrum obtained on a coin of Otton III is presented in figure 20.

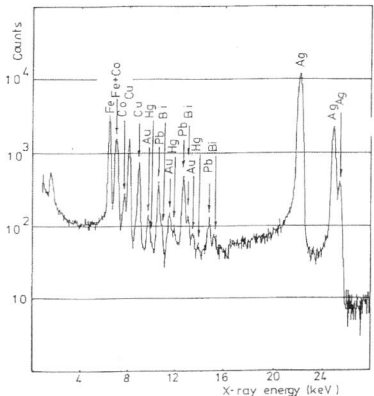

Figure II-1-20. X-ray spectrum induced by 2.8 MeV protons on a silver coin. The signals of Co are induced by fluorescence in the filter inserted between the target and the detector.

No criterion of classification may be extracted from the relative concentrations of Ag and Cu. The four impacts on each coin give elemental compositions whose values often differ by less than 2% for Ag. The variations of the relative concentration of these main elements (Ag and Cu) are given in table 3. Coins showing large variation in one of the 4 irradiated regions are excluded from the data of this table. The general trend indicates that the fineness (silver content) at the surface of the coins decreases slowly from 95,15% to 92,5% for periods extending from 869 to 1300. A decrease of this fineness appears to be very important since the reign of Jean d'Arckel. Except for these last coins, the silver concentrations are always far above the expected concentration (89%), an evidence for the Cu elimination from the surface by wearing.

TABLE II-1-2.   Identification of the analysed coins.

Sovereign or Prince	Dates	Type*	Number of coins
Charles le Chauve	869-870	Tunieras	1
Otton III	983-996	Denier	1
Conrad II	1027-1037	Denier	1
Théoduin	1048-1075	Denier	1
Henri de Verdun	1075-1092	Denier	1
Otbert	1092-1119	Denier	1
Henri de Leyen	1145-1165	Denier	1
Philippe de Heinsberg or Henri de Jauche	1167	Denier	1
Rodolphe de Zaehringen	1167-1191	Denier	1
Albert de Rethel	1191	Denier	1
Unknown provost	1191	Denier	1
Albert de Cuyck	1194-1200	Denier	1
Hughes de Pierrepont	1200-1229	Denier	17
Jean d'Eppes	1229-1238	Denier	1
Robert de Thourotte	1240-1246	Denier	1
Henri de Gueldre	1246-1274	Denier	1
Jean de Flandre	1282-1292	Esterlin	1
Adolphe de la Marck	1313-1344	Volant	1
Englebert de la Marck	1344-1364	Gros bourgeois	27
Jean d'Arckel	1364-1378	Gros bourgeois	1
Jean de Bavière	1389-1418	2 griffons	1
Jean de Heinsberg	1419-1455	Plaquette	1

* The identification on the type is given with its original French name.

TABLE II-1-3.   Variations of the mean Ag and Cu contents with time.

	Mean Ag content in %	Mean Cu content in %
Coins from 869 to 1037 (3 coins)	95.15 ± 1	3.66 ± 1.5
Coins from 1048 to 1200 (8 coins amongst 9)	91.35 ± 2	4.25 ± 2
Hughes de Pierrepont (16 coins amongst 17)	92.1 ± 2	5.65 ± 1.5
Coins from 1129 to 1343 (4 coins amongst 5)	92.35 ± 2	5.5 ± 2.5
Englebert de la Marck (22 coins amongst 27)	92.5 ± 2	7.2 ± 3
Jean d'Arckel (1 coin)	72	25
Jean de Bavière (1 coin)	43	55
Jean de Heinsberg (1 coin)	53	45

We cannot provide any significant information on the variation of the mean concentration for the last three values in Table 3 due to large variations of the measured elemental composition at various impacts.

An attempt of classification is not possible from the Cu and Ag concentration but the computation of the relative concentrations of Au, Pb, Cu is shown in figure 21. The copper contents have been arbitrarily divided by a factor 5 in order to avoid the accumulation of points in the top corner of the ternary diagram.

For the coins of various periods from 869 to 1456, one observes as general trend a loss of Au and a large increase of the Cu content consistent with the debasement after 1300.

The Hg and Bi contents are not completely significant. Hg is only present at high concentration (about several %) in "protected" regions of several coins. A larger proportion of Hg was eliminated by wearing. Zn and As are only present at trace level and cannot be used for any classification.

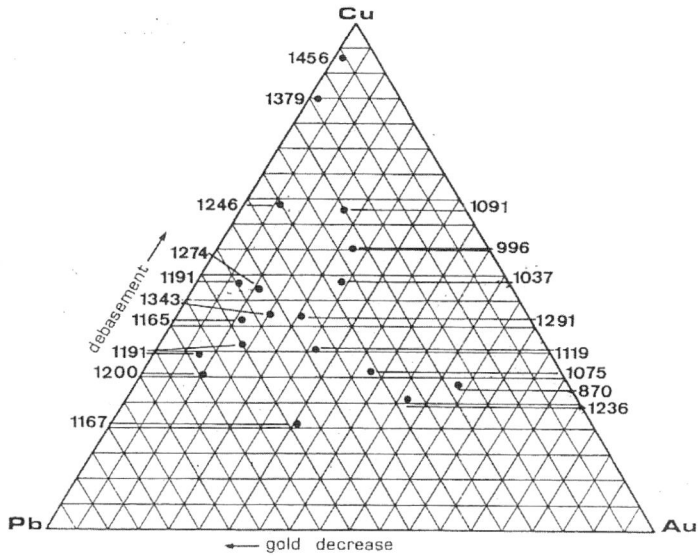

Figure II-1-21. Ternary diagram of Cu, Pb and Au concentrations from the measurements on 20 coins of various periods. The arrows indicate the decrease of gold with time and the important debasement beginning at the end of the 14th century.

5.2 CORRODATED METALS.

Metals provide key evidence of human progress to overcome the environment with increasing skills in manufacturing tools, weapons but also ornaments and currencies. Here metallographic investigations should complete analytical research. Comparisons of the composition of artefact with that of the raw materials cannot resolve the provenance studies as for shards for several reasons:

    a)   the comparative rarity of ores with respect to clays involves trading of materials over considerable distances,

    b)   the extraction by smelting techniques (requiring higher temperatures than for the production of a pottery) induces major changes in elemental composition,

    c)   as many metals are reactive, artefacts become heavily corroded when buried for a long period,

    d)   the recycling of metal artefacts was also frequent.

Since high-energy ion beams induce prompt signals on all the elements of the irradiated material, the simultaneous analysis of the main components is not compatible with the measurement of traces, without selective filters. Furthermore, the analysis on samples "as they come" concerns only the external surface, which is often corroded. Repeated measurements at different positions can quickly reveal the extent to which ratios of elements in an alloy are affected by this corrosion. In most cases, significant results cannot be obtained without abrasion.

Reliable measurements on the composition of corroded metal artefacts with ion beams concern only samples extracted by drilling or analysis after abrasion of a surface layer. The detection of impurities in iron gives information on the skill of ancient metal workers. These impurities might have affected the functional behaviour of the metal, even when they are present in very slight concentrations; phosphorus to harden and/or embrittle the metal but also to inhibit carburization during smelting; and manganese to enhance the carburization during smelting. The Bartol Delaware equipment in complementarity with other microscopic and metallographic examinations by the Lehigh University metallurgists was used to characterize iron artefacts on Bag'ah [14] steel from Transjordan dating circa 1200 B.C. [15]. American iron making processes at the 19th century site of Catoctin in Maryland (16) and the revolutionary area site of valley Forge in Pennsylvania [17].

Even with the best confidence in the provenance of the material, a comparison of concentrations of the main elements cannot always be used to classify the alloys. We have reported [20] results of Cu and Sn determinations by a 14 MeV neutron activation and PIXE with a 30 µm beam diameter on Greek bronze coins from the 3rd century B.C. to the 3rd century A.D. and on blanks coming from one single monetary workshop. The batch was far from homogeneous. Each coin was cut along a diameter in order to display a flat surface of uncorroded material. The comparative results are given in Table 4. We observe that the Sn concentration in the bulk given by NAA is always greater than that given by PIXE analysis of the uncorroded material. The reason cannot be attributed to a systematic error: the same reference materials have been used for both techniques. The Sn and Pb concentrations at the surface are far from constant, as shown in figure 22, but the correlation between Sn and Pb is obvious. The mean concentration of Sn at the surface is always greater than in the bulk, indicating a selective elimination of Cu due to handling and/or burial.

It seems that the problem of copper corrosion was well known by ancient

Chinese metallurgists, as reported in a study involving Rutherford backscattering, PIXE and prompt nuclear reactions [20]. An arrowhead, which was unearthed from Lintong county of Shanxi in 1974, is one of the relics from the mausoleum of Emperor Qin Shihuang (221-207 B.C.). Although 2200 years old, the surface of the arrowhead retains a surprisingly rust-free metallic aspect, its edge being still very sharp. By Rutherford backscattering of α-particles, a group at Fudan University has shown that the substrate of the arrow head contains 82% Cu, 17% Sn and 0.5% Pb, a very hard alloy most suitable for penetrating weapons, as is known today from modern metallurgy. The composition of the sharp edge is completely different; Cu and Sn are in the proportion 42% and 58% respectively. Furthermore, PIXE analysis shows that the edge contains about 3% of Cr, but also Fe, Ni, As at lower concentrations and Al, P, S, Cl, the presence of which might be due to the burial conditions. Nuclear reactions induced by deuteron bombardment of the edge indicate the presence of a thin oxide layer. By careful investigations of all the spectra, the authors attribute the rust-free quality of this well-preserved arrowhead to a layer of CaO + SnO + PbO + Cr_2O_3. They notice that German and US patents of 1937 and 1950 were concerned with similar anticorrosive treatments.

TABLE II-1-4. PIXE and FNAA* of Greek coins (bronze).
Concentrations and precisions are given in per cent.

	FNAA	PIXE Sn	PIXE Sn	PIXE Pb	PIXE Pb	PIXE As	PIXE Ag	
Coins	Sn	Core	Patina	Core	Patina	Core	Core	Importance of corrosion 1- FNAA/PIXE
C5	9.3	7.8	0.3-18	0.62	1.3-5.2	0.04	0.07	0.19
C4	9.6	8.4	8-16	0.53	1.4-3.3	0.08	0.09	0.14
C6	10.1	8.5	0.4-17	0.32	1.3-3.1	0.00	0.06	0.19
C7	12.1	10.1	0.9-13	1.41	2.0-4.1	0.00	0.08	0.20
C1	13.1	9.5	5-16	0.76	1.1-6.2	0.05	0.07	0.38
C3	17.0	10.8	10-21	0.85	1.9-6.1	0.05	0.10	0.57
C2	17.2	11.8	11-23	0.99	1.9-7.4	0.06	0.11	0.46
Precision (%)	5	0.5	1	2	4	10	10	

* FNAA: Fast Neutron Activation Analysis.

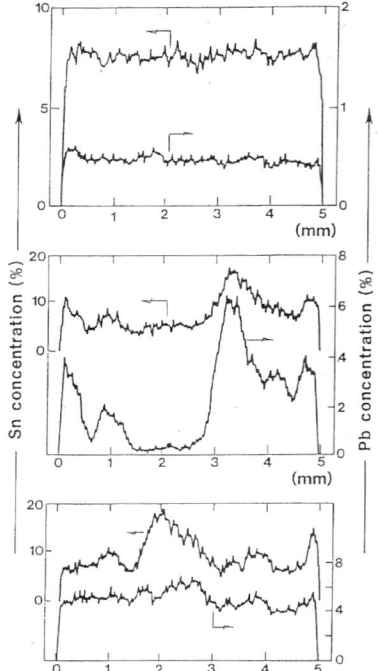

Figure II-1-22. Lateral concentration of Sn and Pb (a) in the core, (b) the patina and (c) the patina (other side) of a Greek coin [19].

.

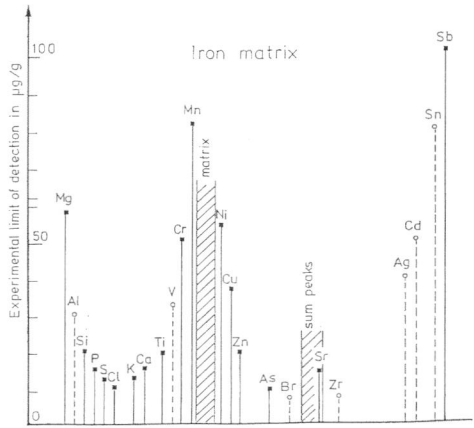

Figure II-1-23. Variation of the limit of detection of elements in an iron matrix by PIXE [15].

Extensive work with a PIXE milliprobe and a collimated PIXE microprobe of Bartol-Delaware and the focused PIXE microprobe of Harwell has been carried out by the MASCA group (University of Pennsylvania, Philadelphia) on numerous tin bronzes of Syrian and Palestinian origin (2000 B.C.) and on more recent American and European bronzes [21-22]. Because corrosion can disturb the elemental profile of the surface [23], polishing of small region was required in order to obtain a clean metal. Repeated polishing procedures were sometimes necessary for quite confident interpretation [24].

Iron artefacts suffer even more than bronze from corrosion and heterogeneity problems. No common analytical technique is therefore especially able to determine the significant alloying elements such as C, Si and P. A small number of analytical studies have been reported [25], but none involves high-energy ion beams. A few (destructive) analyses, carried out after sectioning with diamond, have been performed by the MASCA group [26]. Figure 23 indicates the detection limits of several elements by PIXE. In the low energy region of X-rays, the lack of sensitivity is due to the high absorption of X-rays onto the artefact itself. In the medium part, this lack of sensitivity is due to the presence of a high counting rate in the iron $K\alpha$ and $K\beta$ peaks and the subsequent tails in the experimental peak resolution. Around 13 keV, the lack of sensitivity is due to the sum peak of two simultaneous K signals of iron. In the high-energy region, the Si(Li) detector is less efficient and the cross section for X-ray production is much smaller (see figure 3).

Lead and leady alloys, used extensively by the Romans, have not yet been studied by high energy ion beams [27].

5.3 GOLD JEWELLERY

PIXE has been extensively used at LARN to analyse bulk material and solders on ancient gold artefacts from numerous sites and periods: Neo-Elamite, Suse, Achemenide, Hellenistic, Roman, Byzantine, Gallic and Merovingian jewellery, items of Pre-Columbian civilisations. Our main field of research was to investigate the soldering procedures used to bind various parts of complicated jewellery items and to identify the original processes of surface gilding of tumbaga by south Mesoamerican goldsmiths.

5.3.1 *Byzantine jewellery*.

Let us start this review by the study of a richly decorated gold Byzantine cross (figure 24) containing a piece of wood, said to originate from the Jesus-Christ cross. The front side of the relic is decorated with 48 stones and 68 pearls; on the backside, the number of stones is only 24, but the number of pearls remains 68. The edges are also decorated with pearls, stones, filigrees, … The interior of the box contains many partitions attesting that the jewel has been arranged for several purposes during centuries. This Byzantine cross is known as a piece of a treasure originating from Constantinople, sacked in 1204. It may probably have been constructed in the 7th century A.D. [28]. This cross has been venerated since the 13th century at Tournai (Belgium) and belongs to the treasure of the Notre-Dame cathedral in this town.

The elemental composition at about 70 proton impact sites allows us to appreciate the bulk composition of the basic gold, wires and granulations, repairs and additional elements. Analyses at selected proton impacts on one main side of the jewel are reported in Table 5. The homogeneity of the concentrations in nearly all regions (in flat gold and in wires, into and out of the solders) indicates that no sophisticated procedure of soldering has been used in the construction of the box; the joined parts are sintered (without any additional soldering element) by heating the surfaces to be bound, up to the early melting temperature. We have observed the bad quality of the soldering procedure during the analysis of several elements in the inside of the box. Decorations, stones and pearls on the outside mask the rudimentary procedure of soldering.

Repairs are detected in regions of impacts 9 to 14: copper and silver wires, a silver support (97% Ag – 3% Cu) coated with a Au-Hg amalgam (90% Au – 10% Hg), additional caps enclosing substitution stones.

More elaborated soldering processes appear in jewellery items decorated with granulations, as those worked out by Etruscan craftsmen (North of Italy, 7th to 4th century B.C.). Numerous small granules of nearly pure gold may be soldered together by a "solid state diffusion bonding", a procedure in which copper (from natural ores finely powdered between the fitted surfaces) diffuses at a temperature of about 900°C. This technique was rediscovered in the 1940's by Litteldale at the British Museum. An example of this spectacular workmanship has been identified in an Achemenide pendant (Iran) of the 4th century B.C., which belongs to the Department of Iranian Antiquities of the Musée du Louvre. We have shown that four different procedures of soldering were performed on a distance of 5 mm [29]. This perfection in the use of soldering procedures seems to be unique and never achieved outside the Etruscan world. At the same period, but in the Hispanic Peninsula, one can also find jewelry items finely decorated by granulations and filigrees. They are attributed to the Tartesic period. Several items have been also analysed using the PIXE technique in an external beam assembly. In order to achieve a sufficient spatial resolution to characterise the elements in tiny solders, a collimator has been inserted between the exit foil of the accelerator (see figure 10) and the analysed item in order to limit the beam spot to a circular region of 350 µm in diameter at the target site.

TABLE II-1-5. Composition at several sites on the Byzantine cross (Figure 24). All concentrations are in % by weight [28].

Impact n°	Cu	Ag	Hg	Au	
Original basic gold					
1	1.8 ± 0.1	5.6 ± 0.5		92.6 ± 1.5	Flat part
2	1.7 ± 0.1	6.0 ± 0.5		92.3 ± 1.5	Flat part
3	3.0 ± 0.2	6.6 ± 0.6		89.2 ± 1.5	Edge of a lost pearl
4	3.0 ± 0.2	7.1 ± 0.6		89.2 ± 1.5	Top cover
Original wires					
5	2.3 ± 0.2	7.2 ± 0.6		90.5 ± 1.5	Thick wire between pearls
6	3.4 ± 0.2	7.4 ± 0.6		89.2 ± 1.5	Thick wire between pearls
7	2.8 ± 0.2	9.8 ± 0.6		87.4 ± 1.5	Thin wire across a pearl
8	3.0 ± 0.2	8.7 ± 0.6		87.7 ± 1.5	Thick wire across a terminal pearl
Repairs and additions					
9	8.1 ± 0.4	91.9 ± 1.5		-	Silver wire
10	4.6 ± 0.2	4.5 ± 0.5		90.9 ± 1.5	Gold wire (may also be original)
11	100	-		-	Copper wire to bind a new pearl
12	5.5 ± 0.3	9.4 ± 0.7		85.1 ± 1.5	Cap of an additional stone
13	(0.6)	(20)	(7.2)	(72.2)	Central element in silver coated with amalgam
14	4.2 ± 0.2	94.2 ± 1.5		0.6 ± 1.5	Contains also 1% Pb

5.3.2 *Tartesic gold artefacts.*

The jewellery item (P1 of figure 25.a) originates from a fortuitous discovery in the « El Pedroso » area (Sevilla, Spain). This small piece was found with another one, almost perfectly preserved [30], that helped us to rebuild the original shape of P1 and to date it in the Oriental period of « Valle del Guadalquivir ».

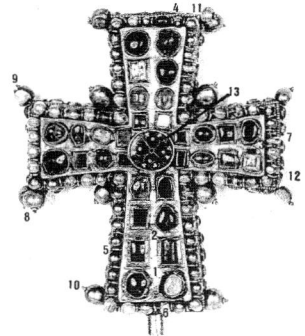

Figure II-1-24. The Byzantine Cross of the Cathedral of Tournai (Belgium).

Because the sample does not have technical details, like filigrees or granulations characteristic of earlier periods, it can be dated to the beginning of that period, possibly in the 5th century B.C.. Each of the 3 elements of the P1 item is almost exclusively made by « repoussé ». The three parts seem to have been soldered together afterwards. This technique enables the modelling of the different elements around the central piece, the shape of which is of typical Oriental style. Although the different components were directly soldered to one another, the solder seems to have been reinforced by adding an inner wire to bind them. The other items (figures 25.b, 25.c and 25.d) are from the « Ebora » treasure and they were found with numerous other Oriental gold items, during agricultural works in the area of « Sanlucar de Barrameda », a village near Cadiz (Spain). These samples are conserved at the Archaeological Museum of Sevilla and have already been widely studied [31-33]. The composition of the alloy in various impact regions on the three jewellery artefacts is given in table 6 [34].

TABLE II-1-6. Copper and silver contents (in % by weight) at impacts of figure 24 [34].

El Pedroso 1 (figure 25.a)			EBORA 3 (figure 25.b)			EBORA 5 (figures 25.c and d)		
Impact N°			Impact N°	Cu	Ag	Impact N°	Cu	Ag
	u	g						
1	3.1	18.5	1	8.1	8.1	1	2.3	14.6
2	2.9	19.4	2	7.5	7.4	2	3.5	16.8
3	2.3	18.5	→ 3	8.2	8.8	3(0)	2.0	13.5
4	3.2	18.7	4	7.6	7.8	4	2.4	16.0
5	2.4	17.5		6.4	7.3	5	2.2	13.3
→ 6	3.0	18.0	6(0)	7.5	7.9	6(i)	4.5	20.5
→ 7	6.6	16.4	7	7.1	7.8	7(i)	4.4	20.4
→ 8	4.3	22.0	8	7.2	7.8	8(i)	4.3	20.1
→ 9	4.2	18.7	9	7.1	7.6	9(i)	4.5	17.6
→ 10	4.0	15.5	10(o)	7.5	7.4	10(i)	3.7	20.1
→ 11	5.2	16.4	→ 11(o)	6.5	7.3	→ 11(g)	3.2	14.8
→ 12	4.7	16.1	→ 12	6.3	8.2	→ 12(g)	2.9	14.7
→ 13	3.6	15.7	→ 13	6.6	8.6	→ 13(g)	3.2	14.5
→ 14	3.9	19.5	→ 14	6.0	8.8	→ 14 (g)	3.0	15.1
→ 15	3.5	15.4	→ 15	7.2	8.1	→ 15(g)	3.0	16.0
16	2.6	17.0	→ 16	7.2	8.3	-	-	-
17	2.8	19.1	17	7.5	6.8	-	-	-
18	2.6	17.9	18	7.5	8.4	-	-	-
19	2.9	19.9	-	-	-	-	-	-
20	3.1	18.6	-	-	-	-	-	-
21	3.2	19.4	-	-	-	-	-	-

The balance to give 100% concerns the gold content. → refer to soldering regions - (i) inner part of the artefact; (g) grazing incidence of incident proton beam.

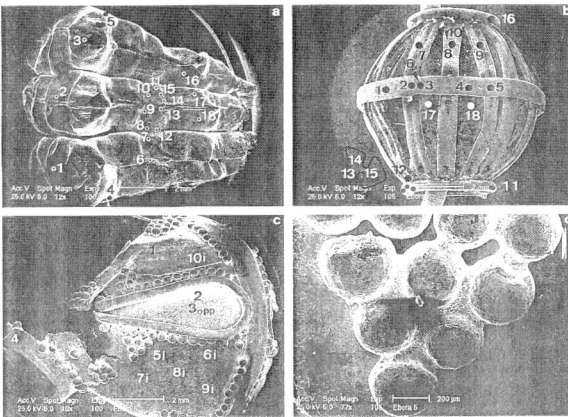

Figure II-1-25. (a) Sample found in "El Pedroso", 5th century B.C. (b) Bead of Ebora made by strips attached on a hollow cylinder. Arrow indicates that the impact corresponds to the symmetric region at the rear. (c) Biconical bead of Ebora (partially broken). Impacts indicated by an "i" refer to inner part of the artefact. Point 3 refers to the opposite. (d) Details of a region of the granulations in the artefact of part C. Brazing is clearly exhibited. The composition at each impact is given in Table 6.

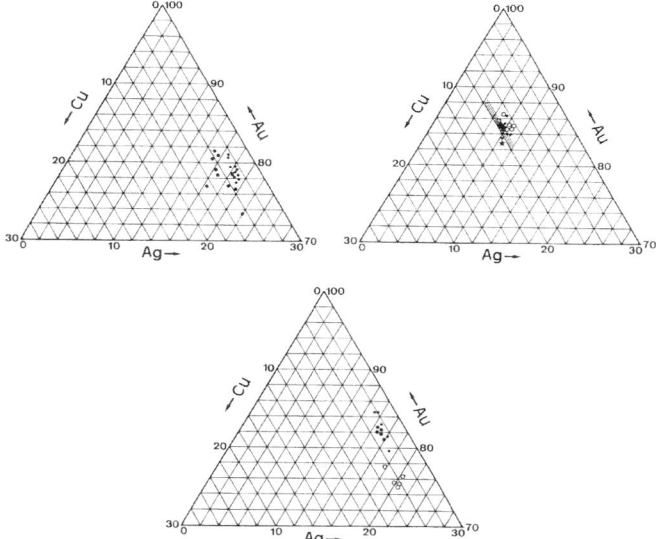

Figure II-1-26. (a) Ternary compositional diagram of sample P1. Small dots indicate impacts in the bulk, large dots refer to soldering regions. (b) Ternary compositional diagram of sample E3. Dots refer to impacts on the strips outside the solders, "open circles" refer to regions of the binding of strips to bottom and upper rings, the star to the soldering of the outer horizontal strip. (c) Ternary compositional diagram of sample E5. Small dots refer to the outside bulk, "open circles" to inside bulk, large dots to granulations.

Highly accurate measurements with a narrow beam (350 μm) have been made in order to be able to suggest an explanation of technological interest. The maximum statistical error on the reported data is ±1% for Cu, ±3% for Ag and ±0.5% for Au. Special attention has been paid to regions of expected solders (indicated by an arrow in the table). It is clear that the concentration of Cu and Ag in those regions is very close to that measured outside the solders. To extract a more pertinent interpretation, compositional data are displayed in the ternary diagrams of figure 26 (a, b, c)

For sample P1, the results outside the solders indicate that the relative Ag/Cu concentration ratio is 6.5, with gold contents ranging from 77 up to 81%. This range of variation clearly indicates that the material (a natural electrum) is not fully homogeneous: the reported values vary by much more than the statistical error. In the solder regions, the Cu content is systematically higher and the results are largely scattered. This situation may be explained by a brazing procedure to attach the elements together: a liquid alloy with a temperature of fusion of 50 to 80° C below the melting point of the bulk material was used. The scattering of the results in the ternary diagram is partially due to the width of the incident proton beam which may extend beyond the soldering regions.

In sample E3, points are closely distributed around 84.5% Au, 8% Ag and 7.5% Cu. A closer look indicates that at impacts outside the solders (dashed area in the ternary diagram) the copper concentration is a bit higher than in almost all soldering regions. The lower Cu concentrations at solders may be understood if we think that some forging procedure has been used to bind strips together. No additional material is used: parts of the item to be bound are simply locally heated at a temperature slightly higher than the solidus temperature. The metal is partially fused and copper, the less noble metal in the alloy, is selectively eliminated by forming oxides that are lost during the final procedure of brushing. At point 3, both copper and silver are slightly more abundant than anywhere and we may think that a brazing procedure has been used to bind both ends of the central circular ring. The central ring is then soldered by a brazing procedure; the strips on the bottom and upper rings are bound by forging.

For sample E5, the experimental values may be distributed in three groups: a) outside any solder in the outer part of the item, b) outside any solder in the inner part the item and c) into the granulations. The gold concentration inside is lower than outside. Furthermore, if we compute the Ag/Cu ratio, we obtain 5 at impacts inside the hollow sphere and 6.1 outside. These observations may be interpreted by some enhancement of both Au and Ag at the external surface produced either by an original surface treatment in order to increase the golden appearance or a chemical treatment to clean (after the excavation) the external surface and to give the object a « better look ». Eliminating the less noble metal was the aim of this cleaning process. In granulations of dimension 200 μm, i.e. lower than the beam size (350 μm), we observe that some brazing procedure involving the melting of a fusible alloy containing more copper was used. The embedded granulations are clearly shown in the micrograph of figure 24.d.

Other Tartesic items have also been studied [35] and the conclusion was that the soldering procedures were forging (or casting) and brazing, but there is no indication of solid state bonding like in Etruscan jewellery.

5.3.3   *Mesoamerican jewellery*

In pre-Hispanic Mexico and Mesoamerica, metallurgy was developed only a few centuries before the Spanish conquest, in a period called "post-classic" (900-1530 A.D.). Indeed, the basic skills of gold metallurgy were imported from South America, where the first methods of copper and gold foiling were developed before the early horizon (about 500 B.C.). Casting, gilding, silvering and soldering, among other techniques, were invented some centuries later. Despite the foreign origin of this technology, the level of performance reached by the goldsmiths of Mesoamerica, and especially in the South of Mexico (Oaxaca) was remarkable. People from this region, the Mixtecs, achieved a skill for casting and gilding gold-based alloy artefacts that was recognized by other Mesoamerican peoples and the Spanish conquerors. Gold alloys artefacts were used by religious people and leaders of the society. Their colour was very important for religious and cultural reasons. In ancient America, metallurgy was developed following the concept of the manipulation of the items colour [37] and this technology was not known in the Old World at that time. The gold alloys used in Mesoamerica (called tumbaga) contained considerable amounts of copper and/or silver and their colours varied from red to white-yellow, depending on the alloy composition.

The most developed techniques in this cultural area were the lost wax and the false-filigree techniques obtained by casting and followed by a procedure of depletion gilding. In South America, soldering was frequently used before the Spanish conquest [38] whereas it appeared to be unknown in Mesoamerica. To manufacture an artefact by the lost wax technique, a core of clay and charcoal was prepared. The shape of this core is close to the final artefact. The core is then covered with wax; the thickness of the wax layer and details in its decoration will determine the thickness of the metal to be casted and the final aspect of the item. Some wax rods are added to the model to provide air vents as well as the casting tunnels through the external cover of the mould made with clay. The hot mold is finally poured with the molten metal trough the tunnels. After cooling, the mold is broken and the core supports are removed. Finally, the artefact is cleaned and polished. False-filigree is based on the application of wax wires to model the artefact. In this way after the lost wax process, very fine and delicate features can be achieved without soldering the wires onto the artefact [39].

The items could undergo successive processes of surface oxidation in order to remove Cu and Ag as oxides and then produce a gilded aspect. To oxidize the surface, the object is heated at a temperature above 500°C. The heated artefact is then dipped in a solution of salts and acids from plants [40]. After several steps of heating and dipping procedures, the surface is "enriched in gold" and appears to be "golden". This process affects a depth up to several microns depending on the number of steps applied and on the original alloy composition. Another gilding technique in ancient America could be the plating of copper with gold alloys using "electrochemical" processes [41].

The objects studied in the present paper include chin and ear ornaments from the Mixtec culture corresponding to the post-classic period from Oaxaca, Mexico. These are all the gold artefacts collection of the exhibition "Mobilier Funéraires des

Zapotèques et Mixtèques", presented during the Europalia Mexico 93 Festival in Brussels (Belgium) [42]. Results of analyses help to determine the metallurgical techniques used at that time and to complement the information on the ancient gold metallurgy of Mesoamerica. Most of the data were obtained using PIXE in the non-vacuum arrangement. We present here only results on several artefacts, all other data may be found in references [43] and [44]. The serpent chin ornament (figure 27.a and Table 8) has almost the same composition at impacts 1 to 4, 7 to 12, and 21, i.e. (65.2 ± 2.6) % Au, (31.5 ± 3.0) % Ag and (3.3 ± 0.7) % Cu, Au-Lβ and Au-Lγ results differ by less than 3%. The measurements have been performed using Lβ and Lγ rays. The reason of this double choice is the possibility to check the surface homogeneity by the way that Lβ and Lγ rays suffer different absorption in copper-rich alloys. The reason is easily understood if we observe that Lγ rays are more efficiently absorbed in gold than its own Lβ rays (Table 7).

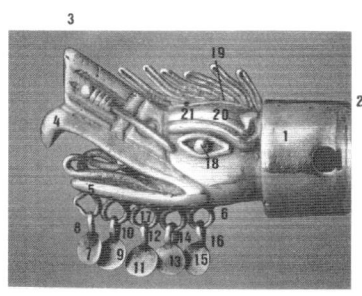

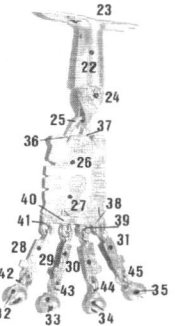

Figure II-1-27.a. Serpent chin ornament (length 4.5cm; height 2.5 cm; width 2.7cm). The analytical results at various impact positions are listed in Table 8.

Figure II-1-27.b. Eagles chin ornament (length 7.5cm; width 2.9cm; thickness 0.7 cm). The analytical results at various impact positions are listed in Table 9.

TABLE II-1-7.   Atomic mass absorption coefficients (μ) of Cu-Kα and Au-Lβ X-rays in a homogeneous Au-Ag-Cu matrix.

X-ray	Energy (keV)	μ (cm²/gr)		
		Au	Ag	Cu
Cu-Kα	8.05	210.5	221.3	52.3
Au-Lβ	11.50	84.5	83.1	152.3
Au-Lγ	13.38	156.8	54.7	101.5

TABLE II-1-8. Elemental composition of the serpent chin ornament. The measuring positions are illustrated in Figure 27.a.

Impacts	Au-Lβ			Au-Lγ		
	% Au	% Ag	% Cu	% Au	% Ag	% Cu
1	63.8	32.9	3.3	61.3	35.2	3.5
2	64.3	32.5	3.2	63.3	33.4	3.3
3	65	31.7	3.3	64	32.6	3.4
4	65.9	30.4	3.7	64.3	31.8	3.9
5	69	28.1	2.9	69.1	28	2.9
6	70.9	25.7	3.4	72.4	24.4	3.2
7	65	32	3	63.7	33.2	3.1
8	65.6	31.4	3	63.3	33.4	3.3
9	65.2	31.5	3.3	63.3	33.2	3.5
10	66.6	30.3	3.1	65.5	31.3	3.2
11	67	29.4	3.6	66.1	30.2	3.7
12	66.3	30.4	3.3	67.2	29.6	3.2
13	67.9	28.4	3.7	68.3	28	3.7
14	67.1	29.4	3.5	67.7	28.9	3.4
15	65.4	31.3	3.3	65.7	31	3.3
16	67.9	28.9	3.2	68.7	28.2	3.1
17	64.3	32.5	3.2	61.7	34.8	3.5
18[1]	64.9	31.8	3.3	64.4	32.3	3.3
19[1]	67.1	29.9	3	66.7	30.2	3.1
20	69.8	27.9	2.3	68.8	28.8	2.4
21	66.7	30.7	2.6	66.3	31.1	2.6

The comparison of results in both parts of Table 8 leads to the conclusion that the surface composition represents also the bulk. The largest variations in comparison with the mean values are observed in positions 5 and 6. The greater gold concentration at positions 5 and 6 (70.4 ± 1.3) % Au, (26.5 ± 1.5) % Ag, (3.1 ± 0.2) % Cu, may be the result of a soldering process without any brazing alloy. In that case local fusion could have been used to fix the rings. During heating, less noble metals like Ag and Cu oxidize as observed also in the Tartesic jewellery presented above. After cleaning the solder, these oxides are removed and the Au concentration is enhanced in the corresponding region. From all those measurements one may deduce that the artefact was completely made by the lost wax technique.

This technique was also used for the eagles chin ornament (Figure 27.b and Table 9). The compositions at positions 22 to 25 and 26 to 45 are quite different. In the upper part of the ornament, the gold content is (78.8 ± 0.4) % with (17.1 ± 1.4) % of Ag and (3.6 ± 0.4) % of Cu; while in the rest of the item one observes a lower concentration in gold: (66.7 ± 2.0) % Au; (28.5 ± 2.9) % Ag; (4.8 ± 1.3) % Cu. The melting point of this last alloy is lower than that used for the upper part. Consequently, we could conclude that two different gold alloys were used for manufacturing this item. No solders were observed, even in regions of rings (impacts 36 to 45). The concentration differences and consequently the melting points of each gold alloy were probably considered instinctively by the goldsmith in order to avoid damages to the fine eagle features during the casting of those additional parts. The ancient American goldsmiths were able to handle melted gold alloys with temperature differences of about 75° C [40].

TABLE II-1-9. Elemental composition of the eagles chin ornament. The measuring positions are indicated in Figure 27.b.
Comparison of results using Lβ or Lγ ray of gold indicates that the material did not undergo surface gilding.

Impacts	Au-Lβ			Au-Lγ		
	% Au	% Ag	% Cu	% Au	% Ag	% Cu
22	78.4	17.7	3.9	78.8	17.4	3.8
23	78.5	18	3.5	77.9	18.5	3.6
24	78.4	18.3	3.3	77.5	19.1	3.4
25	79.9	16.2	3.9	79.8	16.3	3.9
26	67.6	28.8	3.6	66.4	29.9	3.7
27	68.7	27.8	3.5	68.7	27.9	3.4
28	67.3	28.1	4.6	66.9	28.5	4.6
29	66.7	28.1	5.2	66.4	28.4	5.2
30	67.3	26.9	5.8	66.6	27.5	5.9
31	64.6	30.3	5.1	63.5	31.2	5.3
32	66.6	28.8	4.6	65.6	29.7	4.7
33	69.8	25.7	4.5	70.7	24.9	4.4
34	69.6	25.6	4.8	68.7	26.4	4.9
35	67.2	28.1	4.7	65	30	5
36	65.7	29.4	4.9	63.8	31.1	5.1
37	64.4	30.8	4.8	62.2	32.7	5.1
38	66.9	28.1	5	67.2	27.9	4.9
39	65.8	29.5	4.7	65.5	29.8	4.7
40	64.9	30.1	5	65.1	30	4.9
41	65.8	29.4	4.8	63.8	31.1	5.1
42	66	29.3	4.7	65.4	29.8	4.8
43	66.1	27.7	6.2	68	26.2	5.8
44	66.5	27.8	5.7	66.9	27.5	5.6
45	64.9	30.4	4.7	66	29.5	4.5

For flat samples, a sensitive parameter to check the relative homogeneity of bulk compared to the surface is to measure the Cu Kα/Au Lβ yield ratio induced by protons in various geometries of irradiation. This ratio was used as a surface sensitive monitor to characterize depth dependence changes in the gold composition. Some differences in this ratio are expected if a gilding procedure was applied to the item. Dashed lines in figure 28 show the Cu Kα/Au Lβ yield ratios calculated as a function of the beam incidence angle for various homogeneous matrices and for a matrix with a layered structure. As the beam incidence angle θ_1 increases, the projected range for the protons changes in a gold alloy matrix from 24 to 3 µm when we start from normal incidence of the beam to grazing incidence. For a homogeneous matrix the Cu Kα/Au Lβ yield ratio increases with the increase of grazing because the Cu Kα X-rays absorption is stronger. This is clear from the difference of the atomic mass absorption coefficients for Cu Kα and Au Lβ radiation in (mainly) gold (Table 7). The variation of Cu Kα/Au Lβ ratio is more dependent of the choice of the angle of incidence than the Au-Lβ/Au Lγ ratio.

On the contrary, in case of gold surface enrichment, the Cu Kα/Au Lβ yield ratio decreases as the beam grazing increases due to the higher content of gold as shown in the continuous lines in figure 28. It is clear that the behaviour of this ratio depends on the degree of homogeneity of the matrix and this ratio can be used to study the (non-)homogeneity of the target. Figure 29 shows the experimental Cu Kα/Au Lβ yield ratios for various artefacts and the calculated ratios for homogeneous matrices with similar elemental compositions. The alloy compositions were obtained from the results at non-grazing incidence (the reference alloy is also included). For the reference material, the Cu Kα/Au Lβ yield ratio increases with the beam incidence angle and follows the expected behaviour for homogeneous matrices. This behaviour is observed for the serpent chin ornament (figure 27.a), whereas for the ear ornaments (1 and 2) and the butterfly pendant (see figures 30.a, b, c), the corresponding ratios decrease.

The incidence angle dependence of the Cu Kα/Au Lβ yield ratio measurement is a simple method to check the surface bulk homogeneity. Unfortunately, this method cannot be applied to all items because it requires a large flat surface. For irregular items, a differential PIXE analysis (45,46) using various incident proton energies has to be performed to study the homogeneity of the target. A PIXE result at one beam energy represents only the mean elemental concentration within the information depth. In the case of a non-homogeneous matrix, these results can be wrong. At least two very different beam energies (e.g. less than 1 MeV and 2 MeV or more) have to be used in order to decide whether the matrix is homogeneous or not. If an external beam set-up is used, it is not advisable to use low beam energy because the energy loss of the protons in the Al window becomes high and could induce fusion of the foil. In order to change the beam energy, the target-beam distance can be changed, but energy and angular straggling of the beam is a limitation. Nevertheless, non-vacuum analysis provides very useful information about the mean elemental composition of the item. This method is a first step to characterize the artefacts.

Another way to identify a gilding procedure is to use RBS and PIXE simultaneously. The experiment is then made in vacuum. Two items of Columbian origin and belonging to a private collection from France are shown in figures 31 and 32. The corresponding RBS spectra recorded for 2 MeV protons are given in figures 33 and 34. Similar spectra have been also recorded for proton energies ranging from 600 keV to 2.6 MeV. The recorded X-ray spectra indicate that the apparent concentration of copper increases when the incident proton energy is increased. When the energy is increased in that way, the analysed depth concerns deeper and deeper layers. Results at 600 keV yield the approximate concentration of gold at the surface, those at 2.6 MeV are very close to the bulk composition. The computation of RBS and PIXE results at various energies allows us to determine the depth profile as given in Table 11. Developments of that procedure involving PIXE and RBS at various proton energies are continuously improved at LARN (Namur) and at UNAM (Mexico).

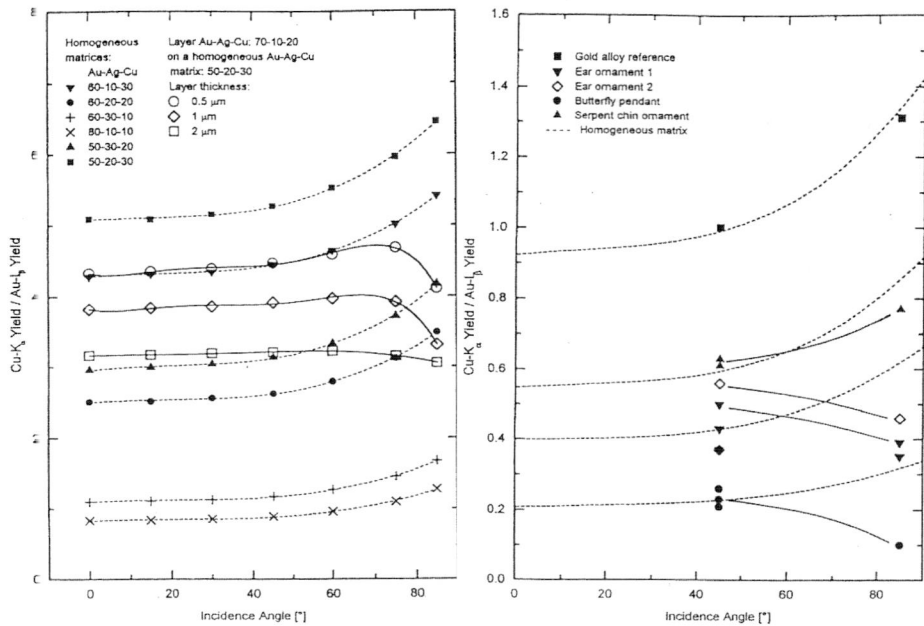

Figure II-1-28. Calculated Cu-Kα/Au-Lβ yield ratios as a function of the proton beam incident angle for various homogeneous matrices and for matrices with a layered structure (2.7 MeV proton beam energy 45° detection angle). The ratio increases as the beam incidence increases for a homogeneous matrix.

Figure II-1- 29. Experimental Cu-Kα/Au-Lβ yield ratios at 45° beam incidence and at grazing beam incidence (85°) for various artefacts compared with the calculated Cu-Kα/Au-Lβ yield ratios for homogeneous matrices with a similar composition to the artefacts. The detection angle is 0° and the beam energy is 2.4 MeV. Some artefact ratios follow the homogeneous matrix (figure 28).

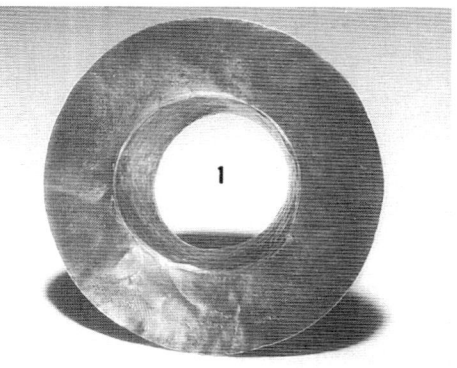

Figure II-1-30.a Ear pendant of Mixtec origin found in Oaxaca.

Figure II-1-31. Solar disk of Columbian origin (1000-1500 A.D) (Private Collection).

Figure II-1-30.b. Sun shape ear pendant of Mixtec origin found in Oaxaca.

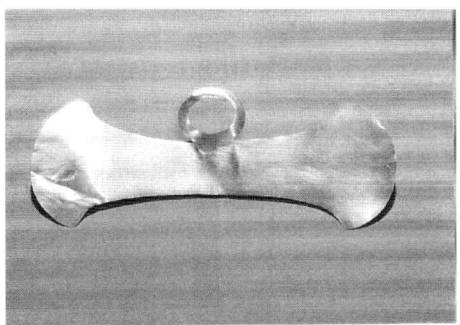

*Figure II-1-30.*c. Nose ornament of butterfly shape origin found in Oaxaca. collection).

Figure II-1-32. Zoomorphic pendant of Columbian (1000-1500 A.D.) (Private

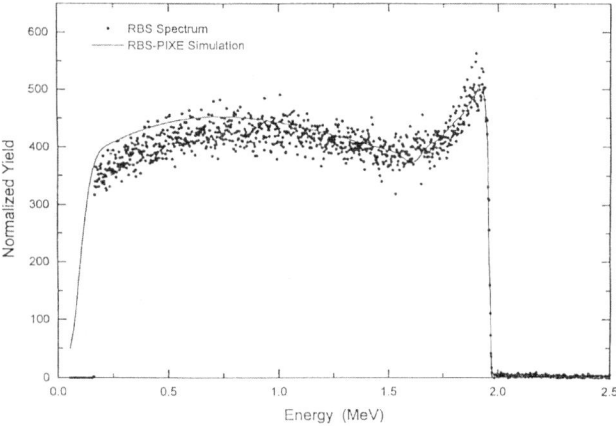

Figure II-1-33. RBS spectrum of 2 MeV protons on the solar disk.

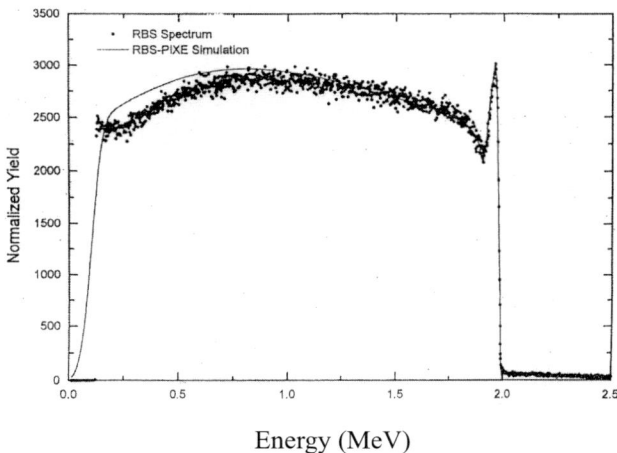

Energy (MeV)

Figure II-1-34. RBS spectrum of 2 MeV protons on the zoomorphic pendant.

References

1.   Deconninck G., Demortier G., Bodart F.: Application of X-ray production by charged particles to elemental analysis, *Atomic Energy Review*, **vol.13, N°2** (1975) 367-412

2.   Kiss A., Elemental Analysis Based on Nuclear Reactions in *"Applications of Ion Beam Analysis Techniques to Arts and Archaeology"*, Eds. M. Respaldiza and J.Gomez Camacho, Universidad de Sevilla (1997).

3.   Lahanier Ch., Amsel G., Heitz Ch., Menu M., Andersen M.H. (ed.): Ion beam analysis in the arts and archaeology, *Nucl.Instr. and Methods in Physics Research* **B14** (1986) 1-167

4.   Demortier G. (ed.): Proceedings of the Second International Meeting on Chemical Analysis by Charged Particle Bombardment. *Nucl. Instr. and Methods* **197** (1982) 1-258

5.   Demortier G.: Applications of nuclear microprobes in Archaeology in *"Principles and Applications of High Energy Ion Microbeams"* (F.Watt and G.W.Grime ed., Adam Hilger, Bristol) (1987) 333-377

6.   Deconninck G., *"Introduction to Radioanalytical Physics"*, Akadémiai Kiado (Budapest), Elsevier Scientific Publishing Company (Amsterdam), (1978)

7.   Deconninck G., Proceedings of the 4th Conf. on the Scientific and Industrial Applications of Small Accelerators, J.L.Duggan and I.L.Morgan (ed.) 76CH 1175-9NPS (1976) 533-542

8.   Cohen D.D. and Harrigan M., *Atomic Data and Nuclear Data Tables*, **33** (1985) 255, and *Atomic Data and Nuclear Data Tables*, **34** (1986) 393

9.   Demortier G.: Differential PIXE analysis of Mesoamerican jewelry items, *Nucl.Instr. and Methods in Physics Research* **B118** (1986) 352-358

10. Ruvalcaba-Sil J.L., Demortier G., Misaledis P. ed., *Application of Particle and Laser Beams in Materials Technology* (1995) 463-470, Kluwer Academic Publ.and :Ruvalcaba-Sil J.L., Demortier G.: ECAART Conference, Zurich 1995, *Nucl. Instr. and Meth*. **B113** (1996) 275

11. Demortier G., *Nucl. Instr. and Meth*. **B49** (1990) 46-51

12. Morrisson C., Barrandon J.N., Brenot C., Callu J.P., Halleux R., Poirier J.: *Numismatique et histoire. L'or monnayé de Rome et Byzance : purification et altération.* Académie des inscriptions et belles lettres. Compte-srendus des séances de l'année 1982, avril-juin (1982), Paris, Diffusion de Broccard, 203-223

13. Meyer M.A., Demortier G. : *Nucl. Instr. and Meth. in Physics Research,* **B49** (1990) 300-304

14. Notis M.R., Pigott V.C., McGovern P.E., Liu K.H., Swann C.P., University of Pennsylvania, *Museum Monographs* **65** (1988) 271

15. Swann C.P., Fleming S.J., *Scanning Microscopy*, **2** (1988) 197

16. Schenk H., *Masca J.,* **2** (1982) 42

17. Notis M.R., McGovern P.E., Moyer H., Pigott V.C., Swann C.P., *Museum Monographs*, **65** (1988), 278

18. Demortier G., Bodart F., Hackens T., Comptes-rendus du 20ème congrès d'archéométrie 26-29 mars (1980). *Revue d'Archéométrie* **3** (1981) 63-72

19. Dion Cl., *Mémoire de Licence LARN-FUNDP* (1978)

20. Chen H.S., Chen J.X., Ren G.C., Xu Z.W., Yang F.C., Zhao G.Q., Zhuo Z.Y., *Nucl. Instr. And Methods*, **191** (1981) 391-396

21. Fleming S.J., *Masca Journal*, **1** (1979) 73

22. Moorey P., Fleming S.J., *Masca Journal,* **1** (1979) 73

23. Mac Clellan J.A., *Masca Journal,* **2** (1983) 114

24. Swann C.P., *IEEE Trans. Nucl. Sci,* **NS-30** (1983) 1298-1301

25. Hughes M.J., Cowell M.R., Craddock P.T.: *Archaeometry,* **18** (1976) 26

26. Pigott V.C., McGovern P.E., Notis M.R., *Masca Journal,* **2** (1982) 35and : Schenck H.R., *Masca Journal,* **2** (1983) 107

27. Bird J.R., Duerden P., Wilson D.J.: *Nuclear Science Applications, section B.:"Ion beam techniques in archaeology and in the arts",* **vol. 1**   (1983) 357-516

28. De Cuyper F., Demortier J., Dumoulin J., Pycke J.: « *La croix Byzantine du Trésor de la Cathédrale de Tournai* », **Aurifex 7**, UCL Press, Louvain-la-Neuve (1987) 1-88

29. Demortier G.: « LARN experience in non destructive analysis of gold artefacts », *Nucl. Instr. and Meth*. **B14** (1986) 152-155

30. Fernandez Gomez F., *"Tesoros de la Antiguedad en el Valle del Guadalquivir",* Cordoba-Sevilla (1977), Figure 101, Caja Sur Publications.

31. Blanco de Torrecillas C.: "El tesoro del Cortijo de Ebora", *Archivo Espagnol de Arqueologia*, **XXXII**   (1959) 50-57

32. Nicolini G.: *"Technique des Ors Antiques. La bijouterie ibérique du IVème siècle »*, Paris (1990)

33. Perea A.: *« Orfebreria prerromana. Arqueologia des oro »*, Madrid (1991), Caja Sur Publications

34. Ontalaba-Salamanca M.A., Demortier G., Fermandez-Gomez F., Coquay P., Ruvalcaba-Sil J.L., Respaldiza M.A.: "PIXE and SEM studies of Tartesic Gold Artefacts", *Nucl. Instr. And Meth.* **B-136-138** (1998) 851-857

35. Demortier G., Fernandez-Gomez F., Ontalba-Salamanca M.A., Coquay P.: "PIXE in an external microbeam arrangement for the study of finely decorated Tartesic gold jewelry items", *Nucl. Instr. and Meth. in Phys. Research* **B158** (1999) 275-280

36. Bray W.: "Gold working in ancient America", *Gold Bull.* **11(4)** (1978) 136-143

37. Bray W.: "Techniques of gilding and surface enrichment in pre-Hispanic American metallurgy", in *Metal Plating and Patination*, S.La Niece and P.Craddock eds., Butterworth Heinemann, Oxford (1993) 182-192

38. Scott D.A., Doehne E.: "Soldering with gold alloys in ancient south America: examination of two small gold studs from Ecuador", *Archaeometry* **32(2)** (1990) 183-190

39. Miller A.M.: "Gold techniques in the new world in pre-Christian Era 2000 B.C.-1500 A.D.", in "Outils et ateliers d'orfèvres des temps anciens", Christiane Eluère (comp.), *Musée des Antiquités Nationales*, **mémoire 2** (1993) 255-258, Saint-Germain-en-Laye, France

40. Scott D.A. : *« Depletion gilding and surface treatment of gold alloys from the Narino aera of ancient Colombia"*, *Journal of the Historial Metallurgy Society* **17(2)** (1983) 99-115

41. Lechtmann H.: "A pre-Columbian technique for electrochemical replacement plating of gold and silver on copper objects", *J.Metals*, **December 1979**, 154-160

42. Solis F., Carmona-Macia M., Carmona-Ortiz E.: "Mobilier funéraire des Zapotèques et Mixtèques", *Catalogue of Exhibition, Europalia 93 Mexico*, Brussels (1993)

43. Ruvalcaba-Sil J.L., Demortier G., Oliver A.: "External beam PIXE analysis of gold prehispanic Mexican jewelry", *International Journal of PIXE*, **vol.5, n°4** (1995) 273-288

44. Ruvalcaba-Sil J.L., Demortier G.: "Non destructive analysis of American gold jewellery items by PIXE, RBS and PIGE" in Applications of Particle and Laser Beams in Materials Technology", P.Misaelides ed., *NATO-ASI Series: Serie E: Applied Science*, vol.**283** (1995) 463-470; Kluwer Academic Publishers, The Netherlands

45. Demortier G., Ruvalcaba-Sil J.L., "Differential PIXE analysis of mesoamerican jewelry items", *Nucl. Instr. and Meth. in Phys. Research* **B118** (1996) 352-358

46. Demortier G., "Complementarity of RBS, PIGE and PIXE for the determination of surface layers of thicknesses up to 25 microns", *Nucl. Instr. and Meth.* **B49** (1990) 46

Chapter II-2

The Origin of Ancient Gemstones Unveiled by PIXE, PIGE and μ-Raman Spectrometry

T. Calligaro
Centre de Recherche et de Restauration des Musées de France,
CNRS UMR-171
Palais du Louvre, 75041 Paris cedex 01
thomas.calligaro@culture.gouv.fr

Keywords: ancient gemstones, PIXE, PIGE, Raman spectroscopy, non-destructive,
      in air, ruby, emerald, garnet

Abstract

Gemstones set on ancient jewels are invaluable witnesses from the past. To obtain these much appreciated items, Men had established long-distance trade routes, the extension and historical evolution of which represent a major archaeological issue. For these studies, chemical characterisation has proved to be very useful. First, the elemental composition directly provides the nature of the gemstone, like for example, SiO_2 for quartz, Al_2O_3 for ruby and sapphire, $Be_3Al_2Si_3O_{18}$ for emerald. Secondly, the trace element content of a gemstone, when compared to reference gems of known origin, may unveil its provenance. However, to perform the chemical characterisation of these valuable items, suitable analytical methods must be non-invasive (neither sampling nor dismounting of the gems from the jewels) and non-destructive (no damage to the gem). Ion beam techniques such as particle-induced X-ray emission (PIXE) or particle-induced γ-ray emission (PIGE) meet these specific requirements. PIXE permits the measurement of all elements heavier than sodium with a sensitivity reaching the μg/g, while PIGE extends the range of elements to lighter ones (Li, Be, B, F). These totally harmless techniques are directly applied to the gem in air with a 20-μm spatial resolution, allowing selecting specific regions of the crystal. The identification of inclusions, a provenance criterion often used in gemmology, can be complementarily performed by Raman micro-spectrometry. Indeed, the inclusions are often located out of reach of the ion beam in the crystal. This paper presents the benefits of the combined use of external beam PIXE/PIGE and Raman micro-spectrometry. The use of these modern analytical techniques is exemplified by the study of ancient gems: rubies from a Mesopotamian statuette, emeralds and garnets set on Barbarian jewels dating from the Dark Ages.

101

M. Uda et al. (eds.), X-rays for Archaeology, 101–112.
© 2005 *Springer. Printed in the Netherlands.*

1. Introduction

Gemstones are defined as precious minerals combining high hardness, bright colour, nice brilliance and good transparency. Among the rare minerals exhibiting these features, the most important ones are diamond, ruby, sapphire and emerald. These gems are called precious stones. Gems have fascinated men since the earliest times and have been sought for their aesthetic and symbolic value. Their superior hardness and hence strong resistance to alteration is interesting from an archaeological point of view as gemstones can travel through time without noticeable weathering, even buried in an aggressive environment. Like spices and other precious and sought-after items, gems used to be recovered from long distance and therefore can be used today as archaeological tracers of ancient trading routes. Each civilisation had its own favourite gemstones: lapis-lazuli and turquoise in Ancient Egypt, amber and emerald in Roman Empire, red garnets for Germanic peoples of the Dark Ages; jade, ruby and sapphire in Far-East cultures. Two major issues concern ancient gems. The first is the identification of the mineral, since its often poor crystal quality (compared to modern ones), rough polish and irregular shape can be misleading. The second issue is the determination of its provenance, often scarcely documented in ancient texts or even completely unknown. For all these reasons, the study of gems conserved in museum collections by modern analytical techniques is a promising research field.

2. Analytical methods suitable for the characterization of gemstones

In gemmology, the identification of gemstones usually relies on the measurement of optical and mechanical properties: refractive index n, birefringence B, visible optical spectrum, and less frequently specific gravity ρ and hardness H [1]. However, since ancient gemstones are frequently mounted in a metallic setting and their surface is seldom flat (often in round shape) and transparent, these simple measurements are often problematic (any dismounting from the setting or sampling being of course prohibited). Under these circumstances, gemstones can still be directly identified by means of their chemical composition. For instance, diamond is a pure carbon crystal; ruby is an aluminium oxide crystal (Al_2O_3) coloured by less than 1% Cr; sapphire is also an Al_2O_3 crystal but its blue colour is due to trace amounts of Ti and Fe; emerald is a beryl of formula $Be_3Al_2Si_6O_{18}$ coloured by traces amounts of Cr. Table 1 summarises the properties of important gems. Note that gems are often composed of light elements (e.g. Be, Li, C, O, F, Al, Si, see for instance the tourmaline); therefore techniques suitable for their chemical analysis must be able to measure a wide range of elements, ideally extending down to hydrogen. The provenance of gems, i.e. the geographical location of their occurrence, is usually determined by the observation of microscopic features inside the crystal such as mineral inclusions, by means of an optical microscope. Raman micro-spectrometry, very effective for the identification of mineral phases, is being increasingly used for this task [2]. Chemical analysis can also be employed to establish the provenance by measuring trace elements present in gems. In fact, trace elements incorporated in the gem crystal during its growth are linked to its geological context (type of terrain and rock-forming history) and can therefore be used as a fingerprint of its occurrence. Recently a new method to determine provenance has been introduced based on the slight variation of isotopic composition, which also depends on the formation conditions of the gem crystal. For example, the concentration of isotope 18 of oxygen ($^{18}O/^{16}O$ ratio) has been successfully applied to determine the origin of ancient emeralds [3]. The isotopic ratio is measured using secondary ion mass spectrometry, a micro-destructive method performed in vacuum that can only be applied to small objects.

Table II-2-1. Properties of a few gems: n is the refractive index, B birefringence, ρ specific gravity, H hardness. In the colour column are indicated the trace elements responsible for the colour of the gems.

	Formula	n	B	ρ	H	Structure	Colour
diamond	C	2.42	0.000	3.52	10	cubic	/
ruby	Al_2O_3	1.77	0.008	4.00	9	rhomboedric	Cr
sapphire	Al_2O_3	1.77	0.008	4.00	9	rhomboedric	Fe, Ti
emerald	$Be_3Al_2(SiO_3)_6$	1.58	0.006	2.71	7.5	hexagonal	Cr, V
topaz	$Al_2(F,OH)_2SiO_4$	1.63	0.010	3.54	8	orthorhombic	Fe, Cr
tourmaline	$Na(Li,Al)_3Al_6(BO_3)Si_6O_{18}(OH)_4$	1.63	0.018	3.06	7.5	rhomboedric	Mn, Fe
quartz	SiO_2	1.55	0.009	2.65	7	rhomboedric	/
amethyst	SiO_2	1.55	0.009	2.65	7	rhomboedric	Fe
garnet	$X_3Al_2(SiO_4)_3$, X=Fe,Mg, Mn	~ 1.8	0.000	4.00	7.5	cubic	Fe, Cr

Table 2 summarises the main features of several analytical techniques used to characterise gems: XRF, scanning electron microprobe, neutron activation, mass spectrometry using ICP/MS and SIMS, Raman micro-spectrometry. Ion Beam Analytical methods have the advantage to combine non-destructiveness, a wide range of measured elements together with a good sensitivity and accuracy. The chemical analysis by IBA can usefully be combined with the structural information delivered for instance by Raman spectrometry. In any case, a careful observation of the gems and their internal features by conventional gemmological means has to be performed prior to proceeding to chemical analysis. However, IBA techniques cannot be used for isotope determination nor does it give information on chemical environment (valence, bonds).

Table II-2-2. Features of several analytical techniques that have been applied for the chemical characterisation of gems. [1] X-ray fluorescence in air, [2] Scanning electron microscope with energy-dispersive microanalysis, [3] Inductively coupled plasma with mass spectrometry and laser ablation, [4] Secondary ion mass spectrometry, [5] μ-Raman spectrometry, [6] Neutron activation analysis, [7] Ion beam analysis.

	any size	non-destructive	no samp preparat.	sensitivity μg/g	lateral resol μm	depth μm	Z range	isotopic analysis	accuracy %
XRF [1]	√	√	√	> 100	~ 1000	1 - 100	> 11	no	10-20
SEM [2]	no	√	coating	> 1000	~ 1	~ 1	> 5	no	5-10
ICP/MS [3]	√	no micro	√	< 1	~ 10	< 10	all	possible	10-20
SIMS [4]	no	no micro	√	< 1	~ 1	< 1	all	√	10-20
μ-Raman [5]	√	√	√	identificat.	~ 1	> 1000	n/a	no	n/a
NAA [6]	no	activation	√	< 1	none	> 1000	selective	no	< 5
IBA [7]	√	√	√	> 1	~ 10	1 - 50	all	no	< 5

2 1. CHEMICAL ANALYSIS BY ION BEAM METHODS

We shall not give a complete description of Ion Beam Analysis (IBA) techniques, as comprehensive reviews are available in textbooks [4,5]. Let us just recall that the principle of IBA techniques relies upon the detection of the products of interaction of an ion beam of a few MeV with the target, that is the gem. Among these techniques, PIXE (*particle induced X-ray emission*) is similar to XRF as it is based on the detection of X-ray emitted by the target atoms, subsequent to an inner-shell ionisation. The only difference stems from the excitation source, which is a charged particle beam instead of an X-ray beam. Because in PIXE the X-ray emission occurs almost without background, this method has a very good sensitivity (reaching the $\mu g/g$ level for the transition elements, which are implied in the colouring mechanism of gems). The lightest measurable element depends on the ability to detect low energy X-rays. With a minimum X-ray energy set to 1 keV, PIXE can measure all elements starting from sodium. For the PIGE technique, acronym for *particle induced γ-ray emission*, there is no such conventional counterpart as for PIXE. PIGE is based on a nuclear reaction between the incident particle and the nucleus of a target atom. Following this reaction, the nucleus emits a γ-ray with a specific energy. This reaction occurs when the incident particle is able to surmount the Coulomb repulsive barrier of the nucleus, a situation mainly met for light target atoms (more specifically Be, Li, B, F when using a proton beam). Therefore, PIGE usefully extends the range of PIXE (from Na to U) to these elements. For the measurement of hydrogen in gems, specific experimental arrangements have been developed to carry out ERDA [6] (*elastic recoil detection analysis*) with a He beam or resonant nuclear reaction with a ^{15}N beam [7].

From the point of view of gemmology, the IBA methods present several interesting advantages. First, in-air IBA techniques performed with an external beam allow *in situ* and non-destructive analysis of gems without any target preparation. This harmlessness is necessary for the study of valuable objects such as historical jewels. Moreover, the availability of a micro-beam with a size of less than 20-μm permits to select inclusions for their identification. Such a small probe is also useful to select an inclusion-free region of the crystal. Indeed, with a broader spot size, there is always the risk that an inclusion incorporated in the analysed area may bias the mean trace element concentration of the crystal (e.g. an ilmenite $FeTiO_3$ inclusion in a ruby might lead to a wrong Fe and Ti mean concentrations).

From the point of view of the IBA techniques, gemstones are ideal samples when compared to other objects of cultural heritage such as paintings or ceramics. They have a simple, homogeneous composition. Moreover, gems exhibit a polished and often flat surface at the scale of the beam spot. Owing to their very stable crystalline structure, gemstones are almost always insensitive to beam damage (no alteration) and charge build-up on these usually non-conductive targets is avoided when IBA are carried out in air.

The experimental set-up used in this study and its successive improvements have been thoroughly described [8], consequently, only the striking features will be recalled here. The PIXE/PIGE system is built upon the external nuclear microprobe line of the AGLAE accelerator facility. In this beam line, the 3-MeV proton beam is focused down to a diameter of 20 μm using magnetic lenses (3 MeV is an optimum energy for analysing geological samples). The beam is impinging on the gem placed in air 2 mm downstream a very thin Si_3N_4 exit foil, which insures the air/vacuum interface. The detection of X-rays is achieved by two Si(Li) detectors located at 45° from the beam in the horizontal and vertical planes. The first one is dedicated to the measurement of light elements, which are often the major constituents of gems (from Na to Fe). With a 10-mm^2 active area, it has a low solid angle, and a minimal filtering using an ultra thin window combined with a helium flow. A magnetic deflector prevents backscattered protons of the beam from entering the silicon crystal. The second detector is dedicated to the measurement of heavy elements (from Ca to U) at the trace level. Its 50-mm^2 active area yields a large solid angle and the absorber is chosen to attenuate X-rays emitted by major elements (typically a 50-μm aluminium foil in case of Si-containing samples). The distance of the two detectors can be adjusted to optimise and balance their counting rates. Gamma-rays are collected using a third detector (high purity germanium with a 20% efficiency). The beam is monitored using the silicon X-ray line emitted by the Si_3N_4 exit foil by means of a compact Silicon-drift detector. Spectra are collected in 1000 seconds with a beam current of the order of 1 nA, yielding an integrated charge of about 1μC.

2.2. RAMAN SPECTROMETRY

Chemical data alone is sometimes insufficient to define the nature of the crystal. Actually a gem is defined by a chemical formula arranged in a specific crystal structure and thus chemical data must sometimes be combined with structural information delivered by complementary techniques, such as XRD (X-ray diffraction) or Raman spectrometry. Raman spectrometry is an optical method similar to infrared spectrometry measuring vibrational bands of structural groups in the crystal like Si-O or O-H. Its probe, a laser beam, allows the direct and non-destructive identification of gems in reflection geometry. The two following examples show how complementary chemical and structural analytical techniques can be for the study of gems. Calcite and aragonite, constituents of coral, shells and pearls have an identical $CaCO_3$ formula, but the two minerals are well separated by their markedly different Raman spectra. Conversely, all pyralspite garnets (gems of formula $X_3Al_2(SiO_4)_3$, X being a divalent element like Mg, Fe, or Mn) have almost undistinguishable Raman spectra whereas they are easily differentiated by their chemical composition.

The Raman measurements have been carried out with a Jobin-Yvon Labram infinity spectrometer, a semi-transportable equipment weighting about 100 kgs. The laser source is a 532-nm green YAG delivering 3.5 mW on the sample. In confocal mode with a 50X objective the analysed volume is approximately 5 μm in diameter by 10 μm in depth. The detector was a Peltier-cooled CCD detector with 1024 channels and the resolution achieved is about 1cm^{-1} in the 100-4000 cm^{-1} Stokes range. The precise positioning of the laser on microscopic inclusions located at any depth inside the gems is performed by a motorized microscope stage. Typical acquisition time is 100 seconds.

3. Application to ancient gems

We describe here three applications of the PIXE, PIGE and Raman techniques to ancient gems from museums performed at the *Centre de Recherche et de Restauration des Musées de France*. It should be emphasised that all jewels studied here are well documented and have traceable history, and there is no doubt about their authenticity.

3 1 RUBY: ISTAR'S EYES

The first illustration of an application of PIXE to gemstones concerns a statuette representing Ishtar, the famous Mesopotamian goddess of Love and Fertility as well as War. The cult of Ishtar was widespread in ancient Babylon and she became under various names the most important Goddess of the Near-East and Western Asia. [fig.1] This remarkable small-sized sculpture carved in alabaster (23-cm high) figuring a naked woman was excavated in the vicinity of Babylon and entered the Louvre collections in 1866 [9]. According to its merged Greek and Mesopotamian style and typology, this representation of Ishtar dates from the Parthian period (2nd c. BC - 2nd c. AD).

Figure II-2-1. Ishtar statuette being analysed using the external beam system.

The issue is to determine the nature of three intriguing red cabochons (rounded shape) inlaid in the eyes and the navel.

The PIXE spectrum shows the lines of the main constituents: 99% alumina (Al_2O_3) associated with less than 1% chromium (fig 2). This confirmed our impression that they are rubies, and not coloured glass or red garnets as previously reported. As shown in the PIXE spectrum of trace elements, slight amounts of titanium, vanadium, iron, copper, gallium concentrations are also measured, all being characteristic of natural rubies.

The similar appearance and chemical composition of the three cabochons suggest the use of a single batch of gems. A very important outcome of the age of the statuette is that these rubies are, up to our knowledge, the oldest ruby found in Middle-East. The fact that Mesopotamia has no source of gems raises the problem of provenance of these rubies, therefore a comparison with contemporary rubies of known provenance has to be undertaken to determine their geographical origins.

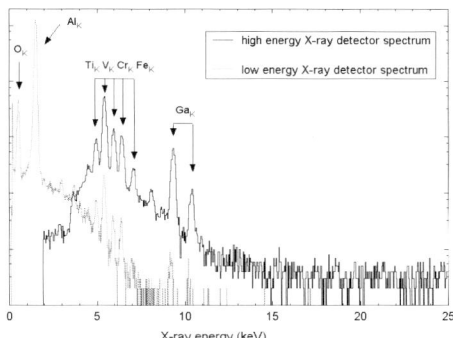

Figure II-2-2. X-ray spectra of major and trace elements recorded on the right eye of the statuette.

Over 500 PIXE analyses have been carried out on rubies from the most important occurrences: Afghanistan, Myanmar (Burma), Cambodia, India, Kenya, Madagascar, Sri Lanka, Thailand and Vietnam [10]. The trace element content of these reference gems appears markedly different for each occurrence and homogeneous within a given deposit, a criterion necessary to use the geochemical data for provenance. Three compositional groups corresponding to three different geological contexts can be identified. As shown in the Fe/Cr plot [fig. 3], Ishtar's gems fall into group I, which has the lowest Fe content and corresponds to deposits from Burma and Vietnam.

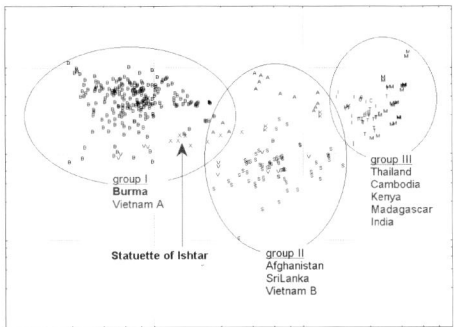

Figure 3. Plot of Cr vs Fe content for the rubies of the statuette and reference rubies.

As Vietnam rubies have only been mined since the 20th century, we can deduce that the eyes of Ishtar were extracted from the deposits of Burma. This result is confirmed by multivariate statistical processing (principal component analysis, hierarchical clustering) involving the entire set of trace elements (Ti, V, Cr, Fe, Ga) and by the observation of mineral inclusions specific to Burmese rubies (aggregates of short rutile needles). The three rubies therefore represent an evidence of a trade over several thousand kilometres between Mesopotamia and Southeast Asia twenty centuries ago.

3.2. EMERALDS: THE   BARBARIAN   TREASURE   OF   THE VISIGOTHS

The second example deals with emeralds set on barbarian jewels from the Dark Ages period. These votive royal crowns and crosses were unearthed in 1858 in the small village of Guarrazar, near Toledo, the capital of the Visigoth kingdom located in the Iberic peninsula. It constitutes one of the most important archaeological remains of the Visigoth period (7th-8th c. AD). The treasure is nowadays divided between the National Archaeological museum in Madrid, Spain and the National Middle-Ages museum in Paris, France [11].

The jewels are made of gold inlaid with gemstones such as sapphire, amethyst, quartz, pearls and emeralds. The archaeological question concerns the provenance of the emeralds, because whereas the origin of emeralds employed during the Roman Empire are well discussed by Pliny the Elder's in the 37th book of *Natural History*, little is known about the origin of emeralds of the early Middle-Ages period.

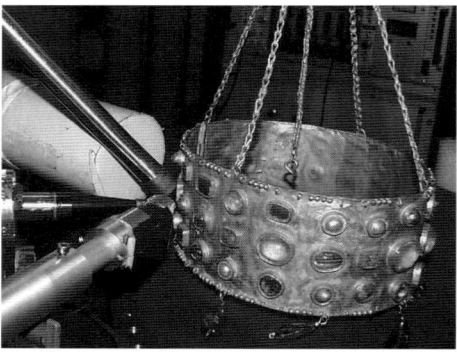

Figure II-2-4. Royal votive crown inlaid with gems positioned in front of the external beam set-up. Note the two Si(Li) detectors and HPGe in the back.

This work aims at comparing these "Barbarian" emeralds to those from various occurrences (Egypt, Austria, Afghanistan, Pakistan, Ural, India, Colombia, Madagascar, Zambia, Zimbabwe, Brazil), on the basis of chemical composition. As shown in fig. 4, the combination of PIXE and PIGE techniques performed with an external beam was used to measure their major constituents (Be, Al, Si) and to determine their trace elements.

In comparison to the rubies in the study presented above, emeralds show a much-extended set of discriminating trace elements (Na, Mg, Ca, Ti, V, Fe, Rb, Ni, Cu, Zn and Cs) including also light ones (Li, F), as illustrated in fig 5.

Emeralds being a hydrated mineral, the OH content was measured by forward elastic recoil technique with an alpha beam (ERDA), so as to give a possible additional provenance criterion. The statistical processing of the entire set of trace elements with discriminant method concluded that ten of the eleven emeralds set on the Visigoth jewels are mined from the Alpine deposits (near Habachtal, in Austria), an occurrence not mentioned in Greek and Roman ancient texts [12].

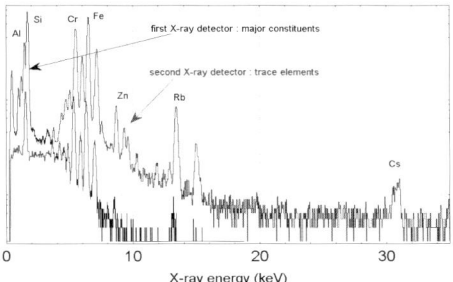

Figure II-2-5. PIXE spectra of an emerald showing the major and trace elements.

By stating that Visigoths had different sources of emeralds than the Romans, this work provides new insights into the trade routes of gemstones at the time of the Great Invasions. Moreover, it reveals that the Alpine deposits were worked much earlier than expected (the most ancient text mentioning these mines was written in 13th century by the Archbishop of Salzburg).

3.3. GARNETS: THE GEMS OF THE FIRST KINGS OF FRANCE

The Germanic tribes, who settled in Western Europe at the fall of the Roman Empire, introduced a very specific type of jewels called *"Cloisonné"*. The main gems used in these jewels are red garnets cut in thin slices (<1mm) and inserted in a honeycomb metallic structure, as shown in fig. 6. Once again, the archaeological issue is the provenance of the garnets used by these nomadic peoples, particularly if we consider the huge quantity of garnets necessary to make these jewels, which were buried in tombs according to the Barbarian custom and not re-used.

Figure II-2-6. Brooch of Frankish queen *Aregonde* with garnets set in "cloisonné" style.

To solve this problem, over 500 garnets set on the jewels discovered in tombs of members of the Frankish court from the necropolis of the Saint-Denis basilica near Paris were studied. The artefacts span the entire Merovingian period (5th-7th c. AD.) and comprise the famous jewels of the Frankish queen *Aregonde* [13]. From the mineralogical point of view, garnets present a highly variable composition. The most common type of garnet is the pyraldine family of chemical formula $X_3Al_2(SiO_4)_3$, where X can be a divalent ion like Fe (almandine), Mg (pyrope) or Mn (spessartite), each combination being called an *end-member*. The situation is actually more complex as natural garnets are a solid solution of end-members in any proportion. Because garnets are relatively widespread and their composition highly variable, it was necessary to cross many criteria to determine their origin [14]. The first criterion is the composition; it was determined by external beam PIXE for major constituents (Mg, Al, Si, Ca, Mn, Fe) and trace elements (Ti, V, Cr, Y). As shown in fig. 7, three groups of garnets were identified. Most garnets belong to the first group (almandine, Fe-rich). The second group (one jewel) consists of intermediate almandine-pyrope garnets sometimes called "rhodolite". The third group comprises pyrope garnets (Mg-rich).

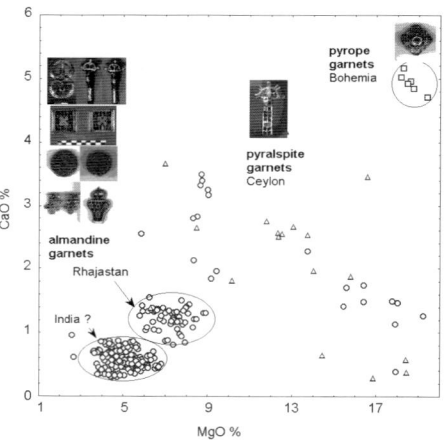

Figure II-2-7. The CaO vs MgO plot for garnets from various jewels shows 4 provenances: *pyropes* from Bohemia, *rhodolites* from Ceylon and two sources of *almandine* in India.

The trace elements content and slight differences in major composition permit to split these three groups in five different sources: two sources of pyrope garnets (with and without chromium) and two sources for almandine garnets (distinctive calcium, magnesium and yttrium contents).

Comparison with published data on garnets suggests that almandine garnets have been mined from India while the "rhodolite" garnets may have been imported from Sri Lanka. The sources of pyrope garnets correspond to the Bohemian deposits (Czech republic).

The second provenance criterion is the inclusions. Micro-Raman spectrometry was used to identify mineral inclusions in almandine garnets: apatite, zircon, monazite, calcite, and quartz. Among these inclusions, two of them were specifically found in archaeological garnets: curved needles of sillimanite (Al_2SiO_5) and small (10 μm) metamict radioactive crystals. Fig. 8 shows the Raman spectra of sillimanite, which is a mineral formed under a high temperature and high-pressure metamorphism.

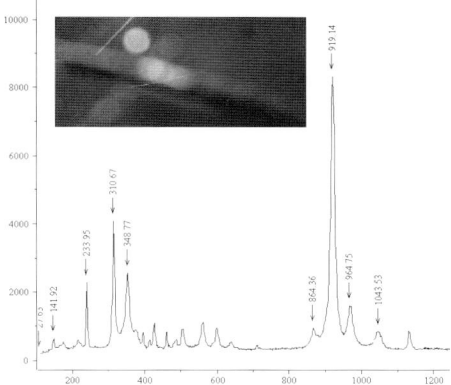

Figure II-2-8. Raman spectra of a *sillimanite* elongated and curved inclusion indicating a high P-T metamorphism of the crystal.

The radioactive crystal was analysed with the microbeam in PIXE mode. Uranium and lead were found as the major constituents of this inclusion. The Pb content being attributed to the radioactive decay of ^{235}U and ^{238}U (0.7 and 4.46 Gy half live, respectively), the Pb/U ratio gives a crystal formation age ranging between 1 to 1.5 billion years. To sum it up, the almandine composition of the garnets, the presence of sillimanite and the very ancient age of the crystal converge towards highly metamorphosed rocks of the Precambrian period. One of the rare parts of earth's crust remaining from that period that has corresponding mineralogical features is the metamorphic belt located in India, a region presenting garnets deposits of gem quality.

The presence of pyrope garnets from Bohemia might also have a historical significance. The pyropes only appear in jewels dated after the end of the 7th century; this indicates that starting from that period Merovingians had to use European garnets instead of Indian ones. This major change in supply is likely a consequence of the closing of the garnet route to India, due to the invasion of the Arabic peninsula by the Sassanids at the end of the 6th century.

4. Conclusions

IBA methods and specially the PIXE and PIGE techniques performed in air appear as tools well suited to the analysis of gemstones mounted on historical jewels, in complement to classical gemmological investigations. This combination usefully provides: 1) a completely non-invasive and non-destructive analysis, 2) the detection of a wide range of elements, 3) the analysis of 10-μm size details, 4) a high sensitivity attaining the μg/g level and 5) highly quantitative results. The identification of mineral inclusions is complementarily achieved by μ-Raman spectrometry. These modern analytical methods were successfully applied to the determination of the nature and provenance of ancient gems, providing new insights into their trade routes. New developments are being undertaken in several directions. One important field is the investigation of the colouring mechanism of gems in relation with geochemistry. For instance, the detailed distribution of cations in complex stones such as jade may provide the understanding of its colour and provenance. The application of X-ray absorption methods (XANES, EXAFS) with a synchrotron facility is being used for study the chemical environment of chromophoric elements in ruby or emeralds. Finally, the measurement of the hydrated layer diffusion building-up with time at the surface of gems by ion beams (ERDA and resonant nuclear reaction with ^{15}N ions) opens new perspectives for relative dating or at least authentication of gems such as quartz.

5. References

1. Anderson, B.W and Jobbins, E.A. (1990) *Gem testing*, Butterworths, London,

2. Schubnel H.-J., Pinet M., Smith, D.C., Lasnier, B. (1992) La microsonde Raman en Gemmologie, Revue de Gemmologie, Association Française de Gemmologie, Paris

3. Gugliani, G., Chaussidon, M., Schubnel, H.-J, Piat, D., Rollion-Bard, C., France-Lanord, C., Giard, D., de Narvaez, D., Rondeau, B. (2000) Oxygen isotopes and emerald trade routes since Antiquity, *Science* 287 631

4. Campbell, J.L., Malmqvist, K.G. (1995) *Particle-induced X-ray spectrometry (PIXE)*, John Wiley & Sons, London,

5. Bird, J.R., Williams, J.S. (1989) *Ion beams for materials*, Academic Press, London,

6. Calligaro, T., Castaing, J., Dran, J.-C., Moignard, B., Pivin, J.-C., Prasad, G. V. R., Salomon, J. ,Walter, P. (2001) ERDA with an external helium ion micro-beam; advantages and potential applications, *Nucl. Instr. and Meth.* B181, 180

7. Dersch, O., Rauch, F. (1999) Water uptake of quartz investigated by means of ion-beam analysis, *Fresenius J. Anal. Chem* 365, 114

8. Calligaro T., Dran, J.-C., Hamon, H., Moignard, B., Salomon, J. (1998) An external milli-beam for archaeometric applications on the AGLAE IBA facility of the Louvre museum, *Nucl. Instr. and Meth.* B136/138, 339

9. Tallon, F. (1999) *Cornaline et pierres précieuses, Actes du colloque-Musée du Louvre 1995*, La Documentation française-Musée du Louvre, Paris

10. Calligaro, T., Mossmann, A., Poirot, J.-P., Querré G., (1998) Provenance study of rubies from a Parthian statuette by PIXE analysis, *Nucl. Instr. and Meth.* B136/138, 846

11. Perea, A., (2002) *El tesoro visigodo de Guarrazar*, A. Perea eds., CSIC editions, Madrid

12. Calligaro, T., Dran, J.-C., Poirot, J.-P., Querré, G., Salomon J.,. Zwaan, J. C. (2000) PIXE/PIGE characterisation of emeralds using an external micro-beam, *Nucl. Instr. and Meth.* B161/163, 769

13. Calligaro T., Colinart, S., Poirot, J.-P., Sudres, C. (2002) Combined external-beam PIXE and μ-Raman characterisation of garnets in Merovingian jewellery, *Nucl. Instr. and Meth.* B189, 320

14. Fleury, M., France-Lanord, A (1998). *Les trésors mérovingiens de la basilique de Saint-Denis*, Gérard Kopp eds, Woippy, France,

Chapter II-3

Investigations of Medieval Glass by a Combined PIXE/PIGE Method

Glassmaking à façon de Venise

Ž. Šmit [1,2] and M. Kos [3,2]
[1]*Department of Physics, University of Ljubljana, Jadranska 19, SI-1001 Ljubljana*
[2]*J. Stefan Institute, Jamova 39, SI-1001 Ljubljana, Slovenia*
[3]*National Museum of Slovenia, Prešernova 20, SI-1000 Ljubljana, Slovenia*
smit@fiz.uni-lj.si

Keywords: Medieval glass, PIXE, PIGE, Ljubljana glass, Venetian glass

The method of proton-induced X-ray emission (PIXE) was used to determine major and several minor elements in glass, produced in the 16th century Ljubljana. The method was combined with the method of proton-induced gamma rays (PIGE) for the analysis of light elements Na, Mg, Al. A specific approach of this technique was normalization according to the argon in the air, based on the realistic experimental geometry, and using silicon as a connecting element between the two types of measurement. The results show clear resemblance of the glasses from Ljubljana to the Venetian white glass.

1. Historic overview

After the Antiquity, the art of making glass in medieval Europe was revitalized in Venice, and followed soon after in Northern Europe. Two distinct Italian centers were Venice (the glassworks on the island of Murano) and Altare near Genova. Due to good commercial relations across the Mediterranean, the Venetian glassmaking developed principally under the Near-Eastern influence, rather than following the local Roman tradition. For the supply of raw materials, the Venetian glassmaking relied entirely on the imported alkalis (Syria, Egypt, Spain), while using the local siliceous sand (reportedly of the Ticino river) [1]. The import of alkalis was Venetian monopoly, jealously looked after, and Venice was a distribution center of alkalis for other European glassworks.

The demand for Venetian glassware was large, and as the Venetian glass-working capacities did not fit the demand, new glassworks were gradually established in the central Europe, attracting the Venetian glass-workers with better

M. Uda et al. (eds.), X-rays for Archaeology, 113–121.
© 2005 *Springer. Printed in the Netherlands.*

income. In 1486, an early glasswork is reported in Vienna, following 1534 in Hall, and 1570 in Innsbruck [2]. Except in Vienna, the glass was produced according to the Venetian recipes and style, and is known today as glass à façon de Venise. With Venetian glass-workers, the Venetian glass working spread as far as Western Europe [3].

Ljubljana, on its way between Venice and central Europe, obtained its first glasswork in the beginning of the 16th century, and operation of at least three glassworks is documented in written sources. An important historical document is the last will of Christoph Prunner, a glasswork leaseholder, which gives a detailed list of contemporary glass products. Excavations within the old city of Ljubljana revealed numerous glass fragments (Fig. 1), stylistically matching the items from the list. This may be regarded as a strong evidence of the domestic glass production, yet a chemical analysis is required for confirmation of this thesis. A similar analytic work is conducted in other European glass-working centers, aimed at distinguishing between the imported and home produced items [3,4,5].

Figure II-3-1
Examples of glass à façon de Venise excavated in Ljubljana (photo T. Lauko, National Museum of Slovenia). (See also Color plate, p. 300)

For the glass objects and fragments kept in the National Museum of Slovenia, sampling was not allowed; investigations with an ion beam in the air therefore proved as an acceptable solution. For control and comparative purposes, a small selection of glasses from Slovenian castles and Late-Roman settlements was involved additionally.

2. The method

2.1 BASIC PRINCIPLES

The analysis of glass generally involves determination of major components, which reveal the glass-making mixture and the type of glass, and the determination of trace elements or impurities, which could be indication of specific raw materials and working places. Further distinction can be made whether analysis can be done *in situ,* or sampling is necessary. An established method for determination of major elements is X-ray fluorescence induced by electrons in the electron microscope. The measured sample of a few mm dimensions is usually cut from the main object and polished to 1µm grade. The analysis, which employs a thin window semiconductor detector, is able to detect low Z elements up to sodium; the polished sample yields the elemental bulk values. Electron bremsstrahlung prevents analysis of minor elements heavier than iron, so the electron microscope measurements have to be combined with some other method. These can be PIXE (using a higher bombarding energy and absorbers to attenuate the low energy X-rays), X-ray fluorescence (XRF), or laser ablation mass spectrometry (LA-ICP-MS). The latter method surpasses the former two in detection limits below ppm; however, sampling is required for the objects larger than a few cm, and the ablation process produces tiny craters or furrows.

Bremsstrahlung is largely reduced in the PIXE method, which is also applicable for the analysis of major elements. Analysis in vacuum or in helium atmosphere can reveal low Z elements up to sodium, though the ion irradiation may produce migration of sodium ions. The technique based on the in-air proton beam is appropriate for the analysis of larger objects where sampling is not desired. Absorption in the air stops detection of elements lighter than silicon. Though helium atmosphere may allow detection of low Z elements, they are more accurately determined by the method of proton induced gamma rays (PIGE), as more energetic protons (required to induce nuclear reactions) penetrate deeper into the target, while the emitted gamma rays are not sensitive to the matrix effects.

Major elements can also be determined by XRF. A large span in atomic numbers requires irradiation with more than a single energy, and absorption of soft X-rays in the air poses similar problems for the detection of elements around sodium. ICP methods can be used for the analysis of major elements as well.

2.2   PIXE/PIGE PROCEDURE

The glasses of Ljubljana were analyzed by the in-air PIXE method that also largely simplified handling with the investigated objects. They appeared in different sizes and forms, from tinny shreds to complete vessels. The proton beam energy was 2 MeV, however, actual bombarding energy was about 1.7 MeV. The measurements were executed in the normal air atmosphere, so air (of 6-7 cm thickness) was also used as an absorber to attenuate the excessive low energy photons. The softest X-rays detected were those of silicon (Fig. 2). For normalization, the signal of argon from the air was used. The argon yield was modelled according to the known geometry, so no empirical proportional factor was introduced. The evaluated concentrations depend essentially on the thickness of air gaps between the exit foil and target and between the target and detector. These two quantities were calibrated with a set of thick elemental and simple chemical compound targets. The exit foil – target distance was kept fixed with the mechanical spacers, yet the calibration procedure was repeated on each measuring day to avoid influences of the air-pressure variation. For control, the SRM glass standards 620 and 621 were used. Systematic differences up to 20% were observed between the deduced and nominal concentrations of the standards, which we attributed mainly to the uncertainties of the proton stopping power database. These differences were then compensated with correction factors for a few elements (Si, K, Ca), differing from unity by less than 20%.

Figure II-3-2

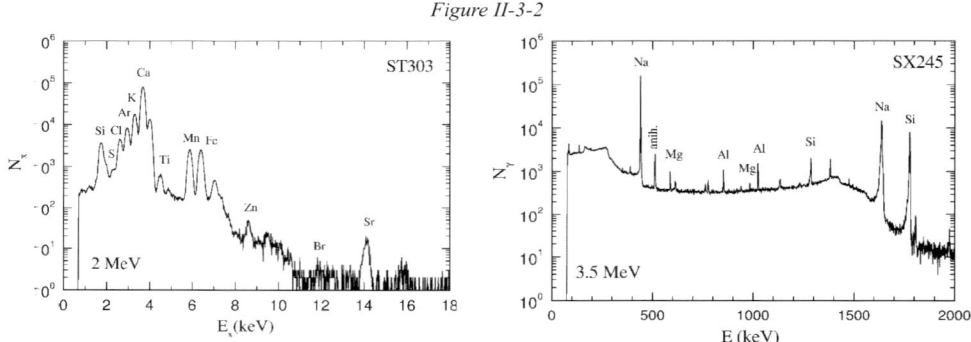

X-ray and gamma ray spectrum of the reference specimen, a piece of the 16th century window glass [6].

The X-ray data already produced useful results for glass characterization [6], yet precise knowledge of low Z elements (Na, Mg, Al) provides us with important information about glass technology. The most convenient IBA method for determination of light elements is proton induced gamma emission (PIGE). These measurements had to be performed at higher proton energy (3.5 MeV), using tantalum exit foil of 2 μm thickness [7]. Among the three elements, determination of Mg appeared most critical as its 585 keV line interferes with the 583 keV line of natural background [7]. Lead shielding of the detector and an intense proton current were used to increase the relative counting rate of the Mg line. The gamma yields originate from greater depths than those obtained with the complementary X-ray methods (PIXE in vacuum or helium atmosphere, XRF), so the impact of surface sensitive effects (i.e. corrosion and surface ion mobility) was largely diminished.

The gamma measurements were calibrated according to the SRM 620 standard, yielding the concentration ratios of Na/Si, Mg/Si, Al/Si. These results were combined with the PIXE measurements in a concentration evaluation algorithm, following the principle of setting the sum of all metal oxides to unity. The oxide concentrations usually converged to a value slightly different from unity, so re-normalization factors were used to force the sum of concentrations to unity. The distribution of re-normalization factors was monitored [7], and it confirms spread of the data to be smaller than 10%. This corresponds to the uncertainty of the proton air-path of about 0.1 mm, and agrees with the expected accuracy of mechanical spacers.

The distribution of evaluated concentrations was studied by the principal component analysis (PCA), using the logarithmically transformed data   [8]. The method further included the redistribution of data according to the multidimensional scaling [9], which improved separation between groups.

3.   Results

The principal-component analysis of the measured glasses revealed several groups [7] that we identify now as shown in Fig.3; the mean concentrations of the groups are presented in Table 1.

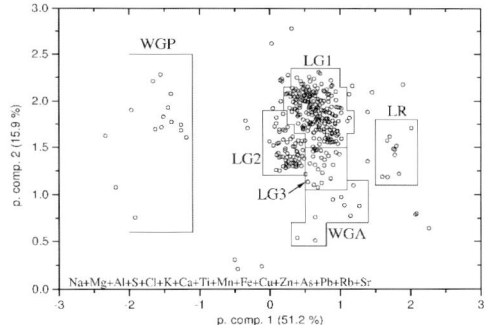

Figure II-3-3
Distribution of the measured glass into groups according to the principal component analysis.

The main group can roughly be divided into three subgroups, denoted as LG1-3. Below them, we find the group WGA, the glasses at the left form a broad group WGP, and the Roman glasses (with one exception) also group closely together. We expect that the grouping is mainly related to the major elements in glass and enhanced further according to the minor and trace elements. These can be brought into the glass either through the alkali medium or by siliceous sand; the potential third component limestone/dolomite may also contain various divalent metals. Fig.4 (top) shows the K/Na ratio of the alkali medium. This distribution essentially reflects structure of the glasses LG1-3, so the differences among these groups are due to the different sources of alkalis. The content of K is characteristic for ash of halophytic plants. Further, the WGP glasses are characterized by the predominant content of K, which reveals them as wood-ash-produced forest glasses. The group WGA was also made of ash, probably of different origins. The Roman glasses exhibit specific potassium content, suggesting a completely different source of alkalis. Our analysis also included a set of plates of colored glass, clearly imported;

tnese data appear separately in Figs. 3, 4, and denote specific workmanship using cifferent types of glasses.

Table II-3-1
Concentrations of major elements in glasses of Fig. 3.

	LG1	LG2	LG3	WGA	WGP	LR
Na₂O	12.3 ± 1.4	12.1 ± 1.3	14.1 ± 1.8	14.3 ± 2.1	1.0 ± 0.7	18.7 ± 1.6
MgO	3.44 ± 0.48	2.77 ± 0.41	2.36 ± 0.48	1.27 ± 0.94	2.24 ± 0.64	0.74 ± 0.19
Al₂O₃	2.64 ± 0.81	2.02 ± 0.47	2.36 ± 0.76	2.3 ± 2.6	2.9 ± 1.8	4.70 ± 1.06
SiO₂	66.3 ± 1.7	67.7 ± 2.3	68.3 ± 3.1	70.7 ± 4.3	64.3 ± 6.1	64.3 ± 2.2
K₂O	2.35 ± 0.34	4.30 ± 0.58	2.83 ± 0.51	2.67 ± 0.84	10.9 ± 4.1	0.72 ± 0.21
CaO	10.4 ± 1.5	8.6 ± 1.4	7.4 ± 1.3	6.5 ± 1.9	16.0 ± 4.6	7.3 ± 1.0
MnO	0.64 ± 0.29	0.75 ± 0.23	0.55 ± 0.15	0.40 ± 0.4	0.69 ± 0.4	0.95 ± 0.5
Fe₂O₃	0.64 ± 0.17	0.54 ± 0.14	0.52 ± 0.16	0.50 ± 0.5	0.86 ± 0.8	0.83 ± 0.4

Strontium is another element that is strongly related to the alkalis. Low Sr data are in the lower part of Fig.3. The absence of Sr denotes that the ash was precipitated into potash, as the procedure of melting, filtering and evaporation removes the insoluble oxides of divalent metals from the product. An independent source of Sr could also be limestone in the form of marine shells, though we estimate this source would contribute Sr of a much lower level. The presence of Sr in practical all our glasses   (except a small part of WGP) then indicates the use of ash rather than potash.

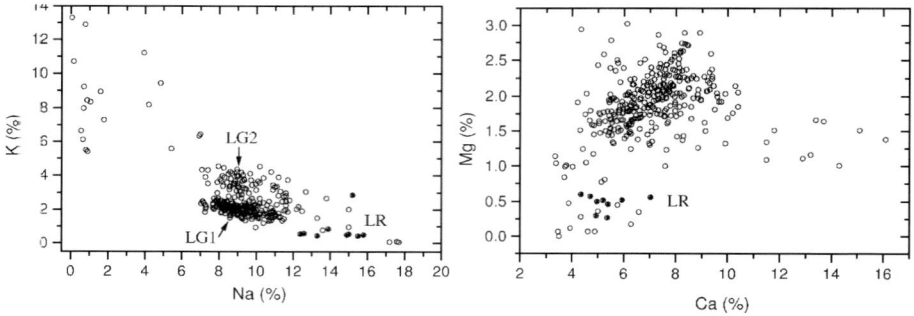

Figure II-3-4
Bivariate concentration plots: K-Na (left), Mg-Ca (right). The Late Roman glasses are marked with dots.

Silicon dioxide, in the form of siliceous sand, is the most abundant glass component, and a rather pure material is required for high quality products. Siliceous sand can be mixed with alumina, calcium and magnesium carbonates, and trace elements such as zirconium. Aluminium can also be contained in ash, and carbonates can be additives to the molten glass mass, so the trace elements only may provide a fingerprint. For the glasses of Antwerp, a distinction between Venetian import and home production was discovered in the concentrations of zirconium and hafnium lower than 25 ppm and 0.7 ppm, respectively [3]. Such low concentrations, attainable by LA ICPMS, were not reached by our PIXE measurements, which exhibited the minimum detection limit for Zr of about 100 ppm. Zr about this level and higher was discovered in about 10% of our glasses. They belonged mostly to the numerous LG1 group, so no definite conclusions about Zr can be made.

Oxides of calcium and magnesium improve the mechanical properties of glass. They are already contained in ash, they may be added separately as limestone/dolomite, or they are admixture in the siliceous sand. Fig.4 (right) shows their respective concentrations. Mg seems to be roughly correlated to Ca for the majority of glasses, suggesting use of a limestone source with a known proportion of dolomite, or ash from a geographically well defined area. Fig. 4 also shows a small group of glasses characterized by a calcium content above 10%. These glasses belong mainly to the WGP (forest glasses) group and denote a high quality potassium glass. A distinct group is also formed in the very low Mg/Ca region. Here we find all our Late Roman glasses: their typical Mg and Ca contents are 0.44% and 5.2%, respectively. Mg was evidently avoided in making Late Roman glasses; it is interesting to compare this finding to the text of Pliny the Elder (1st cent. AD). The text unfortunately seems modified from its original meaning in the part listing raw materials. According to one reading [10], the term 'magnes lapis' could be interpreted as magnesian limestone or dolomite. Due to the low Mg content, this reading now does not seem very probable.

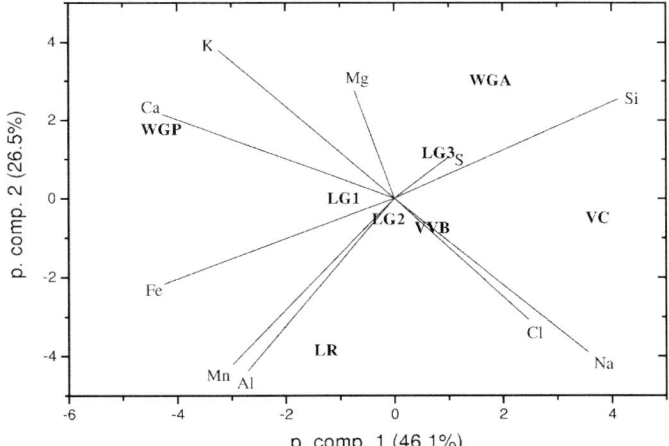

Figure. II-3-5

Relation between particular glass groups according to the principal component analysis. The data were transformed to have zero mean – unit variance. The two Venetian glasses are marked as VB (vitrum blanchum) and VC (cristallo). The eigen vectors were multiplied by a factor of ten.

For the groups of Fig.3 we determined the mean concentrations of the main elements and compared them to the compositions of the two main Venetian glasses, vitrum blanchum and cristallo [3]. Taking our six glass types and the two Venetian glasses as equivalent elements, we inspect their mutual relation in the PCA plot of Fig.5. The closest grouping is obtained for LG1, LG2 and vitrum blanchum (VB). The smallest difference is between LG2 and VB, though the alkali sources - according to K content in LG2 - were clearly different in both cases. These results suggest that the glasses of Ljubljana tried to approach vitrum blanchum as close as possible, using probably the same recipe and experience.

The eigen vectors of Fig. 5 confirm the marine origin of Na through its correlation with Cl. Sodium is inversely correlated to K, reflecting two principally different sources of alkalis (ash of halophytic plants and wood).

The cristallo glass was not reproduced in Ljubljana, while the glass circulation around the city also dealt with imported products, including objects of art made of potassium glass that may have a German origin. Fig.5 also confirms that glassmaking à façon de Venise does not continue the Late Roman tradition: these glasses exhibit distinct properties, summarized in relatively high concentrations of sodium and aluminum, but low of potassium and magnesium.

4. Conclusion

A combined PIXE/PIGE analysis is able to determine all major elements important for glass production, and several trace elements. For the glass circulating in the 16th century Ljubljana, we succeeded to discern the main groups of glasses produced à façon de Venise, and to compare them to the contemporary Venetian products. A high similarity between the glasses of Ljubljana and Venetian products was observed.

References

1.   Toso, G.: *Murano, A History of Glass*, Arsenale Editrice, Verona, 2002.

2.   Kos, M., and Žvanut, M.: *Glass Factories in Ljubljana in the 16th Century and their Products*, National Museum, Ljubljana, 1994.

3.   De Raedt, I., Janssens, K., Veeckman, J., Vincze, L., Vekemans, B, and   Jeffries, T.E.: Trace analysis for distinguishing between Venetian and façon-de-Venise   glass vessels in the 16th and 17th century, *J. Anal. At. Spectrom.* **16** (2001) 1012.

4.   De Raedt, I., Janssens, K., and Veeckman, J.: Compositional distinctions between 16th century 'Façon-de-Venise' and Venetian Glass Vessels, excavated in Antwerp, Belgium, *J. Anal. At. Spectrom.* **14** (1999) 493.

5.   Kuisma-Kursula, P., Räisäinen, J., and Matiskainen, H.: Chemical analyses of European forest glass, *J. Glass Studies* **39** (1997) 57.

6.   Šmit, Ž., Pelicon, P., Vidmar, G., Zorko, B., Budnar, M., Demortier, G., Gratuze, B., Šturm, S.,
     Nečemer, M., Kump, P., and Kos, M.: Analysis of medieval glass by X-ray spectrometric
     methods, *Nucl. Instr. Meth.* **B 161-163** (2000) 718.

7.   Šmit, Ž., Pelicon, P., Holc, M., and Kos, M.: PIXE/PIGE characterization of medieval glass,
     Nucl. Instr. Meth. **B189** (2002) 344.

8.   Duewer, D.L., Kowalski, B.R., and Schatzki, T.F.: *Anal. Chem.* **47** (1975) 2350.

9.   Clayton, E., in W. Ambrose and P. Duerden (Eds.), *Archaeometry: An Australian Perspective*,
     Australian National University, Canberra, 1982, p. 90.

10. Pliny the Elder, *Natural history, Vol. X, Books 36-37*, translated by D.E. Eichholz, Loeb
     Classical Library, Harvard University Press, Harvard, 1962.

Chapter II-4

PIXE Analysis of Pre-Hispanic Items from Ancient America

J.L. Ruvalcada Sil
Instituto de Física, UNAM
Apdo. postal 20-364, México DF 01000, Mexico
sil@fisica.unam.mx

Keywords: PIXE, pottery, bronze, gold, pigment, obsidian, stucco, gemstone, turquoise, Mesoamerica, South America.

Abstract

In this work, a review of applications of Particle Induced X-ray Emission Spectroscopy (PIXE) for analysis of ancient artifacts and materials from various ancient civilizations from different American cultural regions is presented. Obsidians, pottery, metallic items (gold and bronze), teeth, pigments, stucco and decorations of pottery are the most studied materials using PIXE. Artifacts and materials from museums and archaeological excavations have been studied using various analytical approaches. Several methodologies and their results are discussed.

1. Introduction

Most of the ancient civilizations of the American continent attained high levels of development in some regions of North America, Mesoamerica and Western South America in different epochs (figure 1). Many ancient civilizations disappeared centuries before the European arrival, at the beginning of XVI century. With the arrival of European people, mainly Spaniards, the different civilizations were conquered and huge cultural changes occurred. Most of the knowledge of the ancient cultures was lost or forgotten due to the conquest. In other cases, a strong mestization occurred.

The study of archaeological materials and objects from the ancient American civilizations gives rise to new knowledge about cultural and technological development, the use of materials, cultural relationships and their chronology. From biological material it is possible to establish the diet of the populations and their diseases. Most of the analyses may provide useful information to determine the deterioration of the materials and suitable restoration methods.

M. Uda et al. (eds.), X-rays for Archaeology, 123–149.

The historical materials are often heterogeneous and generally, there is not *a priori* information about their composition. For these reasons, when analyzing historical materials it is necessary to use fast non-destructive methods with high sensibility in order to get a maximum of information from their composition, specially when the object is unique or there is a small amount of the sample.

On the other hand, ion beam accelerators are powerful analytical tools for applied research. In arts, archaeology and history, several techniques may be used to induce X-ray emission for rapid non-destructive analysis of the material surface with high sensibility [1]. From this scope, the most useful techniques are Particle Induced X-ray Emission (PIXE) and Particle Induced X-ray Fluorescence (PIXRF). PIXE is the most versatile technique for archaeological applications. This technique is suitable to measure contents of elements heavier than Na from characteristic X-rays emitted by the material with detection limits of µg/g. Full details of the fundamentals of this method are described in ref. [2-3].

Figure II-4-1. Most important cultural areas in ancient America.

2. Analytical approaches

Recently, the use of PIXE in combination with other techniques to study the elemental composition of almost all kind of inorganic materials from ancient America has increased. Metals (gold and bronzes), biological materials (bones, teeth), ancient documents, obsidians, pottery and glazes, amber, turquoises, pigments and paintings, stuccos, have been analyzed. Most of the reported studies are related to the cultural area of Mesoamerica (figure 2).

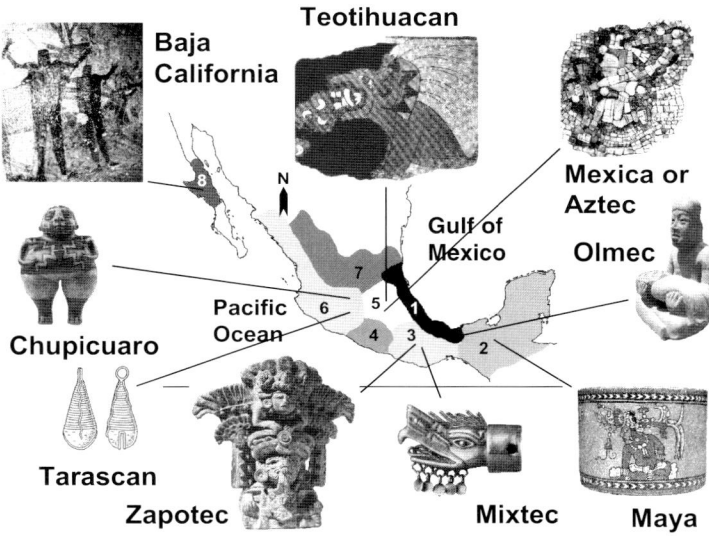

Figure II-4-2. Mesoamerican regions and some artifacts analyzed by PIXE from this cultural area. 1. Gulf of Mexico. 2. Maya. 3. Oaxaca. 4. Guerrero. 5. Central Mexico. 6. West Mexico. 7. North. 8. Baja California.

Different experimental devices and strategies of study have been used. The methodologies depend mainly on local development and available resources of each laboratory [4-7], but there are a reduced number of accelerator laboratories working on ancient American items. Most of the research is carried out on materials of European and Asian origin.

Usually, if the object or sample is small, non-vacuum PIXE is combined with other ion beam techniques, such as Rutherford Backscattering Spectrometry (RBS) for light elements and matrix measurements, and Particle Induced Gamma-ray Emission (PIGE) for specific elements (F, Na, Al) [8].

In some cases, when the beam energy provided by the accelerator is low, a combined analysis may be used to measure major and light elements by PIXE and trace elements by X-ray Fluorescence techniques (XRF).

External beam PIXE is the most suitable set-up for archaeometric applications. There is no limitation on artifact size, form and composition, and the process of irradiation is easier than in a vacuum chamber. The proton beam crosses an exit window and a short path in an atmosphere of air or helium before reaching a region of the artifact. The artifact or material may be set in front of the beam in order to irradiate other regions of the item. X-ray detection is carried out by one or two detectors. When only one detector is available, usually two incident beam energies are employed: a low energy for low energy X-rays and major elements, and a higher energy for X-rays of trace and heavy elements using selective or non-selective X-ray absorbers at the detector window. For two detectors systems, one detector may be used for light elements detection, while the second detector may detect heavier elements simultaneously (figure 3). In this way, the full composition from exactly the same region is obtained. When analyzing heterogeneous materials this is very important, particularly when the irradiation of the same region using several ion beam energies is not guaranteed. Microscopic analysis under vacuum or by external beam is performed by a proton microprobe; the spatial resolution must be larger than $10\mu m$ to get reliable data and to avoid heterogeneity effects of the archaeological material.

The new trends involve the use of portable XRF equipments for diagnostic analysis to select the most relevant and interesting artifacts and materials for ion beam analysis at accelerator laboratories and other techniques such as Raman and Infrared spectroscopy.

PIXE analyses have been done in American countries laboratories where historical artifacts or archaeological remains exist or, when the artifacts are part of the collection of a foreign museum, in local laboratories. The artifacts and materials may be sampled or transported to the laboratory.

In European laboratories (LARN in Namur, AGLAE in Paris, mainly) an important number of ancient American items has been analyzed by PIXE. Only in the case of Louvre Museum, the laboratory AGLAE [9] has carried out directly within the museum facilities the study of ancient American items from the Branly collection, exhibited temporally in this place.

In the United States, PIXE analyses of archaeological materials and artifacts from ancient America have been carried out in some accelerator laboratories (Bartol Research Institute in Delaware, UCL in Davis, Arizona State University) usually in collaboration with museums and archaeologists.

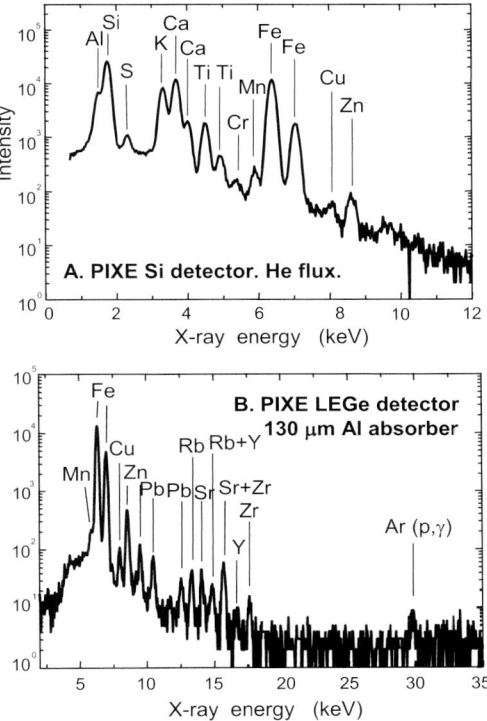

Figure II-4-3. Typical PIXE spectra of a ceramic material obtained by the external beam set-up of figure 4. A. Light and major elements spectrum. B. Heavy and trace elements spectrum.

In Latin American countries, the ion beam facilities are only available in few countries: Mexico, Argentina, Chile and Brazil. For this reason, X-ray Fluorescence Analysis (XRF) and/or electron microprobes are often preferred for archaeometric studies [10-12]. Besides the availability problem of the PIXE technique, another reason for XRF preference over PIXE is the cost of analyses. XRF is often cheaper than PIXE, mainly because most of the experimental devices are not optimized to perform a large number of analyses in one day. PIXE and XRF methods may provide similar information about mean elemental composition but they are really complementary [13-14]. The collections of museums are often analyzed by XRF and others techniques due to the difficulties to bring the item to the accelerators laboratory (expensive insurances, security and transportation).

Argentina and Brazil have big accelerators, but there are very few reports on ion beam applications to archaeology. In the case of the Argentinean TANDAR [15], mainly fundamental research by heavy ion beams has been reported. In the Synchrotron laboratory of Sao Paulo, in Brazil, few analyses of ancient pottery have been done [16]. Recently, an external beam has been developed [17] and archaeometric PIXE applications, also on archaeological pottery, are being started in one of the Tandem Brazilian laboratories [18]. Nevertheless, Chilean and Mexican accelerator laboratories have reported analysis of pottery, obsidians and pigments for archaeometric applications since 1980's [19-20], but only in Mexico in the last years this research has become systematic in collaboration with archaeologists and restorators to study and to preserve the national heritage.

At the National Autonomous University of Mexico (UNAM), special attention was paid to develop a new versatile external beam for all kind of applications (figure 4), in particular for archaeometric research [21]. A large number of analyses may be carried out with this system using two X-ray detectors. PIXE, PIGE and external beam RBS analyses may be performed simultaneously for a complete characterization of the artifact or material. In this way, the analyses have a competitive cost and all kind of materials can be studied in short time. In addition, a portable system is being developed for diagnostic tests at museums or for fieldwork. The use of the accelerator beam time and the laboratory facilities is then optimum.

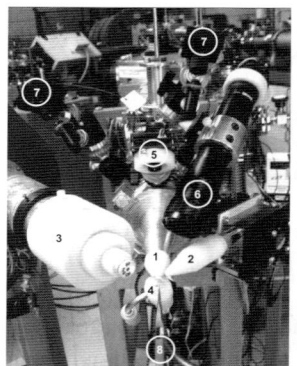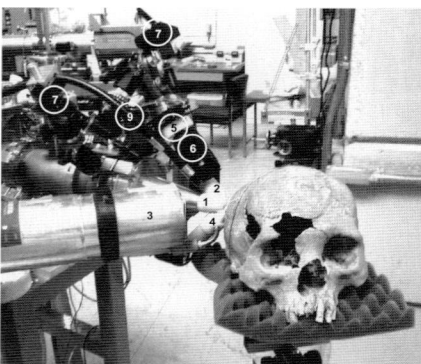

Figure II-4-4. Analysis of an archaeological skull with the external beam set-up of the Pelletron accelerator of Instituto de Fisica, UNAM, Mexico.

1. Exit window and anti-scattering cone.
2. Si detector with helium flux.
3. LEGe detector with selective or non-selective absorber.
4. External RBS detector for matrix measurements.
5. High-resolution color micro-camera.
6. Microscope with color micro-camera (45X).
7. Focusable laser pointers.
8. Sample holder.
9. Illumination by optic fiber.

3. PIXE applications

Due to new electrostatic accelerators, recent developments and several improvements of experimental devices, in the recent years the applications have diversified and practically all kind of materials from ancient America has been analyzed by PIXE: Metals (gold and bronzes), biological materials (bone remains, teeth), ancient documents, pottery and glazes, amber, turquoises, pigments and paintings, obsidians, stuccos. Most important and relevant reports are described. The cultural regions and most important sites related to these studies are shown in figures 1 and 2.

i) Pottery

Pottery is the first stable synthetic material prepared by man. Ceramic materials are abundant in archaeological context and it may be used to determine provenance, technology, trade routes and chronology. Usually, fine clay is mixed with water and another material, called temper, to prepare the paste for pottery. Objects are modeled with this paste and then dried. When these objects are fired at temperatures higher than 600 °C, the mechanical properties of clays are modified to produce a ceramic material. In ancient America, the pottery was fired at low temperature (lower than 1000 °C). High temperature and double firing for glazing were European contributions. Several clays may be used to fabricate pottery. This is one of the most important difficulties for pottery analysis, especially for provenance studies and clay sourcing.

The ceramic objects and sherds may be analyzed directly to study the pigments of decorations, surface finishing and paste composition. This approach is valid when the object is valuable and sampling is not possible. This approach is limited by the grain size of the ceramic paste because the ion beam may impinge in large mineral grains and give rise to non-representative concentrations of the pottery. Usually, sherds are sampled to prepare homogeneous pellets from fine powder. In this way, the homogeneity of the sample is guaranteed.

A Chilean group reported a special sample preparation when small quantities of ceramic material are available [22]. This method was also used to prepare bone remains. Fine sample powder is suspended and deposited on polycarbonate filters to get a mass density of $50\mu m/cm^2$; uniformity of the deposition is verified by transmitted intensity of a He-Ne laser (differences may reach 1%). From 3 fragments from a pre-Hispanic jar (200 A.D.), 15 samples were prepared by this method. Good agreement among the composition of the samples and the corresponding fragments of the jar was found.

In one analysis of six stirrup handled Moche ceramic vessels from Peru (circa 100 B.C. to 700 A.D.), Swann et al. [23] analyzed the matrix and the red and white surface patterns of the pottery by an external beam. In was observed low temperature firing because no glazing or vitrification was attained. Two measurements were done by a 0.5 mm diameter beam at the same region: The first irradiation with a 1.3 MeV proton beam to get signals from major and light elements while the second one at 2 MeV with a 75 μm Al filter measures the trace element contents. Significant differences for copper, zinc and lead contents for one vessel were detected. This vessel has a different reported origin and then this difference is probably due to the clay used to fabricate this object or to a different temporality. High Ca X-rays intensities were observed in white decorations and $CaCO_3$ may be related to this pigment. Red color did not show any difference by comparison to clay composition. A special surface preparation may give rise to this color. Nevertheless, in one case the red pigment was a mixture of phosphorus and iron compounds. Black color may have an organic origin.

One ceramic feminine statue from the culture Chupicuaro (600-100 B.C.) from ancient Mesoamerica was analyzed at AGLAE laboratory [24]. The clay and color decorations of this object, part of the Branly collection, was studied by external beam PIXE and other techniques such as Thermoluminisce dating (TL), Scanning Electron Microscope (SEM), X-ray radiography (XR), and X-ray diffraction technique (XRD). The object was analyzed directly, but sampling was also done under the right foot. XR images indicated that the body is empty, the wall thickness ranged from 5 to 10 mm and several hollows were performed for firing. The pottery paste has a fine texture with small iron oxide inclusions. Elemental composition indicated low calcium clay with 4 % iron. The red color of the decoration is hematite (iron oxide), but black color is a mixture of manganese and barium compounds, called wad (manganite (MnO_2), pyrolosite ($MnO.OH$) and psilomelane ($BaMn_2Mn_8O_{16}(OH)_4$)) with iron oxides and sulphates. Bright white was a kaolin clay rich in iron and calcium compounds (carbonate and sulphate) while dull white had higher amounts of calcium and lower aluminum and sulphur contents. White pigments had different preparations to obtain different finished products. Only by using several types of techniques, it is possible to carry out such a complete analysis. Nevertheless, this is not practical when studying a high number of objects or samples.

PIXE analysis of Salado polychrome ceramics from the North American Southwest shows temporal variation in white pigments [25]. The analysis was carried out using an external beam with 7.6µm of Kapton as exit window. Since there was only one detector for X-ray detection, two incident beam energies (1.46 MeV and 2.81 MeV at the sample surface) were used to detect low energy X-rays elements with no filter and trace elements with a non-selective filter of 300µm Mylar and 22µm of vanadium. Two irradiations were carried out per sample and 38 elements were reported. The Pinto (1270-1325 A.D.), Gila (1300-1450 A.D.) and Tonto (1350-1450 A.D.) phases from the Tonto basin of Arizona were studied. After a multivariate and principal component analysis, aluminum, silicon, potassium, calcium, titanium and iron were the most useful elements to differentiate among samples. Pinto phase (smooth and lustrous) presented less Ca, Ti, and Fe but higher amount of Si than the Tonto phase (dull white to cream color). Silicon and aluminum contents decreased over time during this phase. Differences in raw materials and techniques of preparation were due to site locations and not to choices among geological sources. White color presented weathering by volcanic ash.

In a Chilean study, PIXE was combined with XRF to analyze pottery from one site of the Santiago Valley (El Mercurio) and the coast of Chile (El Peral-C) separated by 150 km [26]. Aluminum to zinc was measured by PIXE while XRF with a [109]Cd source was used to determine rubidium, strontium, yttrium and zirconium contents. Results of the elemental composition indicate that there is a differentiation among the kind of object (pot, mug, vessel) and the decorated pottery of each site. No trade, only local production systems were established.

Aztec (or Mexica) pottery fragments with red decorations from Templo Mayor of Mexico (1325-1521 A.D.) were analyzed using heavy ion beam in vacuum taking advantage of the differences in X-ray production cross sections to excite only elements at the decoration surface [27]. With a proton beam there is no difference among the clay and decoration spectra because the proton beam crosses the thin painted layer. A 5.2 MeV argon beam may reach 7μm in a ceramic material and will provide only signals from elements of the painted layer. In this way, the corresponding X-ray spectrum indicated that the red pigment is an iron oxide with traces of arsenic.

Pottery from Epiclassic and Postclassic periods from Teotihuacan (700-1521 A.D.), in the central high plateau of Mexico, were studied using PIXE, RBS and XRD [28-29]. Teotihuacan was one of the most important civilizations of Mesoamerica during Classic Horizon (100 B.C. to 700 A.D.), but in Epiclassic period Teotihuacan decline started (650 A.D.). During this epoch vast demographic rearrangements occurred while new politic and economic centers were established in all Mesoamerica. The aim of this kind of studies is to determine cultural changes through the ceramic materials. These materials correspond to the cultural phases Xolalpan (400-600 A.D., Coyotlatelco (700-900 A.D.), Mazapan (1000-1250 A.D.) and Aztec 1250-1500 A.D.).

Simultaneous PIXE and RBS analyses in vacuum by 2.6 MeV proton beam were completed with external beam PIXE measurements by 3 MeV protons to determine elemental composition of 38 sherds and several clay sources. RBS was used to measure mainly carbon and oxygen contents while PIXE provided the elemental contents of elements heavier than aluminum. XRD was used to determine the mineralogical phases of the pottery. Sherds fragments were powdered for the analyses. Calcium, iron, titanium, strontium and zirconium were the most relevant elements to distinguish among the sherds. Albite and anorthite were the main mineralogical phases. After cluster analysis of the data, results indicated that several sherds have foreign origin. When using PIXE and XRD the corresponding dendrogram showed a gradual contrast in pottery composition following the cultural sequence between Coyotlatelco and Mazapan phases but higher differences between Mazapan and Aztec phase. This may be due to the exploitation of clays sources or to clay preparation and may be used for relative dating.

The analysis of decorations of the Mazapan phase pottery using the external beam allowed identifying the nature of red and white pigment. Red corresponds to iron oxide while white is a rich titanium oxide. This pigment was observed in pre-Columbian decorations of ancient ceremonial potteries from Chile. Brown color decoration corresponded to a difference in surface oxidation during firing (negative-positive method).

The thickness of layers of red and white decorations was measured by Differential PIXE method (D-PIXE) with the external beam [29] (figure 5). Incident proton energy and the corresponding depth of analysis may be modified by setting 8μm aluminum foils at the exit window, due to energy lost in the foils. The PIXE spectra will change depending on the elemental depth profile. By a comparison of Fe/Mn, Fe/Ca and Fe/Ti X-rays intensity ratios from the spectra of a homogeneous reference material and the decoration spectra, it is possible to estimate the thickness of the paints layers between 75 and 100μm. This method can be used to study the layered structure of paintings and the composition of pigments on sherds.

Following a similar analytical approach by a PIXE-XRD methodology, the pottery from the Caxonos river basin of Oaxaca, Southern Mexico, was studied [30-31]. This river basin is a natural way between the central valley of Oaxaca and the gulf coast of Mexico through the high mountains. Archaeological sherds collected in the Caxonos region sites and the Central Valleys of Oaxaca were analyzed. Visual classification of 13 sherds types was contrasted with the sherds composition in order to establish a basic framework for the study of cultural contacts and the trade routes between these regions. The proposed model for pottery studies includes the data of the site of discovery, ceramic style and typology, sherd composition as well as the geological and mineralogical characteristics of the clay sources. The chronology must be known for archaeological interpretation. It is considered that the process of cultural contact and exchange involves style, finished products, raw material and technology.

For PIXE analysis the external beam was used with two X-ray detectors, a Si detector for light and major X-rays elements and a LEGe detector for high X-rays from trace elements (*e.g.* figure 3). After cluster analysis of PIXE and XRD data, two different groups are formed, each group containing sherds from only one region. Other groups are formed with sherds from both regions. The spatial distribution of these groups indicates short circuits of exchange and contact. Some forms and finisheds from the Central Valley were copied using local clays at the Caxonos basin region, but there are pottery types exclusive of this region. Further studies are in progress using specific diagnostic types of the region.

This PIXE-XRD analytical approach has been tested in isolated communities of Guerrero region of Mexico where modern pottery is still produced following traditional methods and the pattern of production and distribution centers are known [32]. In this case, archaeological parameters and clay sources are controlled. Results of the analysis of sherds and statistical clustering agree with the real data and predict the expected trade pattern of the region.

Other studies have been carried out using PIXE and XRD and other techniques for dating, such as TL. A recent research on authenticity and provenance of samples from Zapotec urns (200 B.C. to 800 A.D), a funerary effigy vessel from the region of Oaxaca, Mexico, revealed that genuine and fake urns have specific and different elemental and mineral compositions [33-34]. Even TL is a unique technique for dating, PIXE and XRD data may suggest probable forgeries and the most probable provenance of the genuine urns. The Zapotec urns (88 pieces) belong to the collection of the Royal Ontario Museum of Canada.

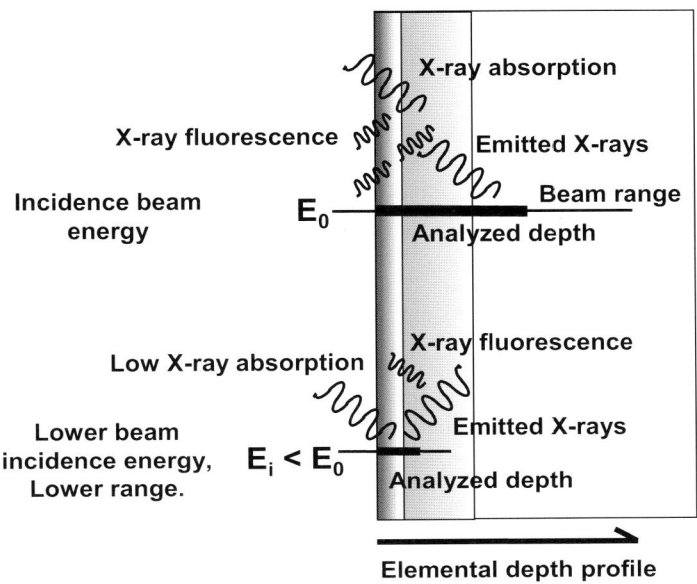

Figure II-4-5. Diagram of Differential PIXE method. By changing the incident beam energy the beam range and the analyzed depth will be modified. The X-ray intensities in PIXE spectra will change as a function of the elemental depth profile. From the X-ratios, the elemental depth profile is measured [47].

ii) Obsidians

Obsidian is an alumino-silicate mineral glass produced by fast cooling of lava. Obsidian beds may be found in volcanic regions easily. In ancient America this material has been extensively used to produce artifacts and for this reason it is a diagnostic material of trade and cultural relationships. This material is very stable and relatively homogeneous. Since this material is not modified by man, sourcing of artifacts is much more easier than clays and pottery. Obsidian is perhaps the most often analyzed material from ancient America. PIXE is preferred to Neutron Activation Analysis (NAA) when non-destructive analysis is necessary or when a large number of samples must be studied.

Obsidians artifacts from the Baja California Sur Caves of Mexico were analyzed by low energy PIXE [35]. The caves are important due to the mural paintings and obsidians may provide useful information about the ancient societies of gathering-hunting groups of this region. A 0.7 MeV proton beam was used only to determine major element concentrations from aluminum to iron. Nevertheless, the results of the San Gregorio cave's cutting tools indicate that potassium, calcium and iron contents are enough to differentiate the obsidians sources of this region. One group clustered most of the artifacts but one sample was scattered, then at least two obsidian sources were exploited.

Other research on obsidian sourcing of central Mexico combined PIXE and Atomic Absorption Spectrometry (AAS) [36]. Fifteen samples from five sites were analyzed using 2 and 3 MeV proton beams for PIXE. 20 elements were measured by these techniques. From rubidium, strontium and zirconium contents, three sources were distinguished.

A very complete ion beam analysis of obsidian sources from Mexico [37] involved Nuclear Reaction Analysis (NRA) for oxygen contents by (d,p) reactions, Particle Induced Gamma-ray Emission (PIGE) for sodium, aluminum and silicon, PIXE for elements ranging from potassium to bromine, and RBS for stoichiometric measurements. The obsidians were irradiated in vacuum for NRA by a 1.7 MeV deuteron beam, then by a 2.62 MeV proton beam for PIGE analysis and by a 2 MeV alpha particles for RBS. An external beam device was used with a 2.6 MeV proton beam for PIXE. Three obsidian sources were identified but a low number of samples were analyzed. In this case, the possibility to carry out a combined and simultaneous analysis by PIXE and PIGE was not exploited, perhaps due to the characteristics of the experimental devices. RBS may be enough for oxygen measurements due to high homogeneity of the material. This kind of analysis cannot be applied to a large number of pieces but it may be considered for a full characterization of diagnostic samples from known sources.

The provenance of pre-Hispanic obsidians from the region of the ceremonial center of Lagunillas, Michoacan, Mexico during early (900-1200 A.D.) and late (1200-1523 A.D.) Postclassic periods were studied by non-vacuum PIXE [38]. This site had an important role of trade and contact between cultural areas in West Mexico. Green and grey obsidian artifacts, mainly projectile points and prismatic blades, and samples from Zinapecuaro, Zinaparo-Varal and Pachuca sources were analyzed with a 2.66 MeV proton beam (2 mm diameter). The obsidians were irradiated 3 times at each region to evaluate the reproducibility of the measurements. Aluminum, silicon, chlorine, potassium, calcium, titanium, manganese and iron contents were determined. The main difference between green and gray obsidian is due to manganese and calcium contents. When comparing the elemental data in ternary diagrams of calcium, manganese and titanium, most of the obsidians clustered with Zinaparo-Varal source samples and few of them overlapped with the distant Pachuca source samples, but none of the artifacts correlated with the closest source of Zinapecuaro. The results suggest that there were two periods for the Lagunillas site, a first one of development and another one of Tarascan domination. During the early period the presence of obsidians from Central Mexican basin (Pachuca) indicated a commercial trade with this region. In late Postclassic period there was a change in the obsidian source exploitation and trade routes.

A set of 38 pre-Hispanic obsidian artifacts and samples from sources from Talca region in Chile and Argentina (Maule River basin) were analyzed at Lucas Heights Laboratory in Australia [39]. PIGE and PIXE were performed simultaneously. X-ray and gamma-ray detectors were set at 45° with respect the incidence beam direction. The study of the sources and artifacts characterization was carried out using nine signal ratios (F/Na, Al/Na, K/Fe, Ca/Fe, Mn/Fe, Rb/Fe, Sr/Fe, Y/Fe and Zr/Fe) in non-parametric cluster analysis. Six major groups were determined but most of the artifacts belong to one group. At least two groups correspond to unknown sources.

PIXE has been combined with fission track dating to determine artifacts trade and the number of pre-Hispanic obsidian sources exploited in Colombia and Ecuador regions [40]. Fission track dating may be used until ages about 10000 years before present. About 142 artifacts from 45 archaeological sites corresponding to three ancient periods were analyzed at AGLAE laboratory. 15 major and trace elements heavier than carbon were measured. It was determined that burial had not significant effect on the elemental concentrations of the surface. Manganese vs. strontium contents clearly revealed clustering. Seven discrete groups were formed but one source presented unusual heterogeneous composition, probably due to incomplete mixing of two magmas. The artifacts composition was similar to known sources (Rio Hondo, Guscatula-Yanaurcu, Mullumica an Callejones), but only three of them have a very different composition, clustering in one unknown source. When fission track dating data were considered, artifacts with the same compositional characteristics sometimes split into different age groups. Then at least three unknown sources were exploited. Solely PIXE data may give rise to erroneous source attribution, as several sources of different ages appear to have indistinguishable chemical composition. In obsidian studies, it is necessary to improve the knowledge about local sources and probable long-distance trade systems.

iii) Gold and bronze artifacts

The use of metals is a characteristic of developed civilizations. Metallic items may be used to establish technology, minerals exploitation, the use of alloys, soldering techniques, finishing methods and trade. Provenance is difficult to determine by PIXE and other X-ray methods because the trace elements signals are strongly absorbed in metallic matrices and usually various metallic artifacts and alloys may be re-melted to fabricate new items.

Artifacts from gold and bronze alloys were produced in several cultural areas of ancient America. Metallurgy was discovered and developed by Chavin civilization around 1500 B.C. in Peruvian area from native metal sources. First alloys were prepared several centuries later (circa 200 B.C.) by Chavin and Mochica cultures. At the beginning of the first century, metallic artifacts already were fabricated by Colombian cultures by lost wax technique. Metallurgy appeared as imported technology in Mesoamerican regions only around 800 B.C., but the lost wax technique reached a high level of development with local improvements. For ancient American cultures, gold artifacts had a symbol of power and hierarchy. In Mesoamerican and South American regions, bronze artifacts were used as utilitarian items before the arrival of European in the 16th century. After de conquest of Mesoamerica and South American regions, many metallic items were smelted and tons of gold and silver were sent to Europeans kingdoms. In the case of Mesoamerica, very few gold and silver artifacts have been preserved in museums or discovered in recent archaeological excavations.

The analytical approach for gold alloys is quite different by comparison to bronze artifacts. Gold alloys usually are not corroded by burial environment and surface analysis may be carried out by direct irradiation of the surface. In the case of bronze, patina is formed and its thickness may avoid direct surface analysis. Depth of analysis by PIXE analysis may reach 10µm for heavy metallic matrices using 3 MeV protons. Depending on analytical conditions, sampling may be carried out to irradiate only the copper or bronze alloy. In other cases small regions of the artifacts surface may be mechanically cleaned and the proton beam may reach non-corroded metal.

A complete description of various ion beam techniques including non-vacuum milliprobe, the use of selective absorbers and XRF induced by proton irradiation for the analysis of gold alloys artifacts and ancient soldering techniques is presented in ref. [41]. In this work, the study of a gold pendant, probably from Central America, revealed that lost wax process was used to fabricate the artifact and no soldering was observed. In the lost wax method, the shape of the jewels is carved in wax and molded in clay. The mould is then emptied by heating and finally refilled with the alloy. The composition of this artifact is quite regular (28% copper, silver<1%, 70% gold) and typical of a rich copper gold alloy, called tumbaga.

In another work, the analysis of a gold Darien pendant from Panama (500-1500 A.D.) was reported [42]. The main body of this pendant represents a seated individual and several rectangular and circular foils decorations hang from the main body. The compositions of the rectangular and the circular foils are different. The rectangular foils presented 26% copper and 23% silver while the circular foils had a mean composition of 40% copper and 14% silver. The composition of the main body is not homogeneous but rich in silver and very poor in copper (0.6%), perhaps there was a gold surface enrichment. Three different gold alloys and a gilding technique may have been used to fabricate this item.

In one of the first works on archaeometry in Mexico, gold artifacts from La Chinantla, in Oaxaca, Mexico, were analyzed by low energy PIXE (0.7 MeV) in vacuum [20]. The artifacts belong to the National Museum of Anthropology and History of Mexico. The four pieces were rich in silver ranging from 6 to 33%, copper contents were low (0.3-2.3%) but the mean composition was different for each object.

A collection of funerary gold artifacts from the Oaxaca region from the National Museum of Anthropology and History of Mexico was studied by external beam PIXE [43]. The thirteen analyzed artifacts were quite homogeneous. In general, the gold alloy contents ranged from 50 to 85% and from 10 to 45% for silver. Copper amounts were in general low (less than 5%), as in many artifacts from Mesoamerica. Only in one component of a bracelet, a 33% copper concentration was observed. The surface homogeneity was tested by grazing beam incidence on objects with flat regions. From X-ray ratio variations as a function of incidence beam angle, a light gold enrichment at the surface was observed maybe due to a depletion gilding technique or a burial natural corrosion.

Special attention received the study of the gilding techniques by ion beam analysis, particularly, the depletion gilding or *mise en couleur* [44]. This original technique was developed in ancient America to provide a golden appearance to rich copper gold alloys. The original copper gold alloy has an orange copper color at the beginning of the process. By oxidizing and pickling in acid plants or corrosive mineral solutions, copper was drawn out preferentially from the artifacts surface because gold does not oxidize. Depending on the original gold alloy composition, oxidizing temperature, pickling process and its efficiency, the original color may be modified. The surface is gradually gold enriched and the enriched thickness may reach until 30μm depth. Usually, the enriched depth ranges from 3 to 15μm [45]. At the end of the process, the rich copper alloy has a golden appearance.

The section of several artifacts from a funerary site of Panama (Sitio Conde, 400-900 A.D.) was analyzed using protons and a microprobe with 50μm of spatial resolution [46]. The gold was used as native metal to prepare thin decorative foils, but rich copper gold alloys were prepared to fabricate artifacts with higher hardness, more wear-resistance, lower melting point and easier casting. When copper content excess 25% in the alloy, high undersurface corrosion may occur and the gilded surface may peel off like a thin foil. Then, the microprobe beam may reach the matrix of the alloy. One decorated plaque and a chisel were irradiated in this way. The mean gold content ranged from 43% in the matrix to 90% at the surface while mean copper contents decreased from 50 % to less than 6%.

Non-destructive analysis of artifacts gilded by this technique may be also studied directly, without sampling, by a milliprobe and a microprobe. The Differential PIXE method (D-PIXE), already described above and fully detailed for gilding analysis in ref. [47-48] (figure 5), was used in vacuum and non-vacuum modalities to measure the elemental depth profile of several artifacts from Colombian and Mesoamerican cultures. RBS is not suitable to measure the elemental depth profile due to its poor elemental resolution for heavy matrices. D-PIXE is then applied: the artifact is irradiated at the same region using several proton beam incidence energies (*e.g.* 1 to 3 MeV) and the X-rays intensities of PIXE spectra will change depending on the elemental depth profile. The method to determine the elemental depth profile lies on the comparison among calculated Cu-Kα/Au-Lβ ratios of a layered structure and measured Cu-Kα/Au-Lβ ratios from PIXE spectra. The thickness and composition of a proposed layered structure is modified following an iterative procedure until the convergence among calculated and measured ratios is attained. RBS may be carried out for qualitative analysis to verify the measured elemental depth profile.

With the D-PIXE methodology, the use of depletion gilding was verified for the first time in a recently excavated Mesoamerican pendant from the Oaxaca region [49]. Within a thickness of 4μm from the surface, gold and copper contents attained at the surface 90% and 5 %, respectively, while the matrix composition was 50 % gold, 40% copper and 10% silver.

For pre-Hispanic pendants from Colombia, D-PIXE combined with RBS allowed to measure the elemental depth profile and determine the use of two gilding techniques [50]. In artifacts corresponding to Quimbaya style (100-1600 A.D.) finished by a depletion gilding technique, a gold enrichment was observed in different depths for each artifact (1.3 and 4 μm). In one artifact of Tairona style (600-1600 A.D.), the RBS and PIXE spectra showed that a rich gold alloy was deposited on a copper rich matrix (10% gold). The measured layer thickness was more than 1.5 μm. In a Tolima style butterfly pendant (0-1000 A.D.), the gold alloy layer thickness was very thin (0.2 μm). These artifacts were finished probably by an electrochemical plating assisted by a thermal treatment to deposit and to bond rich gold-silver layers on the rich copper matrix [51].

Scanning PIXE-RBS analysis of the surface of one of these Quimbaya artifacts by a proton microbeam with a spatial resolution of 10μm showed a surface lateral heterogeneity that may be related to the depletion gilding technique [52]. Analysis revealed gold enriched micro-regions at the surface with different enrichment thickness. During oxidizing by heating and pickling of the surface, the surface dynamic gives rise to gold diffusion and cluster growth by nucleation. The microanalysis results indicated the presence of these microstructures.

A gold enriched surface is not always a synonymous of a gilding procedure. D-PIXE was combined with XRF using radioactive sources to study the elemental depth profile of a gold mask from Colombia [53]. In this case the analysis showed that the surface contains very low amounts of copper, but copper content increases in the bulk (3.3%). The depletion of copper could be due to modern cleaning using acid agents.

On the other hand, smelting furnaces, slag remains, refining crucibles, ingots and finished artifacts were analyzed for the study of ancient copper production at Batan Grande, Peru (1000 A.D.) [54]. The main results indicated that alloying was achieved by deliberated addition of select arsenic rich ores during smelting and slags were reprocessed to obtain higher copper yields.

Concerning bronze artifacts, copper bells are considered as a diagnostic artifact for trade because they were used widely during late Postclassic period (900-1521 A.D.) in Mesoamerica. A set of copper bells from the site of Templo Mayor of Mexico from offerings discovered during archaeological excavations were analyzed by non-vacuum PIXE and metallographic techniques [55]. Several gold artifacts were also studied. RBS technique was applied to determine depth homogeneity. The gold artifacts presented typical low copper concentrations (1-2% copper, 30% silver) and no gold surface enrichment. Several bells had variables amounts of arsenic (1 to 3%), typical of West Mesoamerican items.

Another set of copper bells (or crotals) and minerals (malachite, native copper cuprite and stibnite) from sites and locations from the Great Southwest area of North America were analyzed by a microbeam [56]. Copper bells appear before 1000 A.D. in the Hohokam region and reached a wide distribution after a century. These artifacts were prepared by lost wax casting; there are 10 types exclusive of this region. In the site of Paquimé, in the Casas Grandes region, a workshop was discovered. A random selection of crotals from Paquime collection, a set of bells from Chaco site in Anazasi region and two pieces from Arizona State Museum were sampled using a hand operated twist drill. The samples were mounted on SiO_2 wafers with epoxy and polished before the microprobe PIXE analysis. The spatial resolution was 20μm with a 3 MeV proton beam. A 750μm of aluminum foil was used as non-selective filter to reduce copper X-rays detection, but this choice decreased iron and nickel signal intensities. The metal and its inclusions were irradiated. Crotals contained traces of silver, antimony, arsenic, lead and in some cases strontium, selenium and tin. Several pieces presented no detectable amounts of these elements. Then several workshops may have manufactured these bells. Six groups with different compositional types were obtained (one group of native copper artifacts). From analytical results, early local production sequences and imported artifacts from West Mesoamerica regions were distinguished. By comparison with West Mesoamerica, deliberate alloying of copper with arsenic, tin and silver was not performed.

In a recent study of copper based alloy artifacts from a late Postclassic site of the Mayan region of Chiapas (1500 A.D.), a set of 30 pieces of bells, foils fragments, chisels, one ax, rings and needles were analyzed using an external beam set-up [57]. It was allowed to remove 4 mm^2 of the items patina by a dentist drill. These regions were irradiated by 3 MeV proton beam (1 mm diameter). By comparison with a Si detector, LEGe detector has a higher efficiency for heavy elements X-rays such as silver, tin and antimony. For this reason, a LEGe detector was employed with 15μm cobalt foil filter to decrease the copper X-rays intensity. The following elements were measured: Iron, nickel, arsenic, silver, tin, antimony and lead. Results indicated that few artifacts were smelted from native copper. Most of the artifacts were fabricated with alloys containing arsenic and tin in order to improve the mechanical properties of the chisels and nails. Local production and imported artifacts from West and Central Mesoamerica were determined. Most of the previous analyses on Mesoamerican copper artifacts were carried out by AAS. This method does not provide complete information about elemental composition. For this reason, the analyses of a significant amount of artifacts are uncompleted. For instance, lead contents, usually present in Central Mesoamerican and Mayan artifacts, were not systematically measured until it was considered important. New measurements on collections of artifacts must be considered.

iv) Mural paintings

In ancient America there are beautiful and unique pictorial expressions in mural paintings. Pigments and plasters composition as well as deterioration and restoration of the paintings can be studied on fragments and small samples by PIXE. Heterogeneity of painting support could be a limitation for mural painting studies.

Several painted fragments from La Pintada and San Gregorio caves of Baja California Sur region of Mexico, were analyzed using a low energy proton beam (0.7MeV) in vacuum [35]. Red, black, white and yellow colors were used by the first gathering-hunting peoples of the region to represent human figures, deer and sea animals. Ion oxides were used for red and yellow pigments, but due to the heterogeneity of the wall cave, layer thickness and the nature of other pigments were difficult to determine. Recently, more samples are being studied for restoration purposes. There are very few PIXE studies on this subject and a very important number of mural paintings in caves in all ancient American areas. For this reason, this topic may be considered for future research.

Characterization of mural plaster from the site of Xochicalco, in central Mexico (700-900A.D.) was carried out by PIXE, XRD, Scanning Electron Microscopy (SEM) and Infrared Spectroscopy (FTIR) [58]. Elemental concentration, main crystalline structures, microstructure, morphology and organic components may be analyzed by this set of techniques. Mural fragments with white, black, blue, red, green and orange colors and stucco walls were sampled. In stucco a mixture of calcite and anorthite, albite and sometimes quartz was found. A proton beam of 0.7 MeV was used for PIXE analysis due to short range of the proton beam. In stucco PIXE spectra the silicon and calcium X-ray signals agree with the XRD observations. In red paintings, the presence of high mercury and iron signals suggested a combination of hematite and cinnabar, while orange color is a mixture of iron oxides. Green color presented high sulphur and iron contents, probably due to iron oxides (yellow mineral) mixed with an organic colorant. White color was a mixture of quartz, titanium oxide and clays. Blue color corresponded to a Maya blue type, a mixture of indigo and rich magnesium clay. There was a good agreement among the results of different techniques.

External beam PIXE was used to analyze pigments and stucco wall fragments from different habitational units of Teotihuacan (300-700 A.D.) [59]. Pigments were identified and corresponded to similar materials to those described above. Additionally, the thickness of the paint layer was measured following the differential PIXE method described previously for pottery pigments. The composition of stucco walls (calcite-clay mixture) changed for different constructive periods. This may be used for chronological studies and calcite beds sourcing.

Effects of different deterioration processes produced by underground water level, high humidity, the presence of soil, and water and air pollution on pre-Hispanic painting were studied on samples from the stone benches of the Eagle warriors Precinct at Templo Mayor of Mexico [60]. Analysis of basalts fragments with paint remains were carried out by PIXE and RBS in vacuum and with external beam PIXE. Elemental enrichment factors relative to iron concentrations were defined. Possible contamination by sulphur from reactions with atmospheric SO_2 and chlorine salts dissolved in water as well as Zn scavenged by rain were found. Limonite (hydrated iron oxide) and orpiment (As_2S_3) were observed in yellow color. White color on stucco walls presented high sulphur X-ray intensities probably due to deterioration of $CaCO_3$ by gypsum formation ($CaSO_4.2H_2O$) resultant from dry deposition of SO_2, an abundant atmospheric pollutant in downtown Mexico City. This is a first research to understand the damage in an archaeological site within a city in order to improve future suitable conservation procedures.

v) Biological materials

Teeth, bone remains, hair, animal and plants remains are typical biological materials that may be examined by PIXE. In ancient American sites, environmental conditions strongly determine the deterioration of these materials and the reliability of analyses.

Dry conditions of the Atacama Desert in South America preserve biological tissues. Hair composition from 10 mummies from Archaic (8000 to 2000 B.C.) and Formative periods (200-500 B.C.) were studied using PIXE [61]. Hair samples were divided in two groups of unwashed and washed in acetone and water. Both sample sets were prepared by acid digestion, and then analyzed by PIXE. Results from washed and un-washed samples were compared to determine elements at the hair surface. Calcium to strontium and lead were reported. One reddish sample was found to contain unusual large quantities of arsenic and iron probable due to a mineral dye (hematite and realgar) applied to the hair to obtain the reddish color.

From hair analysis, forensic and nutrition aspects of populations can be investigated. In the case of one mummy of a child, low levels of calcium indicated a disease related to this deficiency. The elemental composition of modern hair is similar to mummy hair but contents of manganese and strontium were higher. This may be related to nutrition differences. Further analysis must be carried out on this topic.

Teeth have been analyzed from different points of view. A possible therapeutic ancient tooth inlay was studied on 5 pre-Hispanic human teeth from La Campana, Colima in West Mexico (2000 B.P.) [62]. Amber colored spots, looked like a dental filling, and enamel regions of the same teeth were irradiated for PIXE and RBS analyses. Dentists' hypothesis suggested that the teeth were prepared to inlay some material. Analytical results indicated that they have similar composition (with some differences in iron contents and trace elements) and suggested that a tooth inlay using healthy tooth grains was glued into the prepared tooth to fill it up for therapeutic purposes.

Teeth have been suggested as dose monitors for the exposure of human to foreign elements. The elemental composition of a set of pre-Hispanic (1300 A.D.), Colonial (1600 A.D.) and contemporary teeth was studied by non-vacuum PIXE [63]. Minor changes were detected for the metal/Ca ratios for manganese, iron, copper, zinc and strontium. In contrast, lead contents showed significant differences among the groups. In pre-Columbian teeth, lead was not found. Colonial teeth showed higher lead levels than contemporary ones. This result suggested that probably the glaze pottery, introduced by the Spaniards after the conquest of Mexico, was the main source of lead. Modern lead sources are environment and pottery. A similar study on a set of Mexican pre-Hispanic teeth and modern teeth of children and adults to get more insights about environmental conditions, diet, teeth health, diseases and post-mortem alteration among populations was carried out [64]. Mean values of manganese, iron, zinc, strontium and calcium contents for each group were reported. Results indicated that metals contents in permanent teeth may be related the kind of diet; low metals contents may be related to anemia. An absence of correlation among the manganese, iron and strontium contents may be considered a criterion of preservation. Nevertheless, further analysis on each specific topic must be carried out for conclusive statements.

Human beings were part of sacrifices and offerings to deities in Mesoamerica. Mexican (or Aztec) culture (1325-1521 A.D.) performed this kind of ritual. Anthropo-physical studies of human remains indicate that most sacrificed children suffered some painful disease, such as brain tumors. Teeth from infants with extremely rare colored enamel regions (white, blue gray and brown) have been analyzed by PIXE and RBS [65]. The objective of this analysis was to determine if a pathology was the origin of teeth coloration since healthy and ill teeth from individuals in the offering were exposed to identical humidity and ground conditions. A comparison among normal, blue and gray colored teeth indicate that colored teeth presented larger mean amounts of phosphorous, manganese and iron, while zinc and strontium did not change too much. Only dark brown regions differed notoriously from hydroxylapatite, the main compound of enamel. Another study from 35 pre-Hispanic colored teeth and soil from the same sites reported variations and differences in zinc, manganese, iron and strontium concentrations [66]. For this group, brown and blue stains on the teeth were due to differences in tooth enamel porosity and post-mortem biogeochemical process. The alteration involves accumulation and digenesis of iron, manganese and organic matter solutions eluviated from soil.

Bone remains have been also studied by PIXE. Disease, paleo-diet and deterioration analyses may be carried out. For dating the elemental composition is very useful. Environmental and burial conditions are relevant for reliable data and interpretation. In the case of ancient American bone remains, there are several researches in progress but there is a lack of PIXE reports on this subject.

vi) Other materials

Turquoise is opaque, crypto-crystalline mineral, composed mainly of hydrated copper aluminum phosphate, $CuAl_6(PO_4)_4(OH)_8 \cdot 4H_2O$, prized throughout the world as a gemstone. Turquoise was used for adornment, pigments and ritual from prehistoric ages and was extensively traded in ancient America. This material may be used to determine social and trade relationships between settlements and resources zones, but also for conservation.

A turquoise disk from the offering 99 of Templo Mayor of Mexico (1502-1521 A.D.), formed by 15000 or small flat pieces, was discovered during excavations. All the fragments were scattered in the offering. A very careful restoration process allowed the reconstruction of the disk. White, light green, light blue and blue colors may be observed in the fragments. Turquoise samples were analyzed by PIXE and XRD to determine the relationship between the fragment composition and its colors, considering the deterioration processes and the environment of the offering [67]. Also, a "paste" sample, material considered as the original base of the disc was included in this study. Four minerals belong to the turquoise minerals family; they have the same crystalline structure but by substitution of aluminum and copper by iron and zinc, the blue color may change gradually to green. Copper substitution by calcium may give rise to a whitish aspect. Results of the analyses indicated that the XRD pattern corresponds to a turquoise structure but the changes in diffraction reflexions intensities pointed out to several replacements in the crystalline structure. PIXE showed that turquoise colors could be related to differences in elemental composition. Blue color presented higher amounts of iron, chromium and cobalt than green color. Whitish turquoise had higher amounts of zinc than the green turquoise. The colors of the fragments may be due to the humid offering environment, temperature changes and the contact with burial soil.

The chemical signatures of turquoise fragments determined by PIXE and XRD can be used to determine mineralogical variability. In a recent study, turquoise artifacts from two large platform mounds communities (Cline Terrace Mound and Schoolhouse Point Mound) from Salado site in the Tonto basin of Central Arizona (1280-1450 B.C.) were analyzed by these techniques to establish whether they shared the same trade network [68]. Results indicated differential access to turquoise sources. This pattern was also observed for obsidians and pottery items.

Amber is an organic fossil resin composed by chains of terpenoid compounds (C_5H_8). This material was used to fabricate in all kind of ornaments and for rituals.

Analytical techniques for organic materials are suitable to characterize the matrix but they require sampling, they are destructive and do not distinguish among beds of the same paleobotanical origin. In amber, there are micro and macroscopic inorganic inclusions of clays, sulphates, calcium carbonates, pyrite, quartz, potassium chloride, etc. These inclusions are related to geological environment and sedimentation conditions of each bed. Non-vacuum PIXE may be used to analyze these inclusions to establish amber provenance of amber artifacts [69]. The analysis is non-destructive and may differentiate among artifacts from different beds of similar or different paleobotanic origin. A set of amber samples from beds of different regions of the world was analyzed by an external beam set-up with 2.95 MeV protons (1 mm beam diameter). Elements from sulphur to iron were detected. There is a correlation between the Fe/S ratios and amber origin. Iron and sulphur contents, and other metallic elements, determine different provenances and beds.

In Mesoamerica, there is only one paleobotanical source of amber. Most beds are found in central Chiapas region. This material was widely traded in all Mesoamerica since ancient times (700 B.C.). In this case, the PIXE analysis applied to two known Chiapas beds and three archaeological artifacts indicated that two of these artifacts have S and Fe contents similar to the Totolapa source, but one of them has a provenance from an unknown source [69]. Further analyses on amber beds samples from Chiapas using an improved experimental set-up are in progress.

Green color stones were also much appreciated in Mesoamerica since ancient times. One anthropomorphic sculpture of 76 cm height of Olmec style from Teotihuacan, part of the Branly collection, polished in green stone was analyzed by non-vacuum PIXE and XRD [70]. Elemental composition indicated that magnesium and silicon are major elements. Nevertheless, this information is not definitive to determine the nature of the green mineral.

Serpentine family of green stones ($Mg_3Si_2O_5.OH_4$) has three minerals with similar composition: chrysotile, lizardite and antigorite. XRD spectrum indicates that the main mineral phase of the green rock is antigorite. Deterioration of the stone was observed due to sulphur and calcium enrichment at the surface.

Paper like-materials and textiles are only preserved in special environmental conditions. Few documents and pieces of textile are available. Most of pre-Hispanic Mesoamerican books or codex were brought to Europe during different historical periods. Nowadays, they are part of collections of the most important libraries and few documents remain in American countries. From the study of these objects, inks, pigments, colorants and technical details of the materials may be determined. In this case, no sampling is required and PIXE may be a unique tool for these analytical purposes. First studies on pre-Hispanic and colonial documents from Mexico have been started recently.

A unique discovery was the pre-Hispanic paper and textiles of offering 102 of Templo Mayor of Mexico (1502-1521 A.D.). The items are in very good conditions of conservation. First PIXE analyses were performed on fragments of decorated paper and painted textile [71]. Results indicated that red color was hematite but for other colors and decorations, the data interpretation is very difficult because there are not enough information on paper, colorants and organic materials used in pre-Hispanic times. Further research and more analyses with other techniques will complete these findings.

4. Final remarks

Several materials and artifacts have been analyzed by different modalities of PIXE technique. The challenge of non-destructive analysis of historical materials gives rise to new methodologies and new developments of experimental devices. Besides the knowledge of cultural and historical heritage, the information obtained from these studies provides objective bases for archaeological and historical hypothesis.

In the case of ancient American materials, a lot of work must still be carried out by PIXE. Some topics require more research: mural paintings in caves, bone remains, ancient documents, colorants and pigments, experimental archaeology and studies related to restoration. Materials and artifacts from colonial periods represent also another practically new research field.

Improvement of methodologies and experimental arrangements are required for more complete and fast analyses. This is necessary for pottery and obsidians studies because a large number of samples must be analyzed with a maximum of information for reliable statistical data. In this way, PIXE methods will become competitive by comparison to other more expensive techniques used traditionally in archaeology (*e.g.* Neutron Activation Analysis).

Collaborations with museums must increase taking advantage of non-destructive characteristics of external beam PIXE.

The new trends of X-ray analysis for archaeological studies, considering the use of portable XRF equipments for diagnostic tests previous to ion beam analysis at accelerator laboratories, represent a suitable strategy for research in American countries where few accelerator facilities are available. In this way, the existing laboratories may become centers for regional studies in arts and archaeology and the use of the experimental facilities will be optimum, especially in regions with rich cultural heritage where highly developed civilizations appeared.

5. References

[1] M.A. Respaldiza M.A. and J. Gómez-Camacho eds., *Applications of Ion Beam Analysis Techniques to Arts and Archaeometry*, Universidad de Sevilla, 1997.

[2] S.A.E., Johansson, J.L. Campbell and K.G. Malmqvist. *Particle-Induced X-ray Emission Spectrometry (PIXE). Chemical Analysis*: Series of monograph on analytical chemistry and its applications, vol. 133, J.D. Widefrodner series ed., John Wiley & Sons, New York,1995.

[3] S.A.E. Johansson and J.L. Campbell, PIXE: *A novel technique for elemental analysis*, John Wiley & Sons, Chichester, 1988.

[4] G. Demortier, Application of nuclear microprobes to material of archaeological interest. *Nucl. Instrum. and Meth.* B **30** (1988) 434-443.

[5] G. Demortier, Review of the recent applications of high energy microprobes in arts and archaeology". *Nucl. Instrum. and Meth.* B **54** (1991) 334-345.

[6] Ch. P Swann., Review of the recent applications of   the nuclear microprobe to arts and archaeology, *Nucl. Instrum. and Meth.* B **104** (1995) 576-583.

[7] Ch. P Swann., Recent applications of the nuclear microprobe to the study of arts objects and archaeological artifacts, Nucl. Instrum. and Meth. B 130 (1997) 289-296.

[8] J.R. Tesmer and M. Nastasi, eds., *Handbook of Modern Ion Beam Materials Analysis*. Materials Research Society, Pittsburg, 1995.

[9] J.C. Dran, Th. Calligaro and J. Salomon, Particle Induced X-ray Emission, in Modern analytical Methods in Art and Archaeology, E. Ciliberto and C. Spoto eds., *Chemical Analysis*: Series of monograph on analytical chemistry and its applications, vol. 155, J.D. Widefrodner series ed., John Wiley & Sons, New York, 2000. p. 135-166.

[10] R. Cesareo, G.E. Gigante, J.S. Iwanczyk, M.A. Rosales M. Aliphat and P. Avila, Non destructive analysis of pre-Hispanic gold objects using Energy Dispersive X-ray Fluorescence. *Rev. Mex. Fis.* **40**. No. 2 (1994) 301-308.

[11] A. Seelenfreund, J. Miranda, M.I. Dinator and J.R. Morales, The provenance of archaeological obsidian artifacts from Northern Chile determined by source-induced X-ray fluorescence, *J. Radioanal. Nucl. Chem.* **251** No.1 (2002) 15-19.

[12] C.R. Appoloni, F.R. Espinoza Quiñones, P.H.A. Aragão, A. O. dos Santos, L.M. da Silva, P.F. Barbieri, V.F. do Nascimento Fihlo and M.M. Coimbra, EDXRF study of Tupi-Gaurani archaeological ceramics, *Rad. Phys. and Chem.* **61** (2001) 711-712.

[13] C. Heitz, G. Lagarde, A. Pape, D. Tenorio, C. Zarate, M. Menu, L. Scotee, A. Jaidar, R. Alviso, D. Gonzalez and V. Gonzalez, Radioisotope Induced X-ray Emisión - A complementary Method to PIXE analysis, *Nucl. Instrum. and Meth.* B **14** (1986) 93-98.

[14] K. Malmqvist, Comparison between PIXE and XRF for applications in art and archaeology, *Nucl. Instrum. and Meth.* B **14** (1986) 86-92.

[15] A. J. Kreiner, M.J. Ozafrán, M.E. Vázquez, A.S.M.A. Romo, M.A. Cardona, M.E. Debray, D. Hojman, J.M. Kesque, J.J. Menéndez, H. Somacal, J. Davison and M. Davison, Heavy Ion induced X-ray emission work at the TANDAR laboratory in Buenos Aires, *Nucl. Instrum. and Meth.* B **99** (1995) 384-386.

[16] http:// www.lnls.br/pop/prop.htm

[17] M.A Rizzutto, M.H. Tabaniks, N. Added, R. Liguori Neto, J.C. Acquedro, M.M. Vilela, T.R.C.F. Oliveira, R.A. Markariana and M. Mori, External PIGE-PIXE measurements at the Sao Paulo 8UD tandem accelerator. *Nucl. Instrum. and Meth.* B **190** (2002) 186-198.

[18] http://omnis.if.ufrj.br/~atomica/

[19] C.M. Romo-Kröger, L. Cornejo-Ponce, J.R. Morales, M.I. Dinator, M.J. Avila, V.H. Poblete, J. Galvez, A. Goldschmidt, A. Trier, J. Valdes, C.M. Rojas, M.E. Cantillano, R. Vera, R. Figueroa and M. Garcia, An overview of X-ray spectrometry at research centers in Chile, *X-ray Spectrom.* **31** (2002) 128-131.

[20] J. Rickards, L. Torres , F. Franco and M.D. Flores, Análisis por la técnica PIXE de artefactos de oro del Museo Nacional de Antropología, *Revista Antropológicas*, IIA-UNAM, No. **15** (1999) 71.

[21] J.L. Ruvalcaba Sil, M. Monroy, J.G. Morales and K. López, The new external beam set-up of the Pelletron Accelerator at the UNAM, Mexico, IX International Conference on Particle Induced X-ray Emission and its Analytical Applications, Guelph, Canada, 2001.

[22] J. R. Morales, M.I. Dinator, F. Llona, J. Saavedra and F. Falabella, Sample preparation of archaeological materials for PIXE analysis, *J. Radioanal. Nucl. Chem. Letters*, **187** (1) (1994) 79-89.

[23] C.P. Swann, S. Caspi and J. Carlson, Six stirrup handled Moche ceramic vessels from pre-Columbian Peru: A technical study applying PIXE spectrometry, *Nucl. Instrum. and Meth.* B **150** (1999) 571-575.

[24] G. Querré, Une statuette feminine en terre cuite de la culture Chupícuaro (Mexique), *Techne* No.11 (2000) 47-52.

[25] D.C. Gosser, M.A. Ohnersorgen, A.W. Simon and J.W. Mayer, PIXE analysis of Salado polychrome ceramics of the American Southwest, *Nucl. Instrum. and Meth.* B **136-138** (1998) 880-887.

[26] J.R. Morales, M.I. Dinator, F. Llona and F. Falabella. *Aplicaciones de PIXE y EDXRF en el estudio de cerámicas prehispánicas de Chile Central, IV Seminario Latinoamericano de Análisis por Técnicas de Rayos X*, 1994, Chile.

[27] D. Tenorio, G. Acosta, M. Bordas, J. Larcher, G- Lagarde, C. Heitz, and A. Jaidar, Thick target X-ray yields and intensity ratios for MeV Br and Kr ion impact and application to the analysis of pottery, *Nucl. Instrum. and Meth.* B **15** (1986) 612-615.

[28] J.L. Ruvalcaba-Sil, M.A. Ontalba Salamanca, L. Manzanilla, J. Miranda, J. Cañetas Ortega and C. López, Characterization of pre-Hispanic pottery from Teotihuacan, Mexico, by a combined PIXE-RBS and XRD analysis. *Nucl. Instrum. and Meth.* B **150** (1999) 591-596.

[29] M. A. Ontalba Salamanca, J.L. Ruvalcaba-Sil, L. Bucio, L. Manzanilla and J. Miranda, Ion beam analysis of pottery from Teotihuacan, Mexico. *Nucl. Instrum. and Meth.* B **161-163** (2000) 762-768.

[30] E. Ortíz, L. Lazos, J.L. Ruvalcaba Sil and L. Bucio, Interdisciplinary Approach for Pottery Analysis of Caxonos River Basin, Oaxaca, *Antropología y Técnica*, IIA-UNAM, **6** (2000) 85-94.

[31] L. Lazos, L. Bucio, J.L. Ruvalcaba, J. Litvak, and E. Ortiz, An Interdisciplinary approach for the study of archeological pottery from Oaxaca, Mexico. Proceedings of the 33rd International Symposium on Archaeometry 2002, Amsterdam, p. 97.

[32] J.L. Ruvalcaba, L. Bucio, G. Gutiérrez and J.G. Morales, Characterization of pottery by PIXE and XRD: The problem of results interpretation in provenance archaeological studies, 66th Annual Meeting of the Society for American Archaeology, New Orleans, 2001.

[33] Y. Martínez, J.L. Ruvalcaba-Sil, A. Sellen, A. Ramírez and L. Bucio, Analysis of the Zapotec Urns from the Royal Onatrio Museum. Proceedings of The Use of Ion beams in Materials Science, Medicine and Archaeology, LARN, Facultés Universitaires Notre-Dame de la Paix, Namur, Belgium, 2003.

[34] A. Sellen, J. Litvak, J.L. Ruvalcaba, L. Bucio, L. Lazos and E. Ortiz D. How to Spot a Fake. Preliminary Results for an Authenticity Study on Zapotec Urns from the Rickards collection at the Royal Ontario Museum. Proceedings of the 32nd International Symposium Archaeometry, Mexico. In CD, IIA-UNAM, Mexico 2001.

[35] J. Miranda, A. Oliver, A. Dacal, J.L. Ruvalcaba, F. Cruz and M.E. Ortiz, PIXE analysis of cave sediments, prehispanic painting and obsidians cuttings tools from Baja California Sur caves,. *Nucl. Instrum. and Meth.* B **75** (1993) 454-457.

[36] D. Tenorio, M. Jiménez-Reyes and G. Lagarde, Mexican obsidian samples analysed by PIXE and AAS. *Int. J. PIXE* **7** No. 1&2 (1997) 17-24.

[37] G. Murillo, R. Policroniades, D. Tenorio, B. Méndez, E. Andrade, J.C. Pineda, E.P. Zavala and J.L. Torres, Analyses of Mexican obsidians by IBA techniques, *Nucl. Instrum. and Meth.* B **136-138** (1998) 888-892.

[38] R. Esparza, D. Tenorio, M. Jiménez-Reyes, G. Murillo and L. Torres-Montes, Provenance of Obsidian Artifacts studied by PIXE from Lagunillas, an archaeological site in Michoacán, México, *Int. J. PIXE* **11** No. 1&2 (2001) 1-9.

[39] A. Seelenfreund, Ch. Rees, R. Bird, G. Bailey, R. Bárcena and V. Durán. Trace element analysis of obsidian sources and artifacts of central Chile (mauler river basin) and Western Argentina (Colorado River), *Latin American Antiquity* **7** (1) (1996) 7-20.

[40] L. Bellot-Gulet, Th. Calligaro, O. Dorigehl, J.-C. Dran, G. Poupeau and J. Salomon, PIXE and Fission track dating of obsidian from South American prehispanic cultures (Colombia, Ecuador), *Nucl. Instrum. and Meth.* B **150** (1999) 616-621.

[41] G. Demortier. Ion beam analysis of gold jewerly, *Nucl. Instrum. and Meth.* B **64** (1992) 481-187.

[42]   G. Demortier, Elementanalyse von Goldschmuck in Physik in unserer Zeit, Nr. 1, 23 Jahrg. (1992) 13-21.

[43] J.L. Ruvalcaba-Sil, G. Demortier and A. Oliver, External Beam Analysis of Gold Pre-Hispanic Mexican Jewelry. *Int. J. PIXE* **5** (1995) 273-288.

[44] W. Bray, Techniques of gilding and surface enrichment in pre-Hispanic American metallurgy, in *Metal Plating and Patination*, S. La Niece and P. Craddock eds. Butterworth–Heinemann, Oxford, 1993, p.182.

[45] D. A. Scott, Depletion gilding and surface treatment of gold alloys from the Narino area of ancient Colombia, *J. Hist. Metall. Soc.* **17** 2 (1983) 99.

[46] S.J. Fleming, C.P. Swann, P.E. McGovern and L. Horne, Characterization of ancient materials using PIXE spectrometry. *Nucl. Instrum. and Meth.* B **49** (1990) 293-299.

[47] G. Demortier and J.L. Ruvalcaba-Sil, Differential PIXE Analysis of Mesoamerican jewelry items, *Nucl. Instrum. and Meth.* B **118** (1996) 352-358.

[48] J.L. Ruvalcaba-Sil, Analyse non destructive par faisceaux d'ions de bijoux anciens d'Amérique, PhD Thesis, Facultés Universitaires Notre-Dame de la Paix, Namur, Belgium, 1997.

[49] J.L. Ruvalcaba Sil, Estudio no destructivo mediante haces de partículas de la técnica de dorado Mise en Couleur, in Tecnologías Metalúrgicas en América Prehispánica. Facultad de Ingeniería, UNAM., D. K. de Grinberg, y F. Franco coord., 2002, p. 129-143.

[50] J.L. Ruvalcaba-Sil and G. Demortier, Elemental concentration depth profile in ancient gold artifacts by ion beam scattering, *Nucl. Instrum. and Meth.* B **113** (1996) 275-278.

[51] H. Letchmann, A pre-Columbian technique for electrochemical replacement plating of gold and silver on copper objects, J. Metals, December 1979, 154.

[52] J.L. Ruvalcaba-Sil and G. Demortier, Scanning RBS-PIXE study of ancient artifacts from South America using a microbeam, *Nucl. Instrum. and Meth.* B **130** (1997) 297-302.

[53] G. Demortier, Y. Morciaux and D. Dozot, PIXE, XRF and GRT for the global investigation of ancient gold artefacts, *Nucl. Instrum. and Meth.* B **99** (1995) 384-386.

[54] S.J. Fleming and C.P. Swann, Pre-Hispanic copper alloy production at Batán Grande, Peru: Interpretation of the analytical data of ore samples, in Archaeometry of pre-Columbian sites and artifacts, D.A. Scott and P. Meyers, The Getty Conservation Institute, Los Angeles, 1994, 199-229.

[55] U. Méndez, D. Tenorio and J.L. Ruvalcaba Sil, Análisis Metalúrgico de las Ofrendas del Templo Mayor de Tenochtitlán mediante Técnicas Nucleares y Convencionales in Tecnologías Metalúrgicas en América Prehispánica. Facultad de Ingeniería, UNAM., D. K. de Grinberg, y F. Franco coord., 2002, p. 71-83.

[56] J.W. Palmer, M.G. Hollander, P.S.Z. Rogers, T.M. Benjamin, C.J. Duffy, J.B. Lambert and J.A. Brown, Pre-Columbian metallurgy: technology, manufacture and microprobe analysis of copper bells from the Greater Southwest, *Archaeom.* **40** 2 (1998) 361-382.

[57] J.L. Ruvalcaba, B.M. Monroy, A. Morales, L. Lowe and C. Alvarez, PIXE Analyses of Mayan Copper Artifacts. Proceedings of the 33rd International Symposium on Archaeometry 2002, Amsterdam, p.18.

[58] V. Rodríguez-Lugo, L. Ortiz-Velázquez, J. Miranda, M. Ortiz-Rojas and V.M. Castaño, Study of prehispanic wall paintings from Xochicalco, México, using PIXE, XRD, SEM, and FTIR, J. Radioanal. Nucl. Chem. 240 no. 2 (1999) 561-569.

[59] C. Martínez, J.L. Ruvalcaba, M.A. Ontalba and L. Manzanilla. Caracterización mediante Haces de Partículas: Estudios Interdisciplinarios de Pintura Mural Teotihuacana, XXIV Coloquio Internacional de Historia del Arte: Arte y Ciencia P. Krieger Ed. , Instituto de Investigaciones Estéticas, UNAM, 2002, p. 239-263.

[60] J. Miranda, M.L. Gallardo, D.M. Grimaldi, J.A. Román-Berrelleza, J.L. Ruvalcaba-Sil, M. A. Ontalba Salamanca and J.G. Morales, Pollution effects on stone benches of the Eagle Warriors Precinct at the Major Temple, Mexico City. *Nucl. Instrum. and Meth.* B **150** (1999) 611-615.

[61] A.Y. Du, N.F. Mangelson, L.B. Rees and R.T. Matheny, PIXE elemental analysis of South American mummy hair,   *Nucl. Instrum. and Meth.* B **109/110** (1996) 673-676.

[62] E. Andrade , J.C. Pineda, E.P. Zavala, G. Murillo, R. Chávez, R. Lazcurain, Ma. L. Espinosa and O. Villanueva, IBA analysis of a possible therapeutic ancient tooth inlay, *Nucl. Instrum. and Meth.* B **136-138** (1998). 908.

[63]   C. Solis, A. Oliver, L. Rodríguez-Fernández, E. Andrade, M.E. Chavez-Lomeli, J. Mansilla and O. Saldivar, Lead levals in Mexican human teeth from different historical periods using PIXE, *Nucl. Instrum. and Meth.* B **118**   (1996) 359-362.

[64] C. Solis, A. Oliver, E. Andrade, R. Macias, J. Mansilla and M.E. Chávez-Lomeli, Trace element analysis of teeth from pre-Columbian Population groups in México, *Int. J. PIXE* **9** No. 3&4 (1999) 501-508.

[65] L. Rodríguez-Fernández, J.L. Ruvalcaba-Sil, M. A. Ontalba, J.A. Román-Berrelleza, M.L. Gallardo, D.M. Grimaldi, O.G. de Lucio and J. Miranda, Ion beam analysis of ancient Mexican colored teeth from archaeological sites in Mexico City. *Nucl. Instrum. and Meth.* B **150** (1999) 663-666.

[66] J. Mansilla, C. Solis, M.E. Chávez-Lomeli and J.E. Gama, Analysis of colored teeth from Precolumbian Tlatelolco: Postmorter transformation or intravitam processes?, *Am. J. Phys. Anthropol.* **120** 1 (2003) 73-82.

[67] J.L. Ruvalcaba-Sil , L. Bucio Galindo, M.A. Marín, A. Velázquez Castro, F. Hinojosa, J. Carrillo and M.A. Ontalba Salamanca, Non-destructive analyses of turquoise samples from Templo Mayor, Mexico, by ion beams and XRD, 32nd. International Symposium on Archaeometry, 2000, México. p. 244-245.

[68] J. Kim, A. Simon, V. Ropoche, J. Mayer and B. Wilkens, PIXE Analysis of Turquoise artifacts from Salado platform mound sites in the Tonto basin of Central Arizona, Proceedings of the IBA conference, Arizona State University, 2003, p1-50.

[69] L. Lowe and J.L Ruvalcaba Sil, Provenance Studies of Amber by PIXE, Antropología y Técnica, IIA-UNAM, 6 (2000) 41-48.

[70] G. Querré, Une statuette anthropomorphe du Mexique, *Techne* No.11 (2000) 84-87.

[71] J.L. Ruvalcaba Sil, L. Gallardo and F. Lucarelli, PIXE analysis on paper-like samples and various items from offering 102, Templo Mayor, Mexico. Bienal Report Book 2000-2001, Dipartimento di Fisica, Laboratorio Van de Graaff K3000. Università degli Studi di Firenze, Italia. In print.

Chapter II-5

PIXE Study on Chinese Underglaze-Red Porcelain Made in Yuan Dynasty

H.S.Cheng and Z.Q.Zhang
 Institute of Modern Physics, Fudan University,Shanghai, China
E.K.Lin
 Institute of Physics, Academia Sinica,Taipei, China
Y.P.Huang
 Jingdezhen Jiayang Ceramics Co.Ltd., Jingdezhen, China
hscheng@fudan.edu.cn

Keywords: Underglaze red porcelain, PIXE, Composition, Trace elements

Abstract

This paper reports the results of the PIXE analysis of ancient Chinese underglaze red porcelain produced at Kuan kiln (Jingdezhen, Jiangxi province) in the Yuan dynasty (AD1206-1368). In China, the firing and glazed red techniques of producing underglaze-red porcelain began early in the Yuan dynasty. So far the published date from the scientific investigation on the ancient Chinese underglaze red porcelain are rather scarce due to the difficulty of sample collection. In this study elemental composition analysis of the obtained samples was carried out using the PIXE facility of the 3MeV tandem accelerator at the Fudan University. The major, minor and trace elements of the clay body, white and red glazes were determined, and details of the results are presented. The obtained data can be used for identification of precious Chinese Yuan underglaze-red porcelain.

1. Introduction

As is known, underglaze-red is one of the most important traditional underglaze colors made for porcelain in ancient China, and the underglaze-red porcelain is a famous variety of ancient Chinese porcelain among the high-temperature glazed pottery [1]. Early attempts to produce the underglaze-red porcelain began in the Yuan dynasty in Jingdezhen, southern China. Since the proper red color transmutation required at Yuan age, only a limited amount of underglaze-red porcelain was successfully produced, and very few specimens were preserved at present. In fact, it is also difficult to collect their potsherds used as pieces of samples

151

M. Uda et al. (eds.), X-rays for Archaeology, 151–158.
© 2005 *Springer. Printed in the Netherlands.*

for scientific investigation. Using the ICP method F.K. Zhang and P.S. Zhang [1] measured the chemical composition of several underglaze-red potsherds, but did not report on the potsherds made in the Yuan dynasty. In this work, we used the PIXE method for non-destructive analysis of Chinese underglaze-red potsherds and we determined the chemical composition and trace elements of the body and glaze materials for white and red colors. Our attention is focused on the variation of trace elements in the glaze materials. The present investigation is of particular interest, since the underglaze-red porcelain made in the Yuan dynasty was examined for the first time with use of modern PIXE technique. Our results will provide very useful information for the characterization of precious ancient Chinese underglaze-red porcelain.

2. Experiment

A collection of underglaze-red potsherds, excavated from the ruins of ancient Kuan klin located in the geological layer of Yuan dynasty, was obtained over years from Jingdezhen. Six specimens were selected for the present investigation, including five underglaze-red potsherds of irregular shape and one monochrome-red glazed bowl with high legs glazed in white color. In addition, two underglaze-red potsherds of Ming dynasty (AD1368-1644) and more than ten modern underglaze-red porcelain were included in this study.

Experiments were performed at the Institute of Modern Physics, Fudan University, Shanghai. External-beam PIXE was carried out for all samples using the 9SDH-2 beamline of the 3.0MeV tandem accelerator. Samples were placed at 10mm outside the beam exit window (7.5μm Kapton). After passing through the Kapton film and air, 2.8MeV protons with beam current of 0.05-0.5 nA hit the sample with a small spot of size 1mm in diameter. The induced x-rays were detected using an ORTEC Si(Li) detector with an energy resolution of 165eV (FWHM) at 5.9KeV. For measurements of trace elements ($Z>23$), an Al absorber (0.125 mm thick) was placed in front of the detector in order to suppress the low energy x-rays, and high beam current (0.5nA) was used to increase the X-ray emission. The obtained PIXE spectra were recorded and analyzed with conventional electronics followed by a multi-channel analyzer. Details of the experimental method and PIXE set-up have been described elsewhere [2-4].

3. Results and discussions

Figures 1 and 2 show the PIXE spectra measured from the glaze for chemical composition and trace elements, respectively. The data analysis was performed using the GUPIX-96 program [5]. Table 1 lists the results for the chemical composition of clay and glaze of six samples obtained from the analysis. For the samples studied here, we found that the clay minerals contain mainly $SiO2$ (73.0%-76.0%) and $Al2O3$ (18.3%-20.7%) with small content of K_2O (2.35%-3.03%), Fe_2O_3 (0.81%-1.33%) and CaO (0.14%-0.32%). As for the glazes, we detected lower contents of Al2O3 and SiO2 and higher contents of K_2O and CaO; the obtained contents of K_2O were 2.2%-4.0%, and the obtained content of CaO amounts to 5.6%-7.8% for the white glaze and were higher (9%-10%) for the red glaze. The result indicates that the glazes used in the Yuan dynasty belong to the Al_2O_3-SiO_2 system with rich Ca, and the main chemical compositions of clay and white glaze are similar to the underglaze-blue porcelain made in the Yuan dynasty [6].

Red is the key color for underglaze-red porcelain. Traditionally the red colorant used by ancient potters in the Jingdezhen region was known as copper bloom, in Chinese term as "tong hua", which is the main material generally adopted for firing underglaze-red porcelain. The "tong hua" is a kind of copper oxide layers, which is made by firing a Cu bar at high temperature. With adding some kinds of clay material the mixture was ground in a grinder to become fine powder and was then used as colorant to paint in an underglaze-red. The development of making copper red glaze in Jingdezhen is not clear. For the samples analyzed, we found in the red color area to have considerable amount of CuO content in a variation with degree of red color from sample to sample. As listed in Table 1, for samples No.1, 2, 3 and 5, the red glaze shows the CuO content in the range 0.23%-5.97%. The CuO content is less for light red color and amounts up to 4%-6% for dark color of samples No.2 and 3, and on the glazed surface of their dark areas show to be somewhat rough and slightly sunken, not being as smooth and even as compared to the other glazed areas in vivid red color. Sample No.4 is believed to be the typical one of failure in firing, part of the colorful pattern turned to brown color. In the cutting edge of this potsherd, a certain amount of CuO material used for the glaze was still in its clay layer, not completely diffused into its glaze surface where the content of CuO was not noticeably observed. In the white glaze part of some samples, it is possible to observe with help of a magnifier certain groups of red spots in different density, arising from the diffusion of CuO grains into the white glaze surface during the firing process. A small amount of CuO content (0.1%-0.2%) showed up in the white glaze for samples No.1 and 3. Sample No.6 is a monochrome–red bowl with high legs. The red color in the glaze surface is uniform and vivid. An amount of 0.208% for the CuO content was observed in the red glaze. From the appearance of samples No.2, 3 and 4, we believe that the underglaze-red industry in the period of Yuan dynasty was still in a primitive stage.

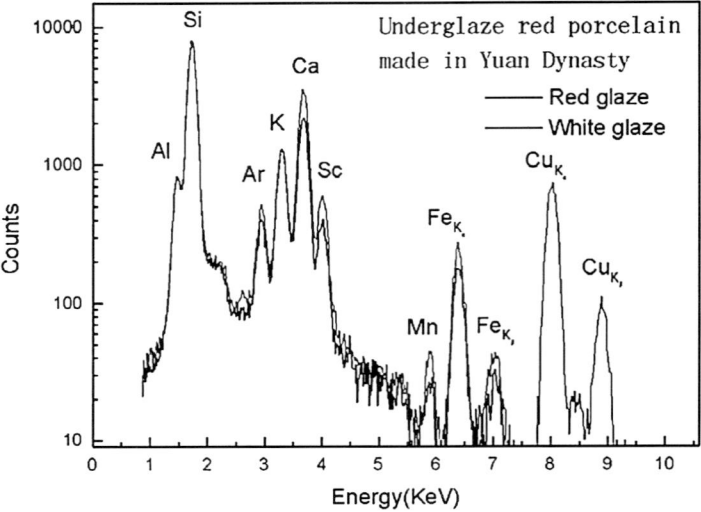

Fig. II-5-1   PIXE spectrum of an underglaze-red potsherd (Yuan dynasty).
Measurements were made on the red and white colored areas
of the glazed surface for chemical compositions analysis.

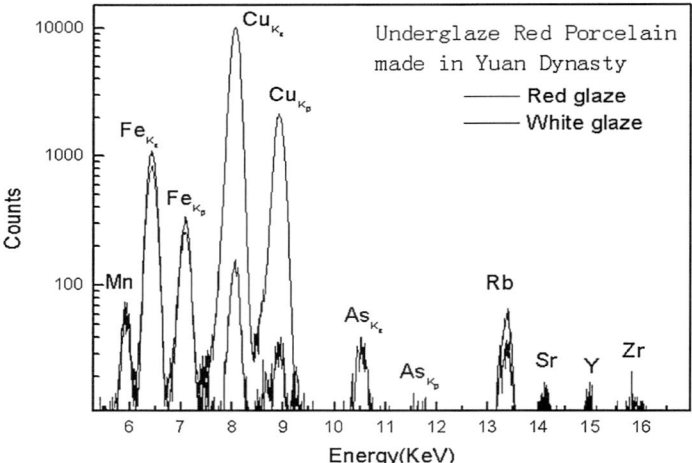

Fig. II-5-2   PIXE spectrum of an underglaze-red potsherd (Yuan dynasty).
Measurements were made on the red and white colored areas
of the glazed surface for trace elements analysis.

Table II-5-1. Chemical composition of body, glaze (white and red) measured from underglaze red porcelain made in Yuan Dynasty (wt.%).

Sample		Al₂O₃	SiO₂	S	K₂O	CaO	TiO₂	MnO	Fe₂O₃	CuO	remark
	body	18.29	75.64	0.25	2.78	0.32	0.072	0.037	1.07	——	
1	white	13.26	72.56	0.27	3.33	7.52	0.022	0.086	1.32	0.100	
	red	13.38	72.09	0.13	3.58	7.11	0.022	0.135	1.80	0.229	
	body	18.54	75.46	0.17	2.87	0.19	0.071	0.06	1.11	——	
2	white	13.52	73.93	0.20	3.28	6.39	0.022	0.085	1.02	0.033	
	red	13.54	72.01	0.16	3.19	7.59	0.021	0.098	1.10	0.774	
	red	12.79	66.72	0.36	2.84	8.48	——	0.130	1.15	5.97	Black
	body	18.83	76.03	0.077	2.35	0.17	0.051	0.066	0.809	0.077	
3	white	12.75	74.10	0.24	2.79	7.02	0.047	0.144	1.18	0.209	
	red	12.69	72.85	0.26	2.48	7.77	——	0.134	1.34	0.928	
	red	12.71	68.77	0.12	2.25	9.10	——	0.138	1.09	4.315	Black
	body	19.48	74.46	0.068	2.63	0.14	0.103	0.062	1.23	0.017	
4	white	13.76	74.37	0.069	3.33	5.88	0.023	0.089	0.879	0.098	
	red	13.46	74.52	0.013	3.62	5.75	0.036	0.090	0.970	0.047	
	body	20.71	73.36	0.03	2.64	0.27	0.081	0.073	1.33	0.008	
5	white	13.77	73.48	0.034	3.68	5.61	0.050	0.138	1.71	0.021	
	red	14.10	73.94	——	4.00	4.46	0.079	0.152	2.35	0.246	
	body	20.60	73.00	0.19	3.03	0.26	0.083	0.080	1.22	0.015	Monochrome red porcelain
6	white	12.52	73.98	0.19	3.47	6.79	0.043	0.155	1.29	0.010	
	red	12.28	70.87	0.24	2.78	10.86	0.011	0.069	1.16	0.208	

Table 2 lists contents of nine trace elements Mn, Ni, Cu, Zn, Ga, As, Rb, Sr and Zr determined for the clay and glaze of the Yuan underglaze-red potsherds obtained from the external-beam PIXE measurement. There is a significant difference between the obtained constituent of the white and red glazes. For the red glaze, the Cu content appears to be considerably high and the element As is present for all six samples. Their corresponding x-ray peaks were clearly observed in the measured PIXE spectra (fig, 2). While for the white glaze, the Cu content is much lower and only a very small amount of As was detected for samples No.1-4, but not for samples 5 and 6. Variation of the As for the red glaze was situated in the range of 150-360ppm. It is interesting to note that after a thin layer of translucent glaze on the outer surface of sample No.4 was removed, the amount of As content in the red colored area inside increased. Apparently As is originated from the red pigment used for the Yuan dynasty underglaze-red porcelain. It was reported [7,8] that there are two kinds of As ore (FeAsS and CuAsS₃) used in ancient China, and many famous bronze artifacts excavated were found to have As admixture. Its appearance in the Cu red pigment is likely from the Cu ore, although the provenance of Cu ore used in Jingdezhen during the Yuan dynasty needs further investigation. For comparison, we examined two underglaze-red potsherds of Ming dynasty appreciable amount of As, even higher than 2000 ppm in some areas. Fig. 3 depicts the PIXE spectrum of one sample of Ming dynasty. Fig. 4 represents the PIXE spectrum obtained from the glaze of modern underglaze-red porcelain, it shows clearly that no As peak appeared there, as the firing and glazed color techniques have been modified in the modern age.

Table II-5-2. Trace element of body, glaze (white and red) measured from underglaze red porcelain made in Yuan Dynasty (ppm).

	Sample	Mn	Ni	Cu	Zn	Ga	As	Rb	Sr	Zr	remark
1	body	285	20	169	128	47	19	297	69	86	
	white	743	50	1107	97	65	12	327	160	29	
	red	876	88	2743	120	89	240	373	195	36	
2	body	377	22	82	97	40	9.3	434	39	31	
	white	755	62	729	62	69	6.0	476	121	31	
	red	758	——	19800	150	86	151	453	121	23	
3	red	952	——	58100	207	108	286	460	159	——	Black
	red	1361	33	39600	294	194	177	438	298	45	
	red	1194	——	50800	285	126	357	349	231	46	Black
4	red	743	37	859	124	65	——	454	148	32	Color not diffuse to surface
	red	813	104	6171	83	45	160	371	128	26	Removed transparent glaze
5	body	1074	181	253	40	71	——	484	109	24	
	red	1665	64	2093	89	77	184	547	142	84	
6	white	1098	109	177	107	70	——	511	123	——	Monochrome red porcelain
	red	847	179	2292	56	66	37	327	263	20	

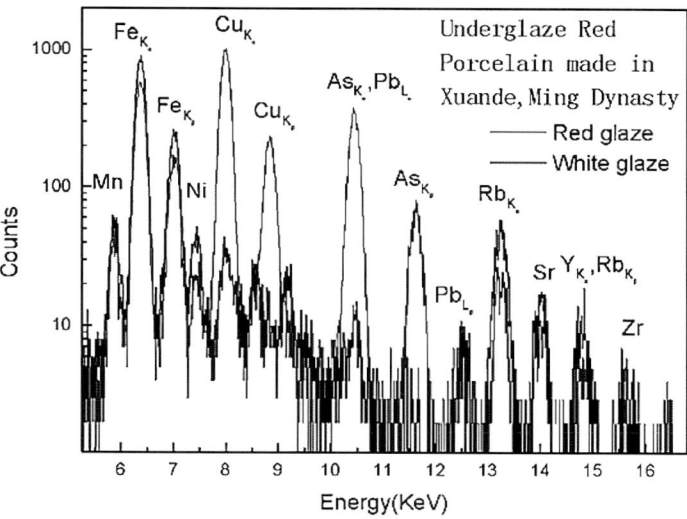

Fig. II-5- 3   PIXE spectrum of an underglaze-red potsherd (Mingdynasty).
        Measurements were made on the red and white colored areas
        of the glazed surface for trace elements analysis.

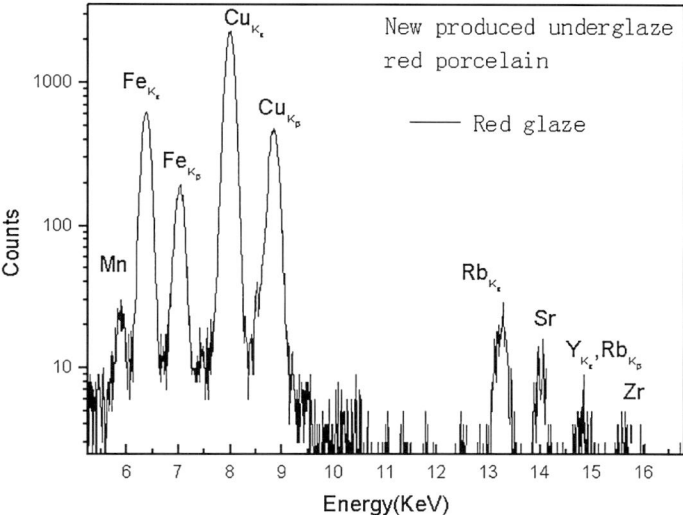

Fig. II-5-4   PIXE spectrum of a modern underglaze-red potsherd.
        Measurements were made on the red and white colored areas
        of the glazed surface for trace elements analysis.

4. Conclusions

We have determined the clay body and glaze composition of six samples selected for the investigation of ancient Chinese underglaze-red porcelain made in the Yuan dynasty. The potsherds studied here were characterized by high purity ceramics, and main chemical compositions of the clay body and white glaze used for underglaze-red porcelain were basically consistent with the underglaze-blue porcelain. Copper was found as an underglaze-red pigment from the red colored area on the glazed surface. The obtained CuO content in the red glaze lies in the range 0.2%-6.0% varying with degree of red color. For monochrome red porcelain as sample No.6, the CuO content in the glaze is low; the measured value of 0.2%(wt) is equivalent to the lowest value obtained for the underglaze-red samples.

Our PIXE analysis confirmed that the copper red pigment was used for making underglaze-red porcelain in the period of Yuan dynasty. The element As was found as an admixture in the CuO pigment. For the six samples analyzed, the As content in the range of 150-350 ppm was obtained. However, no As content was detected in the modern underglaze-red porcelain for the sample studied. Since the amount of major and trace elements may reflect the firing and glazed color techniques of making ancient porcelain, the obtained variation of elemental concentrations will be useful to explore the ages of the production of the Chinese underglaze-red porcelain and to distinguish the products between the true ones produced at Yuan age and the faked ones made with modern techniques. However, our results were obtained from the limited samples collected. Further PIXE analysis of a larger number of samples from the Yuan-dynasty underglaze-red porcelains for a detailed study is necessary.

References

(1) F.K.Zhang and P.S.Zhang (1995), *Proc. Int. Sym on Science and Technology of Ancient Ceramics*, 197

(2).S.Cheng, W.Q.He and J.Y.Tang (1995), *Proc. Int. Sym on Science and Technology of Ancient Ceramics*, .428

(3) E.K.Lin, C.W.Wang, P.K.Teng, Y.M.Huang and C.Y.Chen (1992), *Nucl.Instr. And Meth.* B68, 281

(4) E.K.Lin, C.T.Chen, Y.C.Yu, C.W.Wang, C.H.Hsieh and S.C.Wu (1994), *Nucl.Instr. And Meth.* B85, 869

(5) J.L.Campell, J.A.Maxwell (1996) , GUPIX96: *The Guelph PIXE Program*, University of Guelh, Ontario, Canada.

(6) H.S.Cheng, Z.Q.Zhang, H.N.Xia, J.C.Jiang, F.J.Yang (2002), *Nucl.Instr. And Meth.* B190, 488

(7) Y.X.Li and R.F.Han (1994), *Chinese History of Metallurgy,* 22

(8) S.Y.Sun and T.D.Wang (1994), *Chinese History of Metallurgy*, 180

Chapter II-6

Glassmaking in the Venetian Manner

Mateja Kos[1,2] and Žiga Šmit[1,2]
[1]National Museum of Slovenia, Prešernova 20, SI-1000 Ljubljana, Slovenia
[2] J. Stefan Institute, Jamova 39, SI-1001 Ljubljana, Slovenia
[3]Department of Physics, University of Ljubljana, Jadranska 19, SI-1001 Ljubljana
mateja.kos@narmuz-lj.si

Keywords: Venetian glass, Murano glass, Medieval glass, Ljubljana, PIXE, PIGE

History

In the 15th and 16th centuries, Venice was the largest and most important center of European glass production. The Murano glassworks held a monopoly on the production of glass and on the import of the raw material, of which the most important was soda produced from a special type of plants from the salt marshes of Spain. Venetian glass thus presented the highest criteria for fashion, taste, quality, style and technical perfection.

The Murano glass was desired merchandise through central and west Europe, and the glassmaking in the Venetian manner was soon followed up in many European towns.

Some authors give unquestionable priority to the glassworks in Vienna established in 1486 by Niclas Walch. Others give priority to Ljubljana glassworks. Their argument is that forest glass (Waldglas) and not Venetian glass was produced in Vienna. The next important glass workshop was the one established in 1534 by Wolfgang Vitl in Hall in the Tyrol, while the Royal Glass Workshop in Innsbruck began its operation in 1570.

The owners of the first glassworks in Ljubljana were Andrej Dolenik and Ljubljana's pharmacist Zoan Francisco Catanio[1]. Already in 1526 they acquired the privilege of producing Venetian style glass in Carniola (according to some sources in whole Inner Austria). According to the records, some 14 glassworkers from Murano worked in Ljubljana between 1527 and 1541.

Veit Khisl and Hanns Weilhamer established the second workshop in 1541. In all probability this second glass workshop is the one referred to in the record, which speaks of a dispute between these two owners and the owner of the workshop in Hall, Wolfgang Vitl. The dispute involved the confiscation of a large cargo of ashes.

---

[1] KOS - ŽVANUT 1994, p. 20

M. Uda et al. (eds.), X-rays for Archaeology, 159–162.
© 2005 Springer. Printed in the Netherlands.

The third glassworks owned Ljubljana's pharmacist, Paul Ciriani. He is known a so for smuggling soda from the Spanish city of Alicante. One of his large cargoes was once confiscated because of the Venetian monopoly.

Of key importance is also the will of Ljubljana's master glazier, Christoph Prunner, dated 10 July 1563, and an inventory of glass and other stock, made by the heirs of the late Christoph Prunner and dated 2 February 1564[2].

The Problem

The collections of the medieval glass at the National Museum of Slovenia consist of approximately 800 objects. Medieval glass objects found on the territory of Slovenia rank among those usually classified as Central European. Based on ancient traditions and influences from the Near East, the ingenuity of the local craftsmen helped them create different types of glass vessels. For the major part these were glasses for various beverages (especially wine, beer and water), bottles and equipment for alchemical laboratories. The ornamentation is mostly optical (the glass mass is blown into moulds with different patterns) or applied. The most frequent shapes of glasses are barrels or cylinders with relief ornamentation in the form of molten prunts – the so-called prunt glasses[3] (Fig. 1, 2).

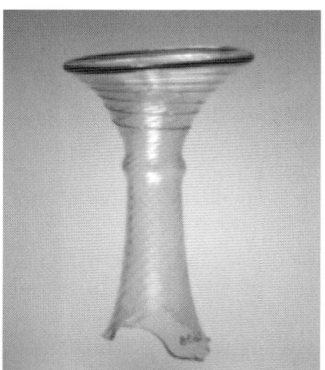
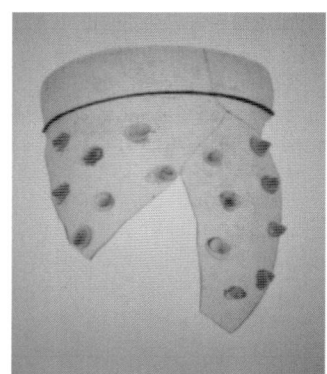

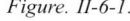

Figure. II-6-1.   Figure. II-6-2.

Fig. II-6-1: Neck of a bottle, 16th Century, found in Ljubljana (photo: National museum of Slovenia).
Fig. II-6-2: Fragment of prunted glass, 16th Century, found in Ljubljana (photo: National museum of Slovenia).

The National Museum keeps some fragments of stained-glass windows, among them the head of St. Catherine, dating from around 1520. The panes for stained glass windows were usually bought ready-made, so it is even more difficult to estimate the provenance (Fig. 3).

---

[2] KOS - ŽVANUT 1994, pp. 46-56

[3] KOS 1995, p. 192

Most objects from National Museum of Slovenia were discovered during archaeological excavations in the center of Ljubljana (Salender Street, Stritar Street [Spittalgasse] and at a location near the Šumi factory [the former dump place]. Among these are a true variety of goblets with differently designed cups, prunted beakers, glasses with trailed glass threads, etc[4]. The location of the finds (at one of the glassworks, and at the city dump), however, is not a sufficient proof that the vessels were actually produced in Ljubljana. So the National Museum of Slovenia, together with the Jožef Stefan Institute started the research project to discover which glass objects were imported from Venice, which originated in Ljubljana, and which were possibly imported from elsewhere (South Germany, for instance).

Fig. II-6-3: Fragment of stained-glass window with St. Catherine, around 1520 (Photo: National Museum of Slovenia).

The Method

The glass material from the excavations in Ljubljana was appropriately studied with a combined art-historian and chemical analysis. The aim of this investigation was to distinguish between home production and import, and to find evidence for production and commercial relations in the 16th century town. The measurements, based on the ion beam analysis (PIXE/PIGE) involved 350 specimens, which provided a sufficient basis for statistical analysis. The glasses from Ljubljana are rather homogeneous, and can roughly be divided into two (or possibly three) groups. It was also possible to identify a few other types of glass made of wood ash or natural soda. The main source of distinction within the main groups can be traced using two different sources of alkalis that were clearly an imported component [5](Fig. 4).

---

[4] KOS 1994

[5] ŠMIT et al. 2002, pp.345-349

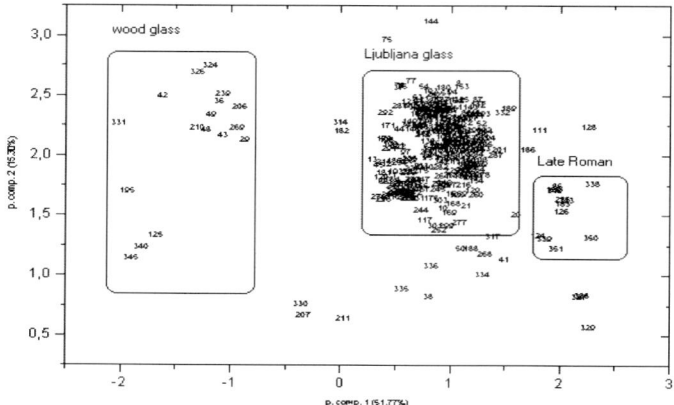

Fig. II-6-4: Graph with three groups of glass: wood, Late Roman and Ljubljana.

Results

The research of the chemical composition revealed that there were at least two main groups of Ljubljana glass, while one of the other groups consists of objects the museum acquired in different ways, that is not through excavations, and late antique objects (a clearly separate group).

This combined method (art-historian and chemical analysis) thus proved very useful for the determination of the origin of the museum objects. Moreover, it is functional for dating objects, but only into clearly limited groups, which differ by the use of raw materials (e.g. Late Antiquity – the Middle Ages – the 17th century), but not within these groups. The method thus fully confirmed our assumptions about the origin of the objects.

References

KOS 1994: Mateja KOS, The Ljubljana (Laibach) Glassworks and its Products in the 16th Century. In: *Journal of Glass Studies* (1994), pp. 92-97

KOS - ŽVANUT 1994: Mateja KOS, Maja ŽVANUT, Ljubljanske steklarne v 16. stoletju in njihovi izdelki – Glass Factories in Ljubljana in the 16th Century and their Products. Ljubljana 1994

KOS 1995: Mateja Kos, Glass in the Gothic Period. In: Gotika v Sloveniji – svet predmetov – Gothic in Slovenia – The World of Objects. Ljubljana 1995

ŠMIT 2002: Žiga ŠMIT et al., PIXE/PIGE characterization of medieval glass. In: *Nuclear Instruments and Methods in Physics Research* B **189** (2002), pp. 344-349

Chapter II-7

Studies on Pigments for Ancient Ceramics and Glass Using X-ray Methods

B. Constantinescu[1], Roxana Bugoi[1], GH. Niculescu[2], D. Popovici[2],
  GH. Manacu-Adamesteanu[3]
[1]*Institute of Atomic Physics, PO BOX MG-6, Bucharest 76900, Romania*
[2]*National Museum for Romania's History, Calea Victoriei 12, Bucharest, Romania*
[3]*Bucharest Municipal Museum, Bd. I. C. Bratianu 2, Sector 3, Bucharest, Romania*
bconst@sun3vme.nipne.ro

Keywords: Cucuteni-Tripolye, ceramics, pigments, glass, XRF, XRD, PIXE, RBS

Abstract

Study of ancient ceramics and glassy objects using nuclear methods can reveal details regarding the provenance of the artefacts and manufacturing techniques. For the present work, several analytical techniques were used: in vacuum 3 MeV protons PIXE (Particle Induced X-ray Emission) at Bucharest Tandem accelerator, 241Am source based XRF (X-Ray Fluorescence), powder XRD (X-Ray Diffraction), external 2.85 MeV protons combined PIXE – PIGE (Proton Induced Gamma-ray Emission) – RBS (Rutherford Backscattering) at Rossendorf Tandem accelerator. Some compositional results on black, brown, red and white pigments of various Neolithical ceramics objects from Romanian sites, especially for Cucuteni culture, in relation with provenance aspects are presented. Some conclusions on the painting procedure for Byzantine (XI-XIII Centuries) glass bracelets found in South-Eastern Romania are discussed.

1. Introduction

As is well known [Demortier G., 1991], the scientific analysis of archaeological objects ideally requires the availability of methods, which are simultaneously non-destructive, fast (so that large number of pieces can be analyzed comparatively), universal (applicable to many materials), versatile, sensitive and multi-elemental. The chemical elements in ores, clays or rocks are characteristic, by their concentrations, of the geology of the area of provenance. Thus, analyses of source materials

M. Uda et al. (eds.), X-rays for Archaeology, 163–171.

combined with analyses of archaeological objects could distinguish from pieces produced in different regions.

Romania has many interesting archaeological sites: Neolithical, Bronze Age, Celtic, Dacian, Greek, Roman and Byzantine. Our purpose was, using elemental analysis nuclear methods, to help Romanian archaeologists to identify objects provenance (workshops, technologies, mines) and, thus, to explain different commercial, military and political aspects. We focused on the composition of Neolithical ceramics pigments, especially on Mn for black colour for Cucuteni culture, and on the painting procedure for Byzantine glass bracelets from XI-XIIIth Centuries discovered in Danube region of Dobroudja.

2. Experimental

For the present work, several analytical techniques were used: in vacuum 3 MeV protons PIXE (Particle Induced X-ray Emission) at Bucharest Tandem accelerator, 241Am source based XRF (X-Ray Fluorescence), powder XRD (X-Ray Diffraction), external 2.85 MeV protons combined PIXE – PIGE (Proton Induced Gamma-ray Emission) – RBS (Rutherford Backscattering) at Rossendorf Tandem accelerator.

For ceramic samples in vacuum PIXE, XRF, and powder XRD were employed, while for the glass bracelets XRF and combined external PIXE – PIGE – RBS were used.

In vacuum PIXE analysis was performed with a 3 MeV proton beam from the Bucharest 8 MV Van de Graaff Tandem accelerator impinging at 45° to the surface. A Canberra GL0110P – Low Energy Germanium Detector (100 mm2 area, 10 mm thickness, 0.075 mm Be window thickness, energy resolution 160 eV FWHM at 5.9 keV, 500 eV FWHM at 122 keV), oriented perpendicularly to the proton beam direction recorded the X-ray spectrum. We routinely used a 0.8 mm diameter beam and a constant proton dose, each acquisition taking roughly 20 min [Constantinescu B., 2002].

XRF measurements were executed with a spectrometer consisting of a 1.1 GBq [241]Am annular gamma-source attached to a support that defines the angle of the incident photons and collimates the fluorescent X-rays in their path to the Si(Li) detector, where they are recorded. A conventional electronic chain, consisting of preamplifier and MCA, is used to accumulate the spectra [Bugoi R., 1999].

A classical XRD experimental set-up with a SEIFERT goniometer, Cu anode X-ray tube (30 kV voltage and 20 mA current intensity) was applied. The acquisition time was roughly one hour, and we used a Ni foil to absorb the Kβ ray of Cu. The detection was made by means of a proportional counter.

As concerning the Rossendorf experimental set-up, the energy of the proton beam on the sample was 3.85 MeV, the whole area around the target and detectors being flooded in a He stream. There are two Si(Li) detectors, one for low-energy X-rays, and another more efficient for high-energy X-ray detection. Backscattered protons are detected at 135° using a Silicon Surface Barrier detector. Gamma-rays were detected using an HPGe detector of 60% relative efficiency [Neelmeijer C., 1996].

3. Results

3.1. CERAMICS PIGMENTS STUDIES

We analyzed the pigments composition of some Neolithic ceramics objects, found on Romanian territory. We studied (chronological classification) ceramic sherds from the following cultures: Cris-Starcevo (6000-4500 b. Chr.), the oldest Neolithical culture, which was covering the whole Balkanic Peninsula and Carpathic Basin, Vinca (4200-3500 b. Chr.), Cucuteni – Tripolye (4500 – 3000 b. Chr.), Petresti (3500-2500 b. Chr.) and Gumelnita (3500-2500 b. Chr.) with its most spectacular archaeological site: the gold cemetery from Varna, Tiszapolgar (4500-3500 b. Chr) (see figure 1). We started with a thorough investigation of Cucuteni culture pieces focusing especially on the presence of Mn in black pigments.

Three analysis methods were used: 3 MeV protons PIXE, 241Am based XRF and powder XRD. For XRF measurements, the elemental intensities data were normalized to their total background spectrum counts – the sum of all characteristic X-rays intensities for the excitation source -, respectively, and subtracted from their corresponding normalized paste composition. Due to the pigments layer's strong heterogeneity and to avoid difficult calibration procedures, we decided to use only the ratios of the main characteristic elements.

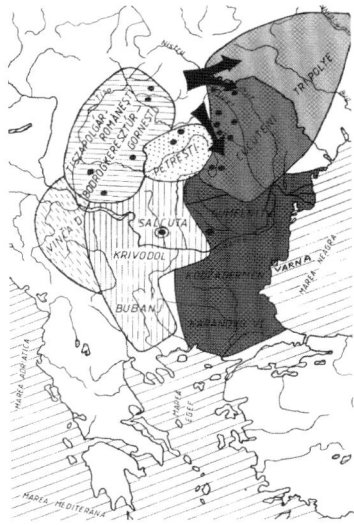

Figure. II-7-1. The map of Neolithic cultures from South – East Europe and the Mn mineral routes (black points represent the archaeological sites where the analyzed sherds were excavated).

3.1.1 *Cucuteni ceramics*

The history of the V-IV Millennia B. C. is marked by the flourishing of great Eneolithical civilizations in the South – Eastern part of Europe: Vinca, Gumelnița and Cucuteni-Tripolye, representing a moment of highest cultural evolution. Spread over a vast territory with a total area of more than 300.000 km2 in Romania, Republic of Moldova and Ukraine, the Cucuteni (in Romania) – Tripolye (in Ukraine) Culture is one of the last brilliant cultural expressions during the Copper Age. "Queen" of the Prehistorical pottery, the Cucutenian ceramics (see figures 2 (a) and 2(b)) represent the most eloquent proof not only of the perfect mastering of pottery (production and temperature control and clay modeling), but also of the extremely developed aesthetic sense that gave birth to genuine and unrivalled prehistorical masterpieces. These represent the proof of a remarkable aesthetic sense and at the same time, of a very complex spiritual life. It is almost certain that the colours had magic significance: red = life (blood), white = the good (light) and black = the evil (darkness).

Figure. II-7-2(a). Cucuteni ceramics cups used for ritual libations.

Figure. II-7-2(b). Cucuteni pottery objects.

(See also Color plate, p. 301)

Some painted (red-white-brown-black) Cucuteni sherds, found in Moldova, in sites situated at the Bistriţa valley down the river (Izvoare, Calu, Caşăria, Ghelăieşti), were studied. For the black (dark brown) colour we found Mn and Fe as the main elements. For provenance studies, the Mn/Fe ratio has a meaning because, due to the very close energy values of the X-rays emitted from both atoms, the matrix effects due to the presence of major (and lighter) elements can be assumed to have the same effect on the numerator and denominator. Three groups of sherds were distinguished: Mn/Fe = 1/20, 1/5, 1/3. In the area of Cucuteni culture there are two main Mn minerals deposits: Iacobeni, up to the Bistriţa river 150 km North from our analyzed sites in Moldova, where the Mn/Fe ratio is under 3/10 and Krivoi Rog (Nikolaev) on the Dnepr river in Ukraine, where the ratio Mn/Fe is from 8/10 up to 10/10 [Treiber I., 1963]. Our results strongly suggest the use of Iacobeni Mn minerals. We also detected a special case for a brown black pigment containing only Fe, identified as magnetite Fe2O4 (also from Iacobeni deposits).

We can presume that the transport from Iacobeni down the river of the Mn and Fe containing clays used as raw materials for pigments had been by rafting (navigation on rafts made of conifers logs tied together using vegetal ropes). In this area, rafting is a traditional way of transportation of timber (wood used in carpentry and in house building), mentioned in documents from the XIVth Century.

Concerning the white colour, Ca is the main element, the pigment raw material being a light clay (type kaolin) relatively frequently detected in the region.

For the red colour, we found Fe and Ti, concluding that the pigment is an iron-rich, clay-based one, with Ti as reinforcing agent.

In the following lines, we will review the previous results obtained using XRF, XRD (X-Ray Diffraction) and some specific mineralogical methods.

By means of the X-ray fluorescence and powder diffraction analysis of a number of Cucuteni sherds from the Poduri settlement (Bacău district – 50 km East from Bistriţa river), Niculescu G., Coltos C. and Popovici D. pointed out that the chromatophorous minerals of the white, red and dark brown pigments were calcium silicate, hematite, and jacobsite (MnFe2O4 – shining black) respectively, which confirmed the results obtained in other Cucuteni-Tripolye settlements [Stos-Gale Z., Rook B., 1981].

In a synthetic study of the Cucuteni-Tripolye culture, Ellis L. made a technological characterization of the pottery. The X-ray diffraction and microscopic analyses were aimed to determine the mineralogical components of the ceramic composition, and the pigments used in the painting of the pottery. The results showed that the sources of clay had been local; the black pigment originated from manganese ores in the region, the red pigment from the alteration of the iron minerals, and the white from marls and kaolins.

A comprehensive study on Cucuteni ceramics from Northern Moldova (including pigments issue) had been reported by Gâţă Gh. in [Marinescu-Balcu S. and Bolomey, 2000]. Summarizing from the above references, especially from Niculescu G., Coltos C. and Popovici D., it can be stated that the pigments used for the Cucuteni ceramics painting were mineral pigments including clay minerals, quartz, and feldspars without iron and manganese oxides for the white chromes, with iron oxides for the red chromes, and with manganese oxides for the brownish-black chromes. These pigments were prepared starting from different coloured clays as raw materials by powdering, dispersion in water, separation of the fine fraction by decantation, and drying. In general, the pigment particle dimensions were of 5-15 μm, based on the width at half intensity of the quartz line at 4.26 Å (Sherrer's law presented by Klug H. P. and Alexander L. E., 1976). The raw materials were cements and concretions with manganese oxides (birnesite and manganite - very often present in Iacobeni deposit) for the brownish-black chromes, iron and manganese concretions, cambic horizons or lehms (goethite - α-FOOH brown-red-brown-yellow colour and lepidocrocite) for the red chromes, and a loam without iron and manganese oxides for the white chrome. Through firing, the iron oxides, goethite and lepidocrocite, were transformed into hematite ($Fe2O3$ - reddish-grey-white colour) for the red chromes, the manganese oxides, birnesite and manganite, into bixbyite, and more rarely into jacobsite, depending on the iron and manganese contents, and rarely into haussmanite. Firing of the clay minerals caused the predominant components to transform into oxides in the process of crystallization, and at a temperature of over 900 °C, γ $Al2O3$ appeared. The fineness of the pigment particles and their mineralogical composition indicated that the raw material for the pigments originated from the proximity of the settlement, or from the region. Experimental preparation of pigments with brownish black and red chrome from cemented sand, iron and manganese concretions, and reddish clays, and the similar composition, chromes and fineness thus obtained, proved that these had been the sources for pigments.

White pigment was prepared from loam deprived of chromophorous oxides; analyses of white and of its sources were identical, when the former had not undergone firing.

In conclusion, our measurements agree with the previous XRD and mineralogical studies, proving the use of local Mn deposits from Iacobeni as black pigments sources for Cucuteni ceramics in this area of Moldova.

3.1.2. *Other Neolithic ceramics*

We extended our study to all Neolithic cultures from Romanian territory, focusing on the black colour especially for the sites from South-East Transylvania (Ariusd zone). The analyzed clay was found to be of local provenances, with great variation in composition (e.g. a lot of Ca for the central region of Transylvania, a lot of Fe for Moldova and Banat, a lot of K in North-East Transylvania). We did not obtain any clue regarding the possible commercial exchanges. When the black colour was close to dark brown, the compositional studies revealed Fe oxides in its content. We found indications of using Carbon from bones (center of Transylvania), but also from graphite (Gumelnita culture, in which connections with the Neolithical graphite mines from North-East Bulgaria were revealed [Dumitrescu V., 1968]). The most interesting case is the one of a site from Cris-Starcevo situated in Oltenia (Salcuta), where Mn was detected as pyrolusite. This mineral was abundant in the North of Greece and largely used in the VII-V b. Chr. centuries to produce the famous attic black glass sherds. We deduced a possible commercial connection South-North in the Balkan Peninsula. We will extend the analysis to other Neolithic sites in the area Carpathian Mountains – Danube in order to confirm this hypothesis.

3.2. GLASSY MATERIALS

During the last years, our team has started a comprehensive study on Byzantine painted glass bracelets. They were excavated from a site near Isaccea, a port on low Danube, which was an important point to cross the river, due to the low level of waters (a ford of Danube). The settlement was around the year 1000 a. D. a lively city, being the residence of a bishop, having its own garrison, a customs office and probably even a Byzantine governor.

There were two kinds of bracelets that were found simultaneously at the same excavation sites. Some of the bracelets were corroded, being covered with a whitish layer, presenting exfoliations and deteriorated painting, as well as a reduced transparency. The other group of bracelets exhibited no corrosion traces, lively colors and no sign of deterioration. These findings were extremely surprising, all bracelets being buried for roughly one thousand years in the same soil and suffering the same kind of weathering process.

The bracelets were twisted and the bulk glass was of different color (black, blue, white, green). On the surface they were painted using red, yellow, golden, silver – grey colors, the stripes coverings only partially the bracelet's surface (see figure 3).

Figure. II-7-3. Fragments of glass bracelets from Constantinople workshops.

Taking into account the great difference in the present status of the glass, it is very likely that the two categories of bracelets (the corroded and the uncorroded ones) were produced in two different workshops, using completely different technologies. One of the workshops was probably a local one, using a primitive technology, while the other was probably an "imperial" one, most likely located in Constantinople, employing a more advanced craftsmanship.

Our study was directed towards the finding of recipes for the pigments used to paint the bracelets, in order to provide evidence to support the hypothesis of different workshops.

We investigated the bracelets employing the [241]Am based XRF method, 3 MeV in vacuum PIXE at the Bucharest 8 MV High Voltage Van de Graaff Tandem accelerator and for certain samples, only uncorroded ones, a combination of external PIXE – PIGE - RBS at the Rossendorf 6 MV Tandem accelerator.

Evaluation of the PIXE spectra indicates the chromophoric elements. Pigments can be deduced based on art, scientific and restoration knowledge [Mora P., 1977].

The PIGE spectra give complementary information regarding especially light elements (Z<15). The presence of an organic overlayer, like varnish or an organic lake, is indicated by the RBS spectra which also provide information on the near surface layer structures of pigments. On the other hand, chemical bond states cannot be identified by these methods.

Our study led to different conclusions depending on the category of bracelets. The local recipes for pigments were very simple, featuring few mineral oxides responsible for each color. For example, red pigment was based on Fe ochre or Pb minium, while the yellow one was made from a mixture of Pb and Sn. The "imperial" bracelets, the ones obtained in advanced workshops impressed through the complicated mixtures used to obtained different hues. For example, the classical red based on Fe ochre and/or Pb minium was nuanced using Cu and Cr as minor elements, while the yellow recipe of Pb and Sn mixture was "spiced" with Sb and As. The grey – silver bands were made of Fe and Ni compounds, while the golden-bronze strips were based on Fe, Pb, Ni, and Cr oxides.

We examined the bracelets with an optical microscope: on the painting layer one could easily see the traces of the brush, i.e. the marks of pig and/or badger hairs. On the "imperial" bracelets a thin protective transparent layer was clearly present. The layer covered the whole bracelet surface, and it could be removed by gently scratching the glass with a knife.

The Rossendorf experiment provided plenty of information, due to the short acquisition time and to the fact that many spectra simultaneously (low energy X-ray, high-energy X-ray, backscattered protons and prompt gamma-rays) were recorded. Low energy PIXE spectra revealed a high Cl content, surprising for a potash glass, as is the case of Byzantine bracelets. RBS data exhibited clear and high C and N signals on the entire surface of bracelets (on the painted and unpainted regions, too). We conclude that the whole glass artifact was covered with a protective organic varnish with a relatively important Cl content.

From the history of painting, it is already known that the mineral pigments were used as a powder mixed with a binding agent in order to be smeared with the brush as paint. There are four categories of natural organic binding media [P. Vandenbeele P., 2000]:

- Proteinaceous media (egg white, casein, gelatin), which contain, beside C, O, H, a certain amount of N;
- Resinous media of terpenoid origin, containing mainly C and H;
- Fatty acid media (linseed oil, poppy seed oil, bee wax), containing only C, H and O;
- Polysaccharide media (potato starch, gum Arabic and cherry gum), containing C and O.

From the elemental analysis methods applied, we could only identify the proteins (through N), and could not distinguish between all the other organic binding media.

Taking into account the transparency of the varnish and the fact that there is no sign of deterioration, we can be quite sure that a resin was used as a binding agent containing proteins, but we cannot exclude the presence of other types of binding media. The resin, most likely turpentine, could originate from the large pine forests around the Mediterranean Sea (there are no conifers in the region of Isaccea, an everglade - plane area). It has been established that the resin, named camphorate, is typical for Mediterranean area that contains plenty of Cl [Beral E., 1961].

The resin protective layer was a very thin one (micron thickness), since the peak intensities for the underlying painting layers (Pb, Fe) were still very high.

Acknowledgements

This work has partially been performed in the frame of a Large Scale Facility Access (LSFA) action of the European Union. Therefore, Christian Neelmeijer and the staff of the Rossendorf Tandem are gratefully acknowledged. For the Bucharest experiments, many thanks are due to the Tandem operationally team and to Florin Constantin for help in data processing.

References

Beral E., Zapan M., (in Romanian) "Chimie organica", Ed. Tehnica, Bucuresti 1961.

Bugoi R., Constantinescu B., Constantin F., Catana D., Plostinaru D., Sasianu A., *Journal of Radioanalytical and Nuclear Chemistry*, Vol. 242, No. 3 (1999) pp. 777-781.

Constantinescu B., Bugoi R., Sziki G., *Nuclear Instruments and Methods in Physics Research* B 189 (2002) pp. 235-240.

Demortier G., *Nuclear Instruments and Methods in Physics Research* B54 (1991) pp. 334-347.

Dumitrescu V., "L'art neolitique en Roumanie", Bucuresti, 1968.

Ellis I. - *Journal for Indo-European studies*, vol. 8 (1-2) (1980) pp. 211-230.

Gâţă G., - "A Technological Survey of the Pottery", in "Draguseni – A Cucutenian Community", Archeologia Romanica II series, Ed. Enciclopedica, Bucureşti and Wasmuth Verlag, Tuebingen, 2000, pp. 111-131.

Klug H. P., Alexander L. E. – "*X-ray Diffraction Procedures*", New York 1976.

Mora P., Mora L., Philipport, "*La conservation des peintures murales*", Ed. Compositori, Bologna, 1977.

Neelmeijer C., *Nuclear Instruments and Methods in Physics Research* B 118 (1996) pp. 338-345.

Niculescu G., Coltos C., Popovici D., Cercetări de Restaurare şi Conservare 2 (in Romanian) (1982) pp. 205-206.

Stos Gale Z., Rook E. – "Analysis of Pigments Used for Decorations of Neolithic Pottery from Bilcze, Zlote Ukraine", in British Museum - occasional paper, no. 19 (1981), pp. 155-161.

Treiber, I. "Petrografia rocilor eruptive si metamorfice" (in Romanian), Bucuresti 1963.

Vandenabeele P., Wehling B., L. Moeno, *Analytica Chimica Acta* 407 (2002) pp. 261-274

Chapter II-8

Compositional Differences of Blue and White Porcelain Analyzed by External Beam PIXE[*]

Y. Sha[‡], P.Q. Zhang, G.G. Wang, X.J. Zhang, X. Wang, J. Liu

Institute of High Energy Physics, Chinese Academy of Sciences,
P.O. Box 918, Beijing 100039, P. R. of China
shayin@ihep.ac.cn

Keywords: Proton induced X-ray emission (PIXE); External beam; Blue and white porcelain PACS: 32.30.Rj; 81.05.Je; 81.05.Mh; 82.80.Ej

Abstract

The elemental contents of two blue and white vases with two holders in the shape of elephant ears (VASEs in brief) and twelve modern imitations of blue and white porcelain (IMITATIONs in brief) were determined by external beam PIXE. Significant differences in the contents of some trace elements, such as Zn and Pb, in the white glaze of the IMITATIONs and the VASEs were detected. It was suggested that this characteristic of the trace elemental contents could be used as a criterion for the judgment of different kinds of porcelain: the VASEs are not the same as the IMITATIONs. The ratios of MnO to Fe_2O_3 in the pigment covered by glaze of the VASEs and the excavated blue and white porcelain pieces produced in the Yuan dynasty are discussed in detail. The VASEs are different from the excavated porcelain manufactured either at the governmental kiln, Jingdezhen of Jiangxi province or at the civilian kiln of Yunnan province during the Yuan dynasty respectively. It is suggested that it was possible that a mixed pigment was used in the VASEs production procedure.

---

[*] This work was financially supported by Chinese Academy of Sciences (KJCX-N04) and the Natural Science Foundation of China (No. 19975046,10075060 and 10135050) and LNAT.

[‡] Corresponding author. Tel:+86-10-88233213; FAX:+86-10-88233186.

M. Uda et al. (eds.), X-rays for Archaeology, 173–180.
© 2005 *Springer. Printed in the Netherlands.*

1. Introduction

The pioneering work on proton induced X-ray emission (PIXE) was published in 1970 by Johansson et al [1]. It is known that the first external beam for proton induced X-ray emission (PIXE) was used in 1972 by Deconninck [2]. Compared with vacuum proton induced X-ray emission (PIXE), external beam PIXE used as nondestructive trace elemental analysis technique has many advantages [3], of which the following aspects are especially favorable for the archaeological study on ceramics (porcelain and pottery).

1) There are no limits of sample size and shape, which is especially suitable for archaeological specimens with big size and irregular shape.
2) Sample loading and positioning is considerably simple.
3) The background resulting from the electric charge build-up on the sample and the sample heating resulting from beam irradiation are suppressed and minimized automatically by the measurements in air. Therefore, it is especially suitable for the analysis of semiconductor and isolator materials.

A facility for external beam PIXE system was installed in our laboratory. In order to protect the accelerator against vacuum breakdown due to rupture of the exit foil, the external beam PIXE facility requires a vacuum safety valve system. However, the commercial vacuum safety systems are complicated and expensive. An improved model with a simple and reliable mechanical vacuum safety valve [4] was developed in our group and installed at the external beam PIXE facility.

Porcelain invented by the Chinese people is famous all over the word. Many ancient porcelain artifacts were excavated after the development of archaeological work in China. At the same time, many modern imitations of the ancient porcelain appeared on the market. Criteria to distinguish genuine porcelain from the fake ones are becoming an urgent task for the collection of cultural relics. It is possible to make the separation of genuine and fake porcelain using nuclear analytical technology besides the traditional method of expert evaluation. For identification of ancient porcelain, it is important to study the "fingerprint" with modern science and technology.

It is essential to find the characteristic fingerprint of trace elements in ancient porcelain. In recent years, the PIXE technique was extensively used for the analysis of archaeological artifacts including ancient porcelain. Some studies [5, 6] were carried out with external beam PIXE for the multi-elemental determination of the major, minor and trace components in glaze, body and pigment of the porcelain. Two VASEs (each blue and white vase with two handles in the shape of elephant ears) and twelve IMITATIONs (twelve modern imitations of blue and white porcelain) were analyzed by external beam PIXE in our laboratory. The evaluation method of different porcelain was studied in the present paper.

2. Experimental

The different kinds of porcelain were analyzed by external beam PIXE. There are two VASEs to be identified, named two unknown VASEs. The two VASEs are about 59 cm height, 15 cm at mouth diameter and 17 cm at base diameter. A photograph of the two unknown VASEs is shown in Fig. 1.

Fig. II-8-1   Photograph of the two unknown VASEs. (See also Color plate, p. 302)

The other twelve porcelains were known to be IMITATIONs sold on the domestic market in China. PIXE spectra of different parts of the porcelain were obtained. The PIXE spectra of the white glaze of one of the IMITATIONs and that of one of the VASEs are shown in Fig. 2 and 3, respectively.

The PIXE spectrum of the pigment covered by glaze of one of the two unknown blue and white porcelain vases with handles whose shapes represent elephant ears is shown in Fig. 4. The chemical composition (major elements, in wt%) and the contents of trace elements (in ppm) of white glaze and pigment covered by glaze of each of the two unknown blue and white porcelain vases were then determined.

In principle, the contents of major, minor and trace elements in the white glaze of the porcelain could be quantitatively analyzed using GUPIX [7] software. However, as there is no certified reference material available today for the analysis of porcelain, and as it is difficult to get reliable and accurate data, the present compositional concentration is only preliminary. Nevertheless, the ratios between various compositions and trace elements are reliable based on the GUPIX approach.

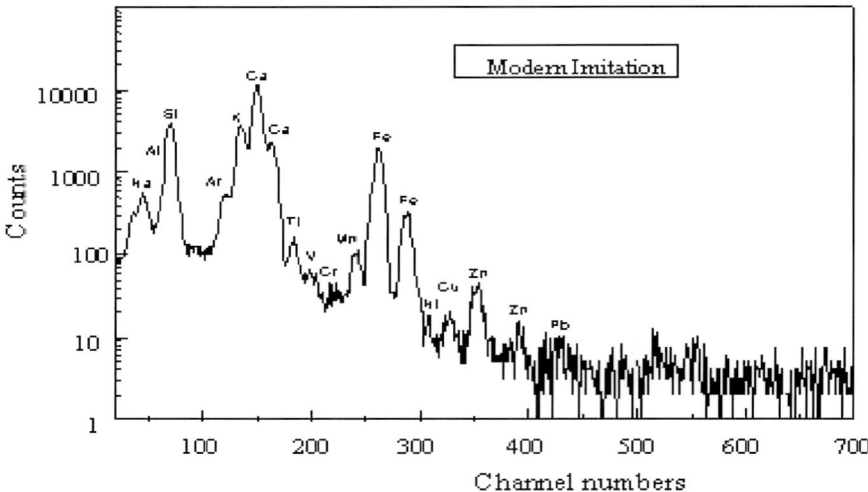

Fig. II-8-2   PIXE spectrum of white glaze of a modern imitation porcelain.

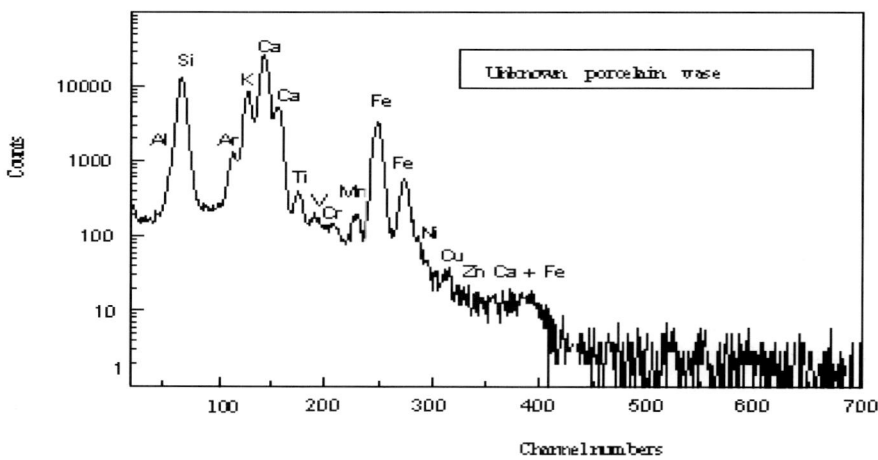

Fig. II-8-3   PIXE spectrum of white glaze of one of the two unknown blue and white vases with two
handles in the shape of elephant ears.

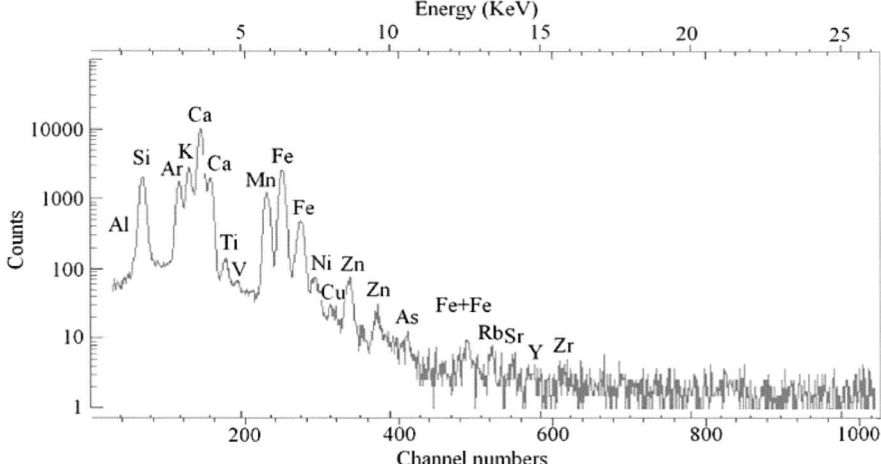

Fig. II-8-4   PIXE spectrum of the pigment covered by glaze of one of the unknown blue and white vases
with two handles in the shape of elephant ears.

3. Results and discussion

Taking into account the determination of "fingerprint" characteristics of elements, especially trace elements, the PIXE technique was used for the identification of ancient porcelain in addition to the traditional methods. Our "fingerprint" characteristics were as follows.

The database for various Chinese blue and white porcelain should be established first. The chemical components (major, minor and trace elements) in various parts of a batch of porcelain of known kind, age, provenance, (original or fake ones) are determined by the PIXE technique. For the blue and white porcelain, the white glaze, the pigment and the body are analyzed. The information on the "fingerprint" characteristics for different kinds of porcelain including originals and imitations could be obtained using statistical analysis of the elemental contents of the specimens.

Finally, the judgment of the kind, age, provenance, original or fake could be made through comparison of the "fingerprint" characteristics with the known porcelain to those to be identified.

It is very difficult to establish a definite and reliable identification of ancient porcelain due to the lack of data of various kinds of porcelain. Up to now only a few results on trace elements of ancient porcelain are available.

A quantitative analysis of a pair of two unknown blue and white porcelain vases was performed by external beam PIXE for the purpose of identifying its originality. The elemental composition of the VASEs is compared with that of several excavated samples with definite kiln, provenance and age of the Yuan dynasty. The concentrations of some elements in the white glaze and pigment covered by glaze of the excavated blue and white porcelain manufactured in the Yuan dynasties of China were determined by Dr. Chen et al [8, 9], in the Shanghai Institute of Ceramics, Chinese Academy of Sciences, using ordinary chemical methods. It was pointed out by Chen that the composition in Fe_2O_3, CoO, MnO and the ratios Fe_2O_3/ CoO and MnO/ CoO play a very important role, but also As as a characteristic trace element in the pigment of blue and white porcelain of the Yuan dynasty. For the elemental characteristics in the pigment covered by glaze of the excavated porcelain manufactured at civilian kiln in the Yunnan province of China using home made pigments (native cobalt materials) during the Yuan dynasty, the composite content of MnO is quite higher than that of Fe_2O_3 based on data from the literature [8, 9]. In our study the average ratio MnO / Fe_2O_3 of the pigment covered by glaze of 5 samples was 1.45 ± 0.27 (see Table 1).

TABLE II-8-1.   The ratios of chemical composition MnO to Fe_2O_3 in the pigment covered by glaze of blue and white porcelain

No. [8, 9]	Mn O/ Fe_2O_3	No. [8, 9]	Mn O/ Fe_2O_3	No.	Mn O/ Fe_2O_3
YU-1	1.77	Y6	0.052	Bg	0.382
YU-2	1.21	Y7	0.045	Sm	0.662
YU-3	1.27	Y8	0.056		
YJ-4	1.72	Y9	0.049		
YJ-5	1.29	Y10	0.033		
		Y11	0.044		
		Y12	0.028		
Average	1.45	Average	0.044	Average	0.522
± sd	± 0.27	± sd	± 0.010	± sd	± 0.140

1, Five samples (No. YU-1, Yu-2, Yu-3, Yj-4 and Yj-5) were the excavated porcelains manufactured at civilian kiln in the Yunnan province, using homemade pigments (native cobalt materials) during the Yuan dynasty.

2, Seven samples (No. Y6, Y7, Y8, Y9, Y10, Y11and Y12) were the excavated porcelains produced at governmental kiln, Jingdezhen in the Jianfxi province of China, using imported cobalt materials as pigment during the Yuan dynasty.

3, Two samples (No. Bg and Sm) are the two unknown blue and white porcelain vases.

In addition, Table 1 shows the elemental characteristics of the pigment covered by glaze of excavated porcelain produced at governmental kiln, Jingdezhen in the Jianfxi province of China, using imported cobalt materials as a component of pigment during the Yuan dynasty. The content of MnO is lower than that of Fe_2O_3 based on data from the literature [8, 9]. The ratio of MnO to Fe_2O_3 of the pigment covered by glaze of 7 samples was calculated yielding an average value of 0.047 ± 0.010. The two VASEs and twelve IMITATIONs were analyzed by external beam PIXE. As to the elemental characteristics in the pigment covered by glaze of the two unknown blue and white porcelain vases, the content of MnO is lower than that of Fe_2O_3. The ratio of MnO to Fe_2O_3 in the pigment covered by glaze of the two unknown blue and white porcelain vases was 0.522 ± 0.140 (average on 2 samples).

We have also measured traces of As in the pigment, which could be considered as a characteristic trace element not present in the pigment of blue and white porcelain untill the Yuan dynasty.

After the comparison using these data the results of analyses may be summarized as follows:

(1) There are some trace elements, such as zinc, barium and lead, with very high traces as impurities in the white glaze of the IMITATIONs.

(2) The contents of some trace elements, such as zinc, barium and lead, are very low in the white glaze of the VASEs, but they are still much lower than those in the IMITATIONs.

(3) The content of MnO is slightly lower than that of Fe_2O_3 in pigment covered by glaze of two the unknown blue and white porcelain vases. The ratio of MnO to Fe_2O_3 of the pigment covered by glaze of the VASEs is about 0.522 ± 0.140.

(4) Arsenic is present at low concentration in the pigment covered by glaze of the VASEs.

Therefore, the results indicate that the two VASEs are different from modern IMITATIONs sold on the domestic market in China. The characteristics of trace elements in white glaze could be used as a criterion for the judgment of different porcelain items. One conclusion could be that the VASEs were not modern imitations sold on the market in China. However, more data are required to confirm this conclusion. A project to establish these databases is currently being carried out by our team under the financial support of the Intellectual Creative Engineering of Chinese Academy of Sciences (KJCX-N04) and the Natural Science Foundation of China.

As the ratio of MnO to Fe_2O_3 of the pigment covered by glaze of the unknown porcelain VASEs is 0.522, representing a value situated between imported cobalt pigment material and native cobalt, it is suggested that possibly a mixed cobal pigment (imported cobalt pigment material was mixed with native cobalt) was used in the VASEs production procedure.

Acknowledgement

This project was supported by Laboratory of Nuclear Analysis Techniques (LNAT) under contract No. K-53 and Intellectual Creative Engineering of Chinese Academy of Sciences under contract KJCX-N04. The authors express acknowledgement to the National Natural Science Foundation of China (NSFC) for the financial support of N0.19975046, No.10075060 and No. [2001] (101020114). The authors also greatly appreciate Prof. Wang Changsui of the Archaeology Laboratory, University of Science and Technology of China for helpful discussion.

References

[1] T.B. Johansson, R. Adelsson and S.A.E. Johansson, *Nucl. Instr. and Meth.* **84** (1970) 141.

[2] G. Deconninck, *Journal of Radioanalytical Chemistry.* **12** (1972) 157-169.

[3]   B. L. Doyle, D.S. Walsh and S.R. Lee, *Nucl. Instr. and Meth.* B**54** (1991) 244.

[4]   A. Anttila, J. Raisanen and R. Lappalainen, *Nucl. Instr. and Meth.* B**12** (1985) 245.

[5]   S.J. Fleming and C.P. Swan, *Nucl. Instr. and Meth.* B**64** (1992) 528.

[6]   Cheng Huansheng, He Wenquan, Tang Jiayong, Yang Fujia and Wang Jiahua, *Nucl. Instr. and Meth.* B**118** (1996) 377.

[7]   J. A. Maxwel, W.J. Teesdale, J.L. Campbell, *Nucl. Instr. and Meth.* B**95** (1995) 407.

[8]   Chen Yao-cheng, Kuo Yeing-yi, Chang Tzi-gong, *Journal of the Chinese ceramic society*, **6** (4) (1978) 225 - 241 (in Chinese).

[9]   Chen Yao-cheng, Kuo Yeing-yi, Chen Hong, *Journal of Chinese ceramics*, **5** (1993) 57- 62 (in Chinese).

Part III: Use of Synchroton Radiation

Chapter III-1

New Trend in Application of Synchrotron Radiation-Induced X-ray Fluorescence Analysis to Archaeology

I. Nakai

Department of Applied Chemistry, Tokyo University of Science
Kagurazaka, Shinjuku, Tokyo 162-8601 Japan
inakai@rs.kagu.tus.ac.jp

Keywords: XRF, XANES, Synchrotron radiation, Chinawares, locality, pottery, firing technique, non-destructive analysis

1. Introduction

X-ray fluorescence (XRF) analysis is a well-established analytical method of archaeological objects enabling non-destructive multi-elemental analysis of the objects. X-ray tubes have been used as a principal excitation source of the conventional XRF spectrometers. Recent development of Synchrotron Radiation (SR) technology induced large advances in the X-ray fluorescence analysis of various materials. Characteristic properties of synchrotron radiation led to a number of new XRF techniques, which could not be realized by using conventional X-ray sources. First, these advantageous properties of SR in XRF analysis of archaeological materials are summarized. Then, principles of two powerful XRF techniques utilizing SR, high energy XRF and fluorescence XAFS (X-ray Absorption Fine Structure), which are very useful in archaeometric analysis are introduced through practical applications in the field of archaeology. Archaeological applications of the SR-X-ray diffraction method are described in a separate section of this paper.

2. Properties of SR as an excitation source of XRF analysis of archaeological materials

Synchrotron radiation is an intense X-ray source produced by an electron storage ring. When electrons, which travel along a ring orbit with a velocity close to that of light, pass the magnetic field produced by bending magnet or insertion device, they are accelerated and emit electromagnetic radiation referred to as synchrotron radiation. SR is highly polarized light and the beam has a high degree of parallelism. The intensity of the SR beam is four to ten orders of magnitude stronger than X-rays from the conventional sealed X-ray tubes. SR has a continuous energy

M. Uda et al. (eds.), X-rays for Archaeology, 183–198.
© *2005 Springer. Printed in the Netherlands.*

distribution so that mono-energetic beams can be produced over a wide range of energies from infrared to hard X-ray regions. Monochromatic X-ray suitable for XRF and XAFS analysis can be easily obtained by crystal monochromator. This, in combination with its polarization in the plane of the synchrotron ring, is extremely important for background reduction in SR-XRF analysis and allows very rapid high sensitive bulk analyses to be obtained on small areas. This factor is particularly attractive for the analysis of archaeological samples. In the case of ED (Energy Dispersive)-XRF analysis, the background due to coherent and incoherent scatter can be greatly reduced by placing the detector at 90 degree to the path of the incident beam and in the plane of polarization. SR-XRF has some other remarkable advantages over the conventional XRF technique. For example, it is easy to obtain a microbeam using slits (beam size: ca.100x100 μm^2) or focusing optics (less than 1x1 μm^2). SR-XRF can be used for the transformation of a small or complicated artistic pattern (e.g. a pattern on pottery or figurine) into a two-dimensional chemical composition image (Nakai 1996). Earliest archaeological applications of SR were reported by the author (Nakai et al. 1991,1992). Excellent reviews of SR applications in archaeology were reported by Janssens (2000) and Pantos et al. (2002). There is a very useful website which introduces publications on archaeological and conservation science related work utilizing SR (Pantos 2004). This text introduces two new approaches utilizing SR in archaeometric analysis: i.e., high energy XRF and fluorescence XAFS. The former technique was used for provenance analysis of old chinaware and the latter for chemical state analysis of iron in potteries to reveal their firing technique.

3. High energy X-ray fluorescence analysis

3-1 PRINCIPLE OF THE HIGH ENERGY XRF ANALYSIS

Trace-element components of archaeological sample often reflect its origin. Heavy elements are particularly useful as fingerprint elements in provenance analyses of archaeological samples, as heavy elements are trace elements in nature and exhibit unique geochemical behavior because of their large ionic radii and relatively high valency. So far, neutron activation analysis has been often used for analysis of the heavy elements. However, destructive sample preparation is necessary for the neutron activation analyses because the sample becomes radioactive after analysis. This precludes the analysis of precious samples such as those preserved in museums. XRF analysis is suitable for non-destructive analysis of precious cultural heritages. However, conventional XRF analysis is not suitable for the analysis of heavy elements. A typical Energy Dispersive (ED)-XRF spectrum of a multi-component material in energy regions of less than 20 keV is usually complicated by an overlap of the L and M emission lines of heavy elements with K lines of the light elements. Therefore, trace heavy elements cannot be measured by conventional XRF technique due to the presence of intense K-line peaks of the light elements. In contrast, the XRF spectrum above 20 keV contains only K line peaks; thus, the spectrum becomes very simple. Therefore, it is thought that the use of the

K-lines would be ideal for the analysis of heavy elements with atomic number Z greater than 45 (= energy of Rh $K\alpha_1$ is 20.12keV). The energy of the K-absorption edge of an element increases with the atomic number of the element, and an X-ray energy of 115.62 keV is necessary to excite the U K- lines.

Third-generation synchrotron light sources such as ESRF in Grenoble, APS in Argonne and SPring-8 in Japan are suitable to obtain high energy X-rays. These light sources incorporate insertion devices, wigglers and undulators. The primary characteristics of the third-generation light sources are their extremely high brilliance and the availability of high-energy X-rays. The 8 GeV storage ring of SPring-8 provide high-energy SR X-rays up to 300keV. We used for the first time 116keV X-ray as an excitation source of XRF analysis and developed a new technique suitable for sensitive analysis heavy elements (Nakai et al. 2001). This chapter introduces XRF analysis of cultural properties utilizing 116keV X-rays, which allows analyzing all the heavy elements utilizing a K series of emission lines. This technique was successfully used to reveal locality of the Old Kutani chinaware.

3-2 ORIGIN OF THE KUTANI CHINA WARES

Kutani chinawares were first produced in the late 17th century in the Kaga Province, which is now Ishikawa Prefecture, Japan. Chinawares from this early period are known as Old Kutani, which are painted chinawares with beautiful colors and is extremely precious because of the high quality art and limited number of production. In 1710, however, after half a century of continuous production of the chinawares, the kiln suddenly closed. Many museums exhibit Old Kutani and we appreciate the quality of the art. Recently, however, it is said that the Old Kutani might originate from Arita, another world famous production place of painted chinawares since 17th century in Japan. Pottery shreds excavated from Old Kunati kilns scarcely contained painted work. On the other hand, Arita produced many beautiful colored painted chinawares until now. Therefore, identification of Old Kutai and Arita is an important and mysterious problem in Japanese art history. The high-energy SR-XRF analysis of porcelain clay bodies is expected to reveal the origin of the source materials of museum grade work.

3-3 HIGH ENERGY XRF ANALYSIS OF CHINA WARES

A schematic diagram of the measurement system for SR-XRF at SPring-8 is illustrated in Fig.1. The high-energy X-rays from an elliptical multipole wiggler were used as an excitation source. Monochromatic X-rays of 116keV were obtained from a doubly bent Si(400) monochromator. Energy dispersive XRF system utilizing, a pure-Ge solid-state detector was used.

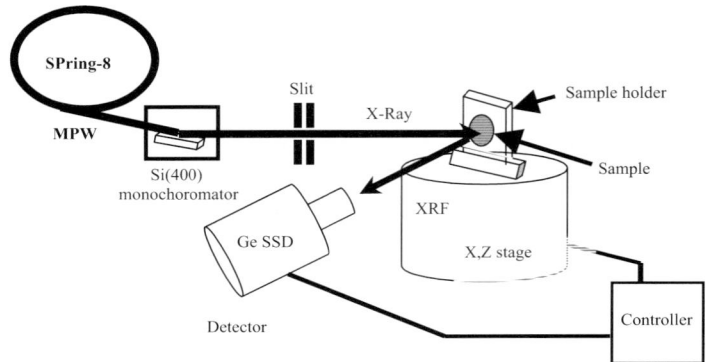

Fig. III-1-1 An illustration of experimental system for high energy XRF analysis at BL08W, SPring-8.

Chinawares excavated from old kilns, dated 17 to 19C in Kutani and Arita districts in Japan, were used as reference samples. Himetani, Hiroshima prefecture in Japan is also known as famous production of painted chinawares during the same periods. Therefore, the excavated samples from Himetani were also included in the samples. Several "original work" of the old dishes in museum grade was also non-destructively analyzed (see Fig. 2). They were very precious so-called Old-Kutani and Arita wares, which were borrowed from several collectors and artists. This was the first non-destructive analysis of museum grade samples of Old Kutani.

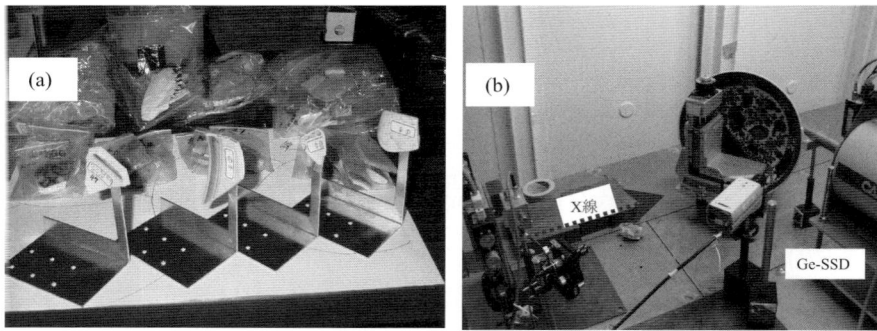

Fig. III-1-2 (a) Samples mounted on a sample holder for XRF analysis. The samples are fragments of excavated porcelain from old Kutani kiln (b) A view of high-energy XRF analysis of "Old Kutani" China ware. (See also Color plate, p. 302)

The analysis confirmed that the irradiation of 116keV X-rays caused practically no damage on the samples. Therefore, this technique is suitable for non-destructive characterization of precious historical samples as is demonstrated in Fig. 2. An example of the high energy XRF spectrum of the excavated fragment of Old Kutani ware is shown in Fig. 3, which demonstrates that tungsten as well as various rare-earth elements produce distinct fluorescence peaks of K-lines and therefore the spectrum is rich in information on heavy elements. It is geologically expected that each porcelain stone has its own heavy element composition characteristic to the production place. Therefore, the heavy element pattern of the chinaware will become an index of production place. The XRF peak intensines of these heavy elements were used as the parameters of some statistical treatments for the provenance analysis. In particular, the Ba/Ce-Nd/Ce plot is the most useful to estimate the origin of the wares. The result is presented in Fig. 4, which shows that Kutani, Arita, and Himetani wares can be clearly distinguished using this plot (Miura et al. 2004).

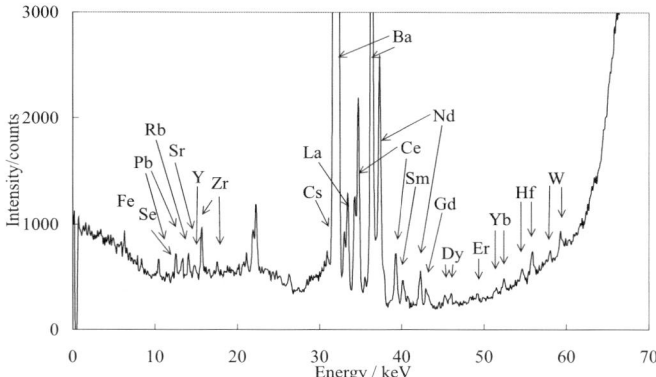

Fig. III-1-3 High-energy XRF spectra of a flagmen of porcelain excavated from the Old-Kutani kiln.

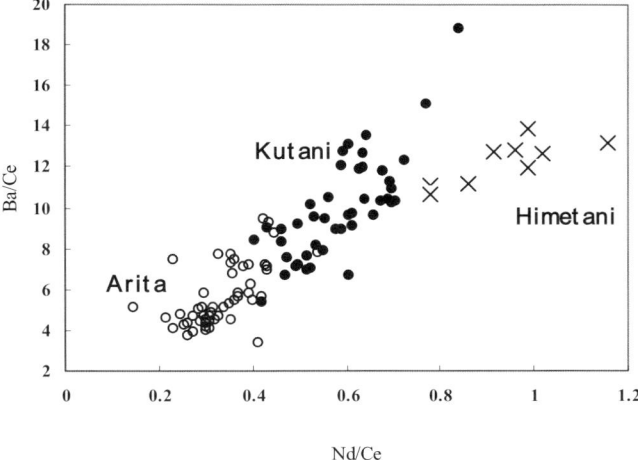

Fig. III-1-4 Classification of porcelain shreds excavated from the old kilns at Kutani, Arita and Fukuyama based on the XRF intensities ratio of Ba/Ce-Nd/Ce.

Fig. 4 illustrates that analytical data of some museum grade samples were located in the region of the Kutani wares in and some in the region of the Arita wares. The data suggest that the former samples were produced using potter's clay of Kutani area truly and the latter were those from Arita areas. Thus, high-energy XRF is able to reveal the original locality of chinawares and is suitable for non-destructive analyses of historical samples. No beam-induced damage of samples was observed after the high-energy XRF analysis, and this technique appears to be truly non-destructive, making it suitable for the analysis of cultural properties. The appropriate samples for high energy XRF analysis will be glass, ceramics including pottery, paints, ink, etc.

4. Chemical state analysis of archaeological materials by fluorescence XAFS

4-1 APPLICATION OF CHEMICAL STATE ANALYSIS TO ARCHAEOLOGY

Chemical state analyses of archaeological materials can provide information on their manufacturing processes. An electronic transition in an atom accompanied by the absorption of visible light of specific energy accounts for the color of substance. The color of pottery, for example, is therefore linked to the chemical state of transition element such as iron, which is affected by the redox condition of their manufacturing processes.

Color of a glass is related to the chemical form (state) of metal ion in it. For example, copper-containing glass becomes red when it is produced under a reducing condition while it becomes blue under an oxidizing condition. XAFS was applied to reveal the colorant elements of the mosaic glasses from Dome of Hagia Sophia in Istanbul (Nakai et al. 1997). A combined use of XAFS and optical absorption spectra revealed the origin of the red color of the copper ruby glass to be metallic copper particles of nanometer size (Nakai et al.1999).

X-ray Photoelectron Spectroscopy (XPS) and Mössbauer spectroscopy have been widely used in the chemical state analysis. XPS was applied to glass in order to investigate the origins of the color (Lambert and McLaughlin 1976) and Mössbauer spectroscopy has been used to study the firing technique of ceramics (Hess and Perlman 1974, Makundi, Waern-Sperber and Ericsson 1989, Wagner et al. 1997, Wagner et al. 1999). The first application of XAFS to archaeology is a chemical state analysis of iron in black glaze of a Temmoku bowl excavated at the kiln site in southern China (Nakai et al. 1991). Ancient iron implement (B.C.3C -A.D. 3C) with zoning of rusts was characterized by micro XANES (Nakai and Iida 1992). In this approach, two-dimensional chemical state imaging of the archaeological sample were made for the first time by selective excitation of iron. In the following section, the principle of fluorescence XAFS is explained through practical examples. Then, this technique is applied to reveal firing conditions of some characteristic potteries excavated at the archaeological site of Kaman-Kalehöyük, Turkey (Matsunaga and Nakai 2004).

4-2 PRINCIPLE OF XAFS AND COMPARISON WITH OTHER SPECTROSCOPIC METHODS

When X-rays pass through matter, some photons will be lost by photoelectric absorption while others will be scattered away. Thus, the absorption spectrum of X-rays by matter shows a sharp increase at a certain energy, called absorption edge, which is caused by the excitations of an electron from a deep core level of an atom to an empty bound state or to a continuum state, as a result the atom is left in a highly excited state. A vacancy has been created in one of the inner shells. The atom will almost immediately return to a more stable electron configuration by emitting fluorescent X-rays or Auger electrons. Such a transition will occur when the energy of the incident X-rays (E) is equal to or higher than the binding energy (Eb) of the electron, the energy required for excitation of the electron. Eb is equal to the absorption edge energy. This process accompanies production of photoelectrons. A fine structure of the absorption spectrum from c.a. 50eV below to 50 eV above the absorption edge is called as XANES (X-ray absorption near edge structure) and more generally as XAFS. On the other hand, well known XPS measures the kinetic energy of the photoelectron (Ek) emitted from an element excited by the X-rays with energy E, and a binding energy (Eb) and is given by the following relation Eb=E - Ek. The energy of the absorption edge or the binding energy reflects the chemical environment of an absorber atom, and therefore XANES and XPS can be used for a chemical state analysis of archaeological objects. The former technique can obtain information from bulk while the latter technique detects that from surface, i.e., less than ten nm from the surface. For archaeological application, XAFS is more advantageous over XPS because non-destructive analysis in air is possible for the former but not for XPS, which requires extremely high vacuum condition because of the detection of the photoelectrons.

Since SR displays a continuous energy distribution, it is ideal for X-rays source of XAFS. XANES provides information on oxidation states and coordination numbers, as well as site symmetry around the absorber atoms (Bianconni et al., 1985). Since the intensity of a fluorescent X-ray is proportional to the absorption coefficient of an element when the sample is thin enough or when it is dilute, the XANES spectrum can be measured using fluorescence detection (Sakurai, Iida and Goshi 1988) instead of absorbance measurement. Fluorescence XANES is far more sensitive than conventional transmission XANES and, therefore is suitable for the chemical state analysis of a trace element in an archaeological sample. Moreover, a combination of an S.S.D. (solid state detector) and X-ray microbeam allow carrying out the nondestructive two-dimensional or spatial measurement of a XANES spectrum (Hayakawa et al. 1991). This is a technique that is unique to fluorescence XANES and cannot be performed by Mössbauer spectroscopy. On the other hand, when the absorber atom in the sample is in a mixture of several chemical states, the resultant XANES spectrum is the sum of the spectra of each state, and therefore only provides limited information about individual components. In this respect, Mössbauer spectroscopy provides more quantitative information about the ratio of the components, such as the Fe^{2+}/Fe^{3+} ratio, though only a limited number of elements such as Fe, Sn and Sb can be measured by Mössbauer spectroscopy.

4-3 APPLICATION OF XANES FOR CHARACTERIZATION OF POTTERY

Since the color of pottery sherd is one feature that is often characteristic of a cultural layer, they are considered as important indexes for dating of the architecture remains at the archaeological site (Matsumura 1994). In the following example, XANES spectra are used to reveal a scientific relation between the colors of pottery and the firing techniques based on experimental data. μ-XANES technique will be applied when two-dimensional resolution was necessary. First, the Fe K-XANES spectra of iron minerals and fired clays were presented to show basic information on the chemical form of iron in the samples. Based on this knowledge, firing techniques of important ceramic classes found at Kaman-Kalehöyük are clarified.

The Fe K-edge XANES spectra were measured in the fluorescence mode using Lytle type fluorescence detector (Lytle et al. 1984) at Photon Factory, High Energy Accelerator Research Organization, Tsukuba, Japan. The XANES spectra were calculated by normalizing the intensity of the fluorescent X-ray (I_f) with that of the incident X-ray (I_0) and by plotting against the X-ray energy. The energy of the absorption edge (E_0) is defined as the mid-point of the edge crest in the normalized absorption curve.

a) XANES SPECTRA OF IRON COMPOUNDS

Figure 5 shows the Fe K-XANES spectra of metallic iron, magnetite (Fe_3O_4), hematite (α-Fe_2O_3), goethite (α-FeOOH) and kaolinite. The characteristic energies for the XANES spectra of each sample are summarized in Table 1. The absorption edge energy (E_0) shifts to a higher energy as the valence state of the absorber atom in a sample increases and that of metallic iron is located at the lowest energy level among the reference compounds (Table 1). Similarly, E_0 of magnetite (Fe^{2+}, Fe^{3+}) is located at lower positions than those in compounds containing trivalent iron.

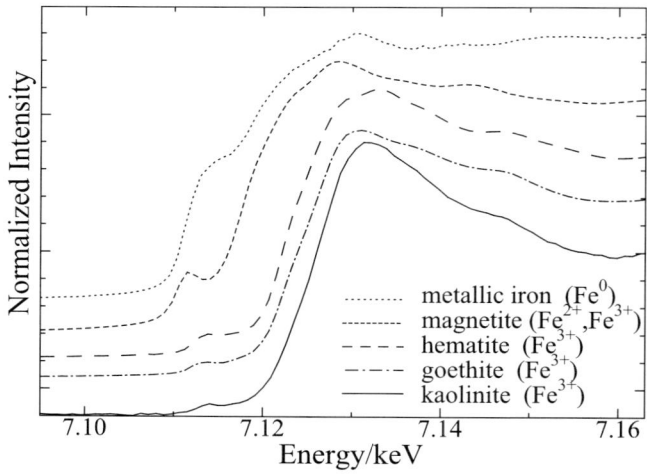

Fig. III-1-5 Fe K-XANES spectra of iron reference minerals.

Table III-1-1 Energy of Fe K-edge features in reference minerals.

sample	energy/keV		
	pre-edge	E_0	edge crest
metallic iron	-	7.115	7.131
magnetite	7.112	7.118	7.128
hematite	7.114	7.122	7.128, 7.133
goethite	7.114	7.124	7.131
kaolinite	7.114	7.124	7.132

Transition elements exhibit a small pre-edge peak on the lower energy side of the absorption edge, corresponding to transitions from 1s to 3d-like levels (Calas, Manceau and Petiau, 1988). This pre-edge absorption is strictly dipole forbidden if the coordination about iron has an octahedral symmetry with a center of inversion. When the symmetry of the ligand is lowered, the pre-edge absorption becomes dipole-allowed due to d-p orbital mixing (Wong et al. 1984). The pre-edge peak also shifts to a higher energy with an increase in the valence state of the absorber atom. The magnitude of the pre-edge peak increases with increasing site distortion or decreasing coordination number (Waychunas et al. 1983). Magnetite has a strong pre-edge peak at 7.112 keV, because it has iron with a tetrahedral coordination due to the oxygen atoms. Hematite also has a rather weaker pre-edge peak at 7.114 keV, which reflects a distortion of the octahedral coordination, and a splitting of the edge crest into 7.128 and 7.133 keV, which originates from the two Fe-O distances (at 2.115 Å and 1.945 Å) at the distorted site. Thus, it is demonstrated that each iron compound has its own characteristic XANES spectrum.

b) CHARACTERIZATION OF FIRED CLAY BY XANES SPECTRA

Fe K-XANES spectra obtained for the unfired clays block (1x1x1.5 cm) and those fired at 300, 600, 900 and 1200 °C in air or with graphite powder in a nitrogen gas flow (50 cm³/min) condition are shown in Figs. 6 (a) and (b), respectively. The characteristic energies for the XANES spectra of each clay sample are summarized in Table 2. The spectrum of the unfired clay is characterized by fine structures with a pre-edge peak at 7.113 keV, an E_0 at 7.124 keV, and the edge crest at 7.132 keV. The oxidation state of iron in this clay was determined to be entirely trivalent, since the energies of the pre-edge peak and the absorption edge are close to those of hematite, goethite and kaolinite, all of which contain trivalent iron. Furthermore, the shape of the spectrum is very close to those of kaolinite and goethite. The iron atoms in goethite and most of the iron atoms in kaolinite are located in octahedral sites in their crystals. Similarly, iron atoms are estimated to be located as well in the octahedral sites in this clay.

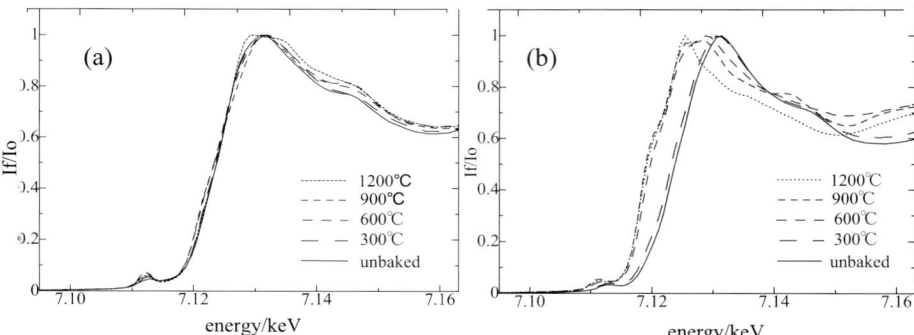

Fig. III-1-6 Fe K-XANES spectra of clays fired at various temperatures (a) under oxidizing conditions and (b) under reducing conditions.

Table III-1-2 Energy of Fe K-edge features in clay samples.

sample	energy/keV		
	pre-edge	E_0	edge crest
unfired	7.113	7.124	7.132
oxidizing condition			
300 $^{\circ}$C	7.112	7.124	7.132
600 $^{\circ}$C	7.112	7.124	7.132
900 $^{\circ}$C	7.112	7.123	7.131
1200 $^{\circ}$C	7.112	7.124	7.129, 7.134
reducing condition			
300 $^{\circ}$C	7.113	7.123	7.131
600 $^{\circ}$C	7.112	7.120	7.129
900 $^{\circ}$C	7.111	7.119	7.128
1200 $^{\circ}$C	7.111	7.119	7.125

Since the chemical shift of the Fe K-edge was negligible between the unfired and the fired clay under oxidizing condition, the iron atoms remained in the trivalent state after firing. On the other hand, the intensities of the pre-edge peaks of the fired clays are stronger than those of the unfired ones. Hematite has a rather strong pre-edge peak because the iron atoms occupy distorted octahedral sites. Probably, the observed increase in the intensity of the pre-edge peak after firing suggests the formation of hematite by heating. At 1200 °C the clay is partly vitrified, and yields a splitting of the edge crest in the spectrum.

Figure 2(b) shows the Fe K-XANES spectra of the clays fired under reducing conditions. Since the spectrum of the clay fired at 300 °C is close to that of the unfired clay, the iron atoms in the clay remain in the trivalent states even after firing. However, above 600 °C, the spectrum showed a chemical shift with the firing temperature. Furthermore, it was found that the clays fired at 600 °C and 900 °C generated broader edge crests at 7.128-7.129 keV, while the unfired clay and the fired clay at 1200 °C produced sharp edge crests at 7.131 keV and 7.125 keV, respectively. This sharp edge crests are often referred to as 'white lines'. These are typical of species in which the iron atoms are located at highly symmetric sites and the valency of the absorber atom is expressed as an integer (Waychunas et al. 1983). On the other hand, broadened edge-crests are often observed in mixed-valence compounds and compounds in which the absorber atoms occupy several different sites. The unfired clay has a trivalent iron and exhibits a XANES spectrum with a sharp edge crest. It is presumed that most of iron in the clay fired at 1200 °C under reducing conditions is divalent, since its edge crest is sharp and its E_0 value is lower than those of reference minerals containing trivalent irons. Accordingly, the continuous change of the XANES spectra was interpreted as follows: iron in the clay changes from a trivalent state to a mixture of trivalent and divalent states, and then finally the divalent iron becomes dominant as the firing temperature increases.

c)CHARACTERIZATION OF POTTERY SHERDS BY XANES SPECTRA
    The color of the pottery is closely related to the chemical form of iron, which is sensitive to firing conditions. The E_0 values obtained from the Fe K-XANES spectra of various potteries are plotted versus their colors in Fig. 7. The samples are thirty-six pottery sherds excavated from Kaman-Kalehöyük, an archaeological site in central Anatolia with Early Bronze Age, Assyrian Trade Colony, Hittite, Phrygian, and Ottoman occupation levels. In addition, the E_0 values of the reference samples are shown for comparison. As a general tendency, it was observed that the E_0 values increased as the colors of the pottery sherds varied from gray to brown to orange. The E_0 value of the orange-colored pottery agrees with those of the silicates that contain trivalent irons. The E_0 values of the grayish pottery agree well with those obtained from the silicates predominantly containing divalent iron. Most of the iron in the pottery occupies the crystallographic sites of silicates. The E_0 values of the black pottery sherds exhibit a wide distribution of energies and cover an area that corresponds to Fe^{2+} and Fe^{3+} in the silicates. Black pottery was produced by various firing techniques (Makundi, Waern-Sperber and Ericsson 1989, Matsumura 1994) and the observed variety in the E_0 values may reflect differences in their firing techniques. Close visual observation of the specimens suggest that the sherds with low E_0 energy, indicating a reduced iron state, displayed a dark gray color at the core of the sherds. On the other hand, the sherds with high E_0 values, indicating an oxidized iron state, had a black surface with red or brown core. The former sherds are produced by firing under reducing conditions, while the latter sherds are produced under oxidizing or not strongly reducing conditions, and their black color is possibly due to the soot during firing or daily use of the vessels in processes that involve heating, such as cooking.

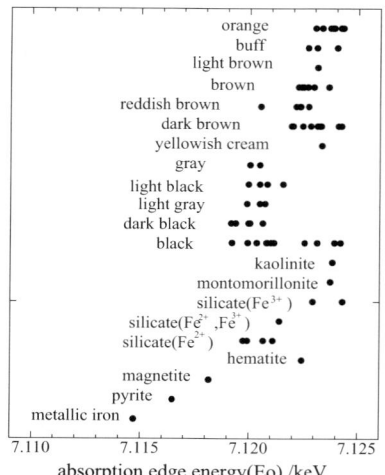

Fig. III-1-7 Relationship between the Fe K-absorption edge energy (E₀) of pottery sherds and their color.

d) FIRING TECHNIQUE OF 'CREAM WARE'

Pottery sherds with cream-colored exteriors were recognized at Kaman-Kalehöyük and are referred to as 'Cream Ware' (Matumura 1992). 'Cream Wares' found at the IIc level of the site, dated to the 7th or 8th century B.C., have the following characteristics: the exterior surfaces of the pottery sherds are cream colored whereas their interiors are light gray. The bodies are coarsely textured and highly tempered. They have no slip layer and they do not contain any organic temper (Tsu 1996). Some of the sherds show color variations from cream to red on the exterior surfaces (Matsumura 1992). Both red- and cream-colored regions of the pottery sherd contain K, Ca, Ti, Mn, Fe, Zn, and Sr in comparable quantities, and no detectable compositional differences were observed among them (Tsu 1996). A cross-section photograph of a typical sherd of the 'Cream Ware' found at the IIc Level is shown in Fig. 8(a). Figure 8(b) plots the Fe K-XANES spectra measured at each point of the sherd. As can be seen from Fig. 8(b), it is found that the absorption edge shifts toward the low-energy side from the surface of the pottery sherd towards the core (cream (A) →dark orange (B)→gray (C)). The XANES spectrum measured in the orange-colored region (point B) agrees with those of the clays fired under oxidizing conditions, and the spectrum measured at the core (point C) exhibits a good resemblance to those of the clays fired under reducing conditions at 600-900 °C (Fig. 8(b)). On the other hand, the spectrum measured at the surface (point A) exhibit a slightly higher edge energy (E₀) than that of clays fired under oxidizing conditions.

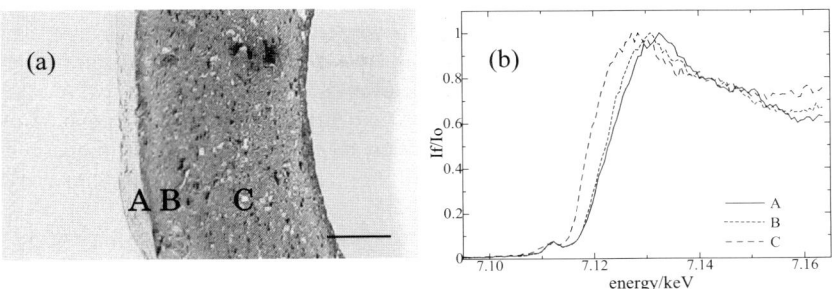

Fig. III-1-8 Photograph of a Cream Ware and its XANES spectra: (a) a cross-section of the pottery sherd. Bar = 3.0 mm. (b) Fe K-XANES spectra of the pottery sherd measured at the points shown in Fig.8(a).

The red color of pottery is associated with the presence of ferric compounds such as Fe_2O_3 (hematite). If a clay is free of iron and organic materials, it will usually be white when fired. Furthermore, it is known that some ceramics made from calcareous clays containing iron exhibit a pale cream color (Rice 1987, Jacobs 1992). When calcareous clay is fired at temperatures above 750 °C, calcium carbonate in the clay decomposes to lime. The lime then reacts with the clay and, at temperatures over 1000 °C, calcium silicates or calcium ferrosilicates with a white or pale yellow color are formed, depending on the composition. It is expected that a similar reaction occurred during the firing of the 'Cream Ware'. A question remains concerning the firing temperature of the 'Cream Ware'. A key to solve this question might be the differences in the oxidation state of the iron measured in each part of the pottery sherd. The analysis of the XANES spectra of the 'Cream Ware' suggests that the pottery was first fired under reducing conditions, and then it was oxidized (Fig. 8(b)). It has been reported that melting of calcium compounds is accelerated under reducing conditions when the calcium carbonate is in the form of very fine particles (Tite and Maniatis 1975, Maniatis et al. 1983). It is probable that the cream-colored substance observed on the surface was produced at temperatures less than 1000 °C at this stage of firing. A microscopic observation of the cream colored sherds suggest that the color is made up of a combination of the red-colored body and the whitish, cream-colored surface. Oxidation in the final stage of the firing would oxidize the remaining iron compounds that had not formed calcium silicates in the clay, and red ferric compounds would be formed. This would contribute largely to make the surface lighter and more reddish in color. We can conclude that the 'Cream Ware' from the site of Kaman-Kalehöyük was probably produced by an oxidation-reduction technique as mentioned above, and that the firing temperature was less than 1000 °C.

e)FUTURE PROSPECT OF XAFS AND SR-X-RAY ANALYSES
    Another important information obtained from XAFS is the local structure about the absorber atom. Fourier transformation of oscillation structure of XAFS spectrum and subsequent curve fitting analysis yield information of the coordination structure, which includes bond distance, coordination number, and Debye-Waller factor, which describes thermal vibration and statistical disorder of the atom. Polette et al.

(2002) reported the local structure of Fe in Maya blue, a famous blue pigment composed of palygorskite clay and indigo. They studied the role of iron in Maya blue using XAFS and high-resolution transmission electron microscopy and reported that iron oxide and the amorphous phase of FeO(OH) may contribute to the optical properties of the pigment or in the characteristic brilliant color. It is the advantage of XAFS that structural information can be obtained, even from an amorphous phase. XAFS spectra can be measured in air for all elements with atomic number heavier than Ti and in vacuum down to S at ordinary SR beam line.

The new trend of archaeological application of SR is a combination of synchrotron micro-beam techniques such as μ-X-RF, μ-XANES and μ-XRD, which allows the identification one to identify trace amounts of samples non-destructively. An example is the synchrotron X-ray micro-beam study of ancient Egyptian make-up reported by Martinetto et al. (2001). The beam size was 2x5 μm^2. Cosmetics dated between 2000 and 1200 BC were examined and galena (PbS), cerussite (Pb$_2$O$_3$) and some lead-based compounds were identified. In order to reveal the origin of the source materials, the impurity of galena was analyzed by μ-XRF, using monochromatic X-rays with energy below the L absorption edge of Pb, and nearest neighbor environment was analyzed by XAFS technique. X-ray micro-diffraction patterns were collected with a high-resolution CCD camera to identify crystalline phases. They revealed high-proportion of Sb in galena, which cannot originate from an Egyptian mine indicating that galena or the make-up was imported. XANES analysis revealed that Zn impurity in galena existed as substituting for Pb and not as an inclusion of ZnS.

5. Conclusion

This article has demonstrated that utilization of SR in X-ray analysis of archaeological samples offers many advantages over conventional X-ray source. μ-XRF and μ-XAFS allow performing two-dimensional analysis, whose spatial resolution is down to 1 μm. Since these techniques use hard X-rays, they are truly non-destructive and can be executed in air. They require no special vacuum chamber like XPS and there is practically no restriction in sample sizes. These characteristics are particularly important in archaeological application, where samples are precious cultural heritage and have various shape and size. Any materials, metals, ceramics, cloths, glasses, paper etc. can be analyzed by these techniques. Trace and major chemical component can be analyzed by XRF, which has fg and sub-ppm level sensitivity. XAFS provide information about local structure and chemical state of a particular element with concentration down to ~100 ppm. XRD can identify crystalline substance as small as 1 μm. Internal structure can be measured by X-ray computer tomography. Probably isotope ratio of an archaeological sample is the only material information, which cannot be obtained by SR X-ray analysis. With these advantages, the future of SR application to archaeology is very promising and SR facilities are waiting new exciting applications of the techniques.

References

Bianconi, A. et al., 1985, *Journal de Physique*, **46**, C9, 101-106.

Calas, G., Manceau, A., and Petiau, J., 1988, Synchrotron Radiation Applications in Mineralogy and Petrogy, 77-96.

Hayakawa, S., Gohshi, Y., Iida, A., Sato, K. and Aoki, S., 1991, *Review of scientific instruments*, **62**, 2545-2549.

Hess, J., and Perlman, I., 1974, *Archaeometry*, **16** (2), 137-152.

Jacobs, L., 1992, *Newsletter, Department of Pottery Technology*, **10**, 7-21.

Janssens,K., Vittiglio,G., Deraedt,I., Aerts,A., Vekemans,B., Vincze,L., Wei, F.,Deryck,I., Schalm, ,O.,Adams,F., Knochel,A., Simionovici,A., and Snigirev,A. , X-Ray Spectrometry 29, 73-91 2000

Lambert, J.B. and McLaughlin, C.D., 1976, *Archaeometry*, **18**, 169-180.

Lytle, F.W., Greegor, R.B., Sandstrom, D.R., Marques, E.C., Wong, J., Spiro, C.L., Huffman, G.P., Huggins, F.E., 1984, *Nuclear Instruments & Methods in Physics Research*, **226**, 542-548.

Makundi, I.N., Waern-Sperber, A., and Ericsson, T., 1989, *Archaeometry*, **31**(1), 54-65.

Maniatis, Y., Simopoulos, A., Kostikas, A., and Perdikatsis, V., 1983,*Journal of the American Ceramic Society,* **66** (11), 773-781.

Martinetto P, Anne M, Dooryhee E, et al., NUCL INSTRUM METH B 181: 744-748 JUL 2001.

Matsumura, K., 1992, *Anatolian Archaeological studies* (Jpn), **1**, 21-45.

Matsumura, K., 1994, *Anatolian Archaeological Studies* (Jpn), **3**, 5-20.

Matsumura, K., 1999, *Anatolian Archaeological studies*, **8**, (in press).

Matsunaga, M., and Nakai, I. 2004,Archaeometry,46, 103-114.

Miura,Y.,Yamato,S., Nakai,I., Terqada, Y.Yamana,K.,Terai, N.,2004,Archaeology and NaturalScience,46,1-21.

Nakai,I., Taguchi,I., and Yamasaki,K.1991, Anal. Sci. 7, Suppl. 365-368 .

Nakai, I. and Iida, A., 1992, *Advances in X-ray analysis*, **35**(B), 1307-1315.

Nakai, I., 1996, *International Symposium on the Conservation and Restoration of Cultural Property Spectrometric Examination in Conservation*, 125-135.

Nakai, I.,.Matsunaga, M., Adachi, M., and Hidakda,K.,1997, J. de Physique IV C2,1033-1034

Nakai,I., Numako, C., Hosono, H., YamasakiK.,1999, J. Am. Ceram. Soc., 82(3),689-695.

Nakai,I., Terada,Y.,Ito,M., and Sakurai,Y., 2001,J. Synchrotron Rad.,8, 360-362

Pantos, E., et al. 2002,Modern Trends in Scientific Studies on Ancient Ceramics, Edited by V. Kilikoglou, Hein, A., and Maniatis,Y., BAR International Series1011, 377-384.

Pantos,E., http://srdweb2.dl.ac.uk/srs/arch/feasibility.html,2004.

Pclette, L.A.,.Meitzner, G, Yacaman, M.J., Chianelli, R.R.,2002,Microchemical J.,71,167-174.

Paris, E., Mottana, A., and Mattias, P., 1991, *Mineralogy and Petrology*, **45**, 105-117.

Sakurai, K., Iida, A., Gohshi, Y., 1988, *Advances in X-ray Chemical Analysis Japan* (Jpn), **19**,57-70.

Tite, M.S. and Maniatis, Y, 1975, *Nature*, **257**, 122-123.

Tsa, M.C., 1996, , *Anatolian Archaeological studies*, **5**, 275-282.

Wong, J., Lytle, F.W., Messmer, R.P., Maylotte, D.H., 1984,*Physical review B*, **30**(10), 5596-5610.

Wagner, U., Gebhard, R., Murad, E., Grosse, G., Riederer, J., Shimada, I., Wagner, F. E, 1997, *Hyperfine Interact.*, 110(1,2), 165-176.

Weychunas,G.A., Apted,M.J. and Brown,G.E.,Jr.,1983,Physics and Chemistry of Minerals,10,1-9.

Chapter III-2

Synchrotron Radiation in Archaeological and Cultural Heritage Science

E. Pantos
CCLRC. Daresbury Laboratory,
Keckwick Lane, Warrington WA4 4AD, UK.
e.pantos@dl.ac.uk

Keywords: Synchrotron Radiation, Cultural Heritage, painting pigments, pottery
glazes

1. Introduction

The use of materials science techniques has long been exploited to address questions posed by archaeologists, particularly those of provenance of ancient material and technological aspects of production and how they fit in the general scheme of extracting relevant information that improves our knowledge of ancient civilisations. X-ray techniques in particular have been employed widely throughout the course of advancement of Archaeological and Cultural Heritage (ARCH) science. Synchrotron Radiation (SR) x-rays have only rather recently been utilised for measurements on archaeological material (Pantos et al (2002), Kockelmann et al (2000)) and references therein) with initial publications in the early 1990s (Nakai et al (1991, 1992)) following the seminal paper by Harbottle et al (1986) suggesting the use of Synchrotron Radiation in Archaeometry.

SR sources are, in the main, large multidisciplinary research facilities supporting a broad research portfolio in physics, chemistry, biology and engineering. The SR spectrum covers a wide spectral range from the far infrared to hard x-rays (see SPring8 website at www.spring8.jp). Advances in electron storage ring and beamline instrumentation have led to the development of third generation sources. In the context of archaeomaterials science the three key SR features, of immediate interest are

a. Brilliance: this permits fast sampling of small amount of sample.
b. Small beam footprint: it enables area mapping at millimetre to micron length-scale.
c. Wavelength tunability: the energy can be selected to suit the problem at hand.

M. Uda et al. (eds.), X-rays for Archaeology, 199–208.
© 2005 *Springer. Printed in the Netherlands.*

The high brilliance combined with fast detectors permits the rapid processing of many samples, or of samples from which only tiny amounts can be extracted (Salvado et al. (2002)). The small beam cross-section combined with high photon flux enables the study of localised details in materials, and can be used to follow changes in composition across a sample down to micro-crystallite scale. The high collimation of SR radiation permits the examination of objects non-destructively (Tang et al (2001)). In addition, unlike conventional laboratory X-ray equipment used for archaeological science research, continuous wavelength tunability allows XANES/EXAFS (X-ray Absorption Near Edge Structure / Extended X-ray Absorption Fine Structure) measurements to be made over small sample areas for the determination of oxidation state or co-ordination environment of the absorbing atomic species (Matsunaga and Nakai (2000)). Finally, the high penetration of X-rays at high photon energies (>30keV) also enables the non-destructive study of the interior of archaeomaterials (Kockelmann et al (2000)).

At the time of giving this lecture, over twelve SR sources have been used for SR-ARCH applications (SRS, ESRF, LURE, HASYLAB, BESSY II, APS, ALS, NSLS, PF, SPRING-8, BEJING) with new users entering the field at a rapid pace. The first concerted effort to bring the archaeological science and SR community together was at the workshop on SR in Archaeometry, held at Daresbury Laboratory in November 1999, followed by the second workshop at SSRL in October 2000 organised by H. Winick.

References to published work and conference presentations can be found at http://srs.dl.ac.uk/arch/feasibility.html, updated by the author as new activity is reported.

Before moving on to specific details of applications, it is worth perhaps keeping in mind that ARCHAEOLOGY IS ABOUT PEOPLE NOT ABOUT THINGS *PER SE*! Because archaeologists study the past, they are unable to observe human behaviour directly. Unlike historians, they also lack access to verbally encoded records of the past. Instead, they must attempt to infer human behaviour and beliefs from the surviving remains of what people made and used before they can begin, like other social scientists, to explain phenomena. A detailed list of archaeological material would be a long one, ranging from human remains (bone, teeth, hair, soft tissue), inorganic (pigments, stone, pottery, metal, glass) and organic remains (animal bone, wood, textile). Traditional questions asked by archaeologists are

- What?   Identification of material e.g. pigments.
- How?   Manufacturing technology, e.g. alloy or glaze composition.
- Where? Chemical fingerprinting of raw material source.
- When?   Chemical/technological typologies to assist dating/provenance/ 'authenticity'.
- Why?   Technological choices practical/conservative/'ritual'.

The majority of early applications of SR in the characterisation of archaeological and cultural heritage material have dealt with problems of identification of elements or mineral phases (pigments, glazes, soils) and manufacturing technology (metals, glasses and glazes). The techniques most suitable for this type of work are X-ray diffraction (XRD), X-ray fluorescence (XRF) and X-ray absorption spectroscopy (XAS). Some work on bone and textiles has also been carried out using XRD and small angle scattering (SAXS). These are not the only techniques or regions of SR radiation that are applicable. It is early days yet and time is needed for the partnerships to mature and new researchers to enter the field. In particular, SR-FTIR and luminescence properties of materials in the 1-1000eV region (VIS - VUV- XUV- Soft X-rays) are expected to play a significant role in the future.

In what follows, some examples are highlighted from work at SR sources carried out by the author and collaborators. Most of this work is at the stage of being prepared for publication so only a summary outline will be given.

1. Pigments

1.1 PAINTING PIGMENTS

The study of ancient paintings is of high interest to scientists and conservators as well as museum curators. Some examples of such interest are: the evolution of painting techniques through time; the characterisation of the painting materials used by important Masters in order to find out correlations with other painters, schools; the determination of alteration phases and layers to provide advice on the correct ways to act in the restoration and conservation of the work of art. Such a detailed study across small paint fragments has been reported recently by Salvadó et al. (2002). It deals with the identification of the green pigments used by one of the most important Catalan Masters in Gothic times, Jaume Huguet. His works fall between tradition (international Gothic style) and innovation (Flemish Renaissance) characteristic of the second half of the 15th century. In his time, Huguet was well known and he obtained some of the most important contracts for painting many religious artworks mainly in the form of altarpieces. Figure 1-(a) shows part of a Huguet altarpiece from which a small fragment of green paint was extracted, embedded in epoxy resin and thin sectioned. A grid of 100-micron areas was examined by SR-XRD, the mineral phases identified and the results described in detail in Salvadó et al. (2002).

In summary, Jaume Huguet used a green pigment, which was synthesised following a procedure similar to that described by a medieval monk, Theophilus the Presbyter. This green pigment was mixed with egg and drying oil and was applied by forming different layers over the plaster surface. Normally, the first layer contained white pigment (a lead carbonate compound) mixed in some cases with some green; the second layer contained the green pigment mixed with yellow (a lead tin oxide). In fact, a number of copper compounds were also identified, chlorides, carbonates and acetates. All these phases are the result of the Theophilus' procedure of green pigment synthesis. Moreover, in all the paintings studied, weddellite (a calcium oxalate) is found especially on the outside surfaces of the painting but also in the plaster. This is probably a mineral phase due to weathering related to the alteration of the organic binding media (egg and drying oils for the chromatic layers or animal glue for the plaster). The painting technique, colours and their alterations were the same for all the altarpieces studied.

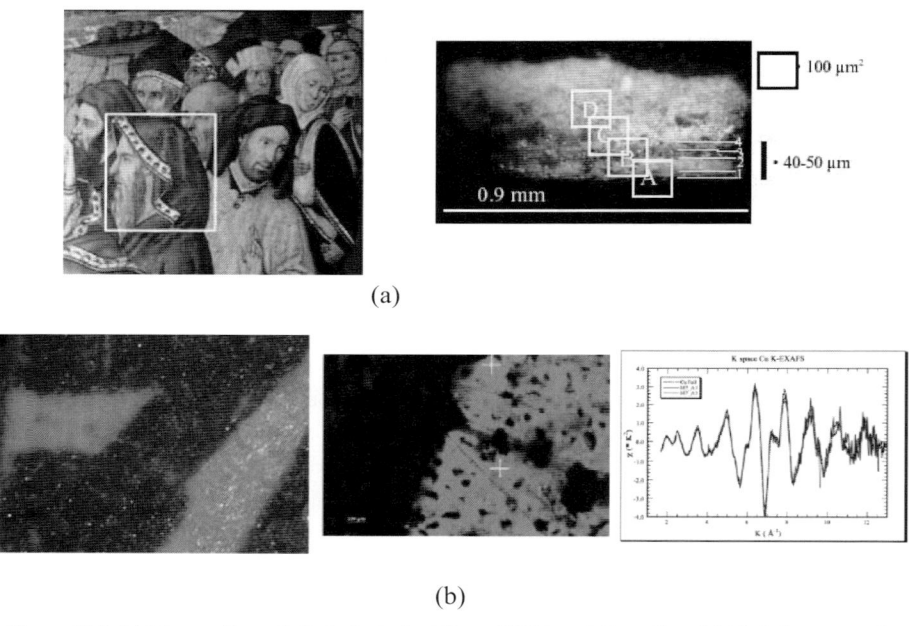

(a)

(b)

Figure.III-2-1.(a) Jaume Huguet's Retaule de Sant Bernadí i l'Angel Custodi, and (b) left: lustre sample, centre: Tricolour elemental map of part of the glazed area where copper is rendered as green, and right: XAS spectra in two positions compared with metallic copper. (See also Color plate, p. 303)

1.2 PHOENICIAN COSMETICS

A number of pigments were discovered during excavations in archaeological sites of Tunisia (Carthage, Kerkouane, Bekalta, Bouaarada and elsewhere). There is little material evidence on the Carthagenian society and their civilization compared to other contemporary cultures in the Mediterranean. Amongst the finds several types of pigments were found, in small quantities and in receptacles such as sea-shells or as small cones. This suggests that they were used for specialist purposes, e.g. as cosmetics. Identification of the mineral nature of these materials may confirm their function and possibly indicate cultural or trade exchanges with other cultures using similar practices.

Natural antique colorants include red pigments such as cinnabar and ochre and pink pigments such as madder. They have been used in different periods and regions of the world from the palaeolithic and bronze age to relatively modern times even in some places in the world. Such pigments have served as ritual and cosmetic make-up. We need to differentiate between make-up used on living persons (cosmetic make-up) and make-up used to vivify the dead (replace the dead person's blood). An unction was applied on the face and the forehead of the dead using cinnabar or ochre. These red materials can be called mourning reds. Traces of red pigmentation of bones and particularly skulls imply that this ritual was extremely frequent in Punic and Roman periods. The study of red pigments found isolated or on bones will allow us to differentiate them from cosmetic make-up (madder-based make-up) based on consistency, purity and texture as well as mineralogical content.

2. Glazes

2.1 HISPANO-MORESQUE LUSTRE WARE

X-ray spectroscopy can only be performed with SR. Of particular interest are cases where two-dimensional mapping is needed to reveal molecular distribution information at a high spatial resolution. MicroXAS was used in an investigation of the lustre finish applied to ceramic ware first developed in the Islamic world of the 10th to 12th centuries. The use of lustre decorated ceramics is one of the most exciting developments in pottery production during medieval times resulting in a decorative finish in which drawings and patterns in a metallic shine layer were applied on a glaze. Subsequent developments in the Hispano Moresque world of the 13th to 15th centuries following the Islamic tradition created highly prized items, illustrated by archaeological findings, which indicate that the majority of the Islamic and Hispano Moresque productions of these centuries were widely exported. Different recipes allowed potters to obtain several colours and hues, such as those exhibited by the Italian maiolica (16th and 17th century) where the lustre shows colours from copper-like to gold-like or greenish.

This distinctive finish was obtained by a third firing (first firing for the ceramic body, second firing for the tin glaze and third for the lustre decorations) to allow the reduction of copper or silver salts to the metallic state. The complexity of the manufacturing process is demonstrated in the number of "non-metallic" lustre layer specimens that have been found, implying that although the potter had intended to produce a metallic lustre, he had no success. One of the questions arising about this pottery was to determine why the colour of a particular piece did not reach the metallic shine. Two possibilities were considered: firstly, that the reduction was not enough to obtain metallic Cu, or alternatively that the metal has formed into crystals too small to exhibit metallic optical properties.

Figure 1-(b) shows the elemental distribution of copper on the surface of a glazed sample. XAS spectra were obtained in the positions indicated. Comparison with copper standards has shown that metallic copper is responsible for the metallic lustre effect. Reproduction pieces for different firing times were also examined. XAS spectra have been fitted by a model that considers a mixture of metallic copper and copper oxides (Cu_2O and CuO) in different proportions (Smith et al (2003)). An increase in the proportion of metallic copper in front of the oxides is obtained for longer firing times.

2.2 ETRURIAN BLACK GLOSS

Non-destructive examination of ancient objects is of course to be preferred whenever this is possible. Mineralogical phase composition by XRD normally requires preparation of powders or of thin sections the measurements being performed usually at a fixed wavelength. The possibility at SR instruments of being able to vary the wavelength, within a reasonable range, provides us with the opportunity to carry out, non-destructively, the mineralogical characterisation of pottery by controlling the penetration depth of the X-rays as a function of the wavelength. Computation of the dependence of the attenuation length (1/e of intensity transmitted) for typical aluminosilicate glazes indicates that the penetration of X-rays varies significantly, approx. between 1 and 100 microns in the wavelength range 0.5Å to 2Å and an angle of incidence between 1 and 5 degrees. This means that if diffraction patterns are obtained at a low angle of incidence, the data would show any variations in mineralogical composition as the depth of penetration varies by changing the wavelength.

A study of Etrurian black gloss pottery using microscopic, elemental and mineralogical analysis has investigated the classification of the various types that had been assigned on archaeological grounds (typology, finding site) and on the visual appearance of the black gloss which ranged from a type with a distinctive blue hue to shiny metallic black colour. Diffraction patterns at different wavelengths for a black gloss pottery sample of the blue hue type show that the dominant diffracting phase at high wavelength (low penetration) is hercynite. As the wavelength is decreased, the X-rays penetrate further into the ceramic paste and phases such as hematite, feldspars and quartz are sampled as the gloss-body interface is approached at about 1.2 Å (thickness approx. 20 microns). Figure 2 shows two such diffraction patterns collected at different wavelengths showing the changes observed as the penetration depth changes. This approach has allowed the rapid characterisation of several samples in the collection. The extreme cases are characterised by hercynite or magnetite as the dominant iron oxide phase.

X-ray spectroscopy using SR can extend to the soft X-ray region where the L-edges of elements such as iron absorb. A study was carried out at the Fe L-edge to identify the iron compound present at the very top layer of the surface of the gloss (Gliozzo et al (2003)). The penetration depth in this energy region (around 710eV for the Fe L-edge) is of the order of 0.1micron. Figure 3 shows the XAS spectra for two of the samples examined. Comparison with standards containing iron in the FeII and FeIII oxidation states showed that the classification of the samples according to visual appearance followed closely the classification according to oxidation state.

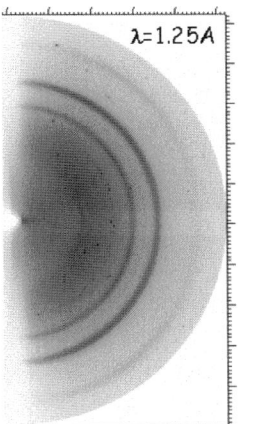 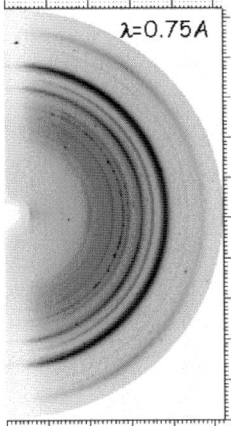

Figure III-2-2. Diffraction patterns from the gloss of a blue-hue black gloss sample at 1.25 Å and 0.75 Å. Hercynite is the dominant iron oxide phase throughout the gloss.

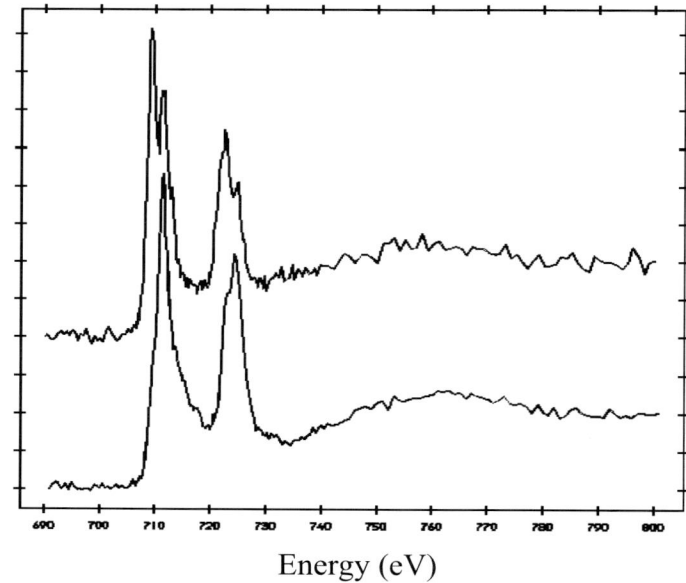

Energy (eV)

Figure III-2-3. Iron L-edge XAS spectra of two of the black gloss samples from Etrurian sites. The spectrum at the top resembles closely that of Fe^{II} compounds such as hercynite ($Fe^{II} Al_2O_4$) while the one at the bottom that of magnetite ($Fe^{II} Fe_2^{III}O_4$).

3. Summary

Archaeological and Cultural Heritage science utilising synchrotron radiation has come of age. The spectrum of applications has varied from ceramic material and glazes to pigments, glasses, metals and corrosion products, stone, wood and soft tissues or fibrous material as found in bone, textiles and paper. The examples highlighted above demonstrate how SR modalities can provide unique opportunities for the molecular speciation of archaeomaterials from the macroscopic (mm) to the microscopic length scale (1micron). There is uncharted territory to be explored and new frontiers to be opened. Of particular promise are the full utilisation of key SR properties, such as brilliance and energy range, for spatially- and time-resolved studies.

Acknowledgements

The work described in these notes is the product of collaborative work with a number of colleagues whom I would like to acknowledge: Nati SALVADÓ, Trinitat

PRADELL from ESAB- CEIB Universitat Politècnica de Catalunya, Barcelona Spain. Judit MOLERA, MariusVENDRELL Departament de Cristal·lografia i Mineralogia. Universitat de Barcelona, Barcelona. Mathew MARCUS Advanced Light Source, Lawrence Berkeley National Laboratory, Berkeley, USA. Elisabetta GLIOZZO, Isabella MEMMI-TURBANTI, Earth Sciences Department, University of Siena, Siena, Italy. Ashfia HUQ, Peter STEPHENS, Department of Physics and Astronomy, State University of New York at Stony Brook, Stony Brook, USA. Naceur AYED, Housam BINOUS, Institut National des Sciences Appliquées et de la Technologie, Tunis, Tunisia. Lucian BURGIO, Victoria and Albert Musem, UK. Andy SMITH, Ian KIRKMAN, Miroslav PAPIZ, Daresbury Laboratory, UK. The encouragement and support of the Daresbury Laboratory managment is gratefully acknowledged.

References

Harbottle, G., Gordon, B.M. and Jones, K.W., (1986). Use of Synchrotron Radiation in Archaeometry, *Nucl. Instr. and Meth.* B **14** 116-122.

Gliozzo, E. , Kirkman, I.W., Pantos, E. and Memmi-Turbanti, I., (2003). Black gloss pottery: production sites and technology in ancient Tuscany. Part II: gloss technology. *Archaeometry*, in press.

Kockelmann, W., Pantos, E., and Kirfel, A. (2000), Neutron and synchrotron radiation studies of archaeological objects Neutron and synchrotron radiation studies of archaeological objects. In *Radiation in Art and Archaeology*, Ed. D.C. Creagh & D.A. Bradley, Elsevier.

Nakai, I., Taguchi, I. and Yamasaki, K., (1991). Chemical Speciation of Archaeological Objects by XRF/XANES Analysis Using Synchrotron Radiation, *Analytical Sciences* Vol **7**, supplement.

Nakai, I. and Iida, A., (1992). Aplications of SR-XRF Imaging and Micro-XANES to Meterorites, Archaeological Objects and Animal Tissues, *Advances in X-ray Analysis*, **35**, 1307-15.

Pantos, E., Tang, C.C., MacLean, E.J., Cheung, K.C., Strange, R.W., Rizkallah, P.J., Papiz, M.Z., Colston, S.L., Roberts, M.A., Murphy, B.M., Collins, S.P., Clark, D.T., Tobin, M.J., Zhilin, M., Prag, K., and Prag, A.J.N.W. (2002). Applications of Synchrotron Radiation to archaeological ceramics, *Modern Trends in Scientific Studies on Ancient Ceramics*, Edited by V. Kilikoglou, A. Hein and Y. Maniatis, BAR International Series1011, 377-384.

Salvadó, N., Pradell, T., Pantos, E., Papiz, M.Z., Molera, J., Seco, M. and Vendrell-Saz, M., (2002). Identification of copper based green pigments in Jaume Huguet's Gothic altarpieces by Fourier transform infrared micro-spectroscopy and synchrotron radiation X-ray diffraction, *J. Sync. Rad.* **9** 215-222.

Smith, A.D., Pradell, T., Molera, J., Vendrell, M., Marcus, M. and Pantos, E., (2003). MicroXAFS studies into the oxidation states of different coloured glazes originating from the early Islamic world, *J.Phys.IV* Vol. *104* pp.519-522.

Tang, C.C., Mclean, E., Roberts, M., Clarke, D.T., Prag, J.and Pantos, E., (2001). A study of attic black gloss sherds using synchrotron radiation, *J.Arch.Sci.* **28**, 1015-1024.

Chapter III-3

Study of the Elemental Distribution in Ancient Chinese Porcelain Using Synchrotron Radiation X-ray Fluorescence

Y. Y. Huang[1, 2]*, P. L. Leung[2], W. He[1]
[1]*Beijing Synchrotron Radiation Facility, Institute of High Energy Physics, Chinese Academy of Sciences, P. O. Box 918, Beijing 100039, China*
[2]*Department of Physics and Materials Sciences, City University of Hong Kong, Kowloon, Hong Kong, China*
huangyy@mail.ihep.ac.cn

Keywords: porcelain, synchrotron radiation X-ray fluorescence, elemental
　　　　　distribution, suitable spot size

Abstract

In this paper, the scanning results of the elemental distribution of an ancient porcelain potsherd in a 200 x 200μm^2 area with a 20 x 20μm^2 micro X-ray spot were obtained by synchrotron radiation X-ray fluorescence microprobe technique. The elemental distribution of the same porcelain potsherd in the different areas was also measured by using different spots from 100 x 100μm^2 to 2 x 6mm^2 respectively. The X-ray spot size, the number of measured point and the associated angular distribution of intensity of synchrotron X-ray were carefully investigated in order to find a suitable spot size by which the average elemental distribution of a few measured points can represent the elemental distribution of a whole porcelain ware.

1. Introduction

The homogeneity of elemental distribution is an important factor for X-ray fluorescence measurement with a small spot size. As we know, the ancient porcelains were made of several raw materials of ceramics clay following high temperature firing; therefore, the elemental distribution in porcelain is usually not very homogeneous. It becomes an interesting research project to specify how to choose an appropriate X-ray spot size for X-ray fluorescence analysis and how many minimum numbers of measuring points are required in order to acquire the elemental distribution showing a whole porcelain ware. Although some researchers had realized the importance of inhomogeneous element distribution in their

M. Uda et al. (eds.), X-rays for Archaeology, 209–216.

analytical areas [1, 2, 3, 4], no detailed study related to ancient porcelain measurement by X-ray fluorescence analysis was reported until recently.

In the work of reference [1], it was pointed out that there is non-homogeneity of the element distribution in and under the surface of porcelain. In the surface, the solution is to choose a bigger size of X-ray spot or take a mean of the different measurement points by using a smaller X-ray spot. However, it is difficult to solve the problem under surface. In this paper we mainly focused on the surface distribution of elements.

The microbeam scanning analysis is an important way to solve this problem. In this study, the elemental homogeneity measurement of two-dimension distribution of ancient porcelain was performed by synchrotron radiation X-ray fluorescence (SRXRF) analysis. The results of homogeneity for a typical ancient porcelain sherd are presented.

2.   Experimental facility

Synchrotron radiation is a powerful advanced light source compared to conventional X-ray tube radiations, and has many unique properties, such as high intensity, natural collimation, well-defined polarization, wide spectral range and energy tunability. SRXRF technique is a non-destructive and free contact analytical method, and especially large specimen testing can be performed in air environmental condition. It is very suitable for valuable ancient porcelain analysis. Measurements for this paper were carried out at the XRF experimental station at the Beijing Synchrotron Radiation Facility (BSRF) [5, 6].

3.   Experimental results and discussion

3.1 MICROBEAM SCANNING ANALYSIS OF MAJOR AND TRACE ELEMENTS

The experimental specimen used here is a sherd of celadon porcelain made in the South Song Dynasty. In the experiment the scanning area is $200 \times 200 \mu m^2$, the micro beam size is $20 \times 20 \mu m^2$. The white spectrum of synchrotron radiation was used; the energy range is 3-30 KeV at Beam line 3W1A of BSRF. The scanning distributions of the major element Ca and Fe, as well as the trace element Cu, Rb, and Zr of glaze of the porcelain sherd are shown in the Fig.1. X and Y-axis indicate the relative position of the scanning beam; Z-axis indicates the relative elemental content. Fig.1 shows that the elemental distribution of the porcelain sherd is inhomogeneous, especially for the element Zr in the analytical area. This special non-homogeneity distribution of element Zr was little mentioned; this may be because the sensitivity of the convention XRF and electron probe analytical techniques is usually lower than SRXRF in such a small analytical area. Similar results of non-homogeneity distribution of Zr element from other Chinese porcelain sherds made in the different dynasties were often found in our research work. For the other major and trace elements, similar results of inhomogeneous elemental distribution were also observed, here only a few typical results are shown.

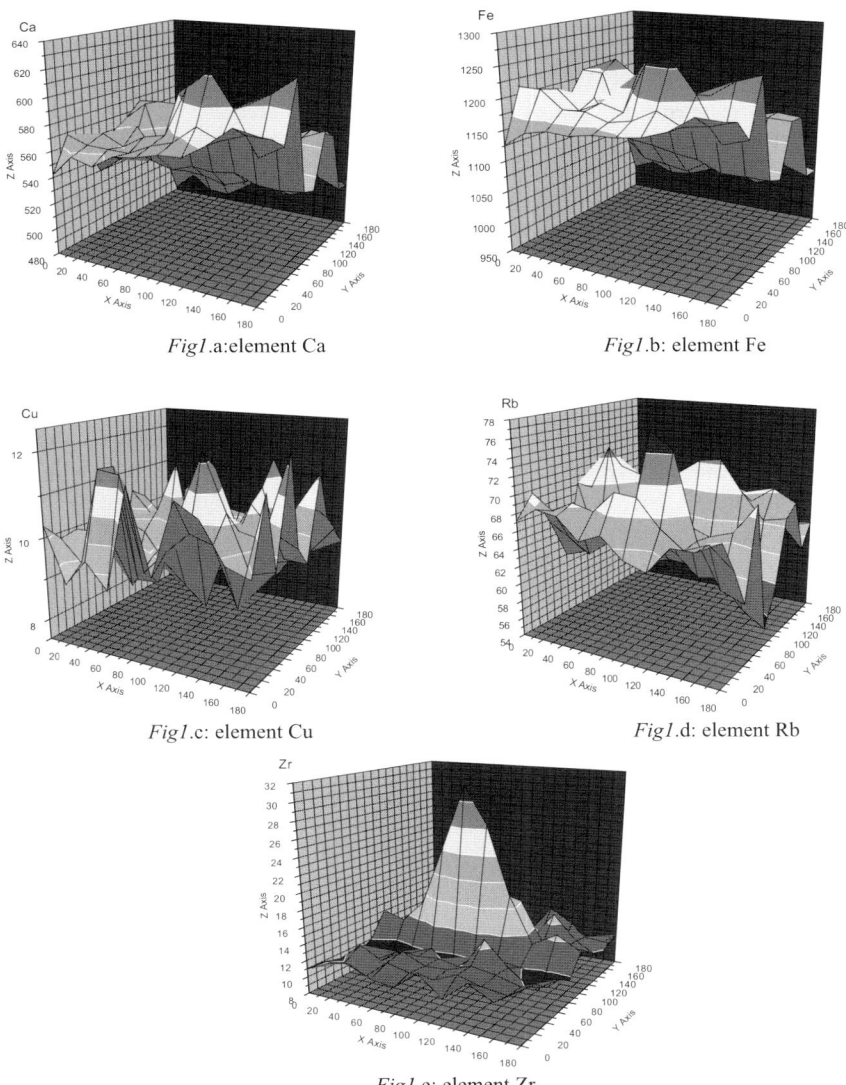

Fig1.a:element Ca

Fig1.b: element Fe

Fig1.c: element Cu

Fig1.d: element Rb

Fig1.e: element Zr

*Fig. III-3-1.*The scanning distributions of the major element Ca and Fe, as well as the trace element Cu, Rb, and Zr of glaze of the porcelain sherd, a: element Ca, b: element Fe, c: element Cu, d: element Rb, e: element Zr. X and Y-axis indicate the relative position of the scanning beam; Z-axis indicates the relative elemental content.

3.2 BEAM SIZE AND ANGULAR DISTRIBUTION EFFECT OF SYNCHROTRON RADIATION INTENSITY

In the experiment we measured five different positions of the glaze of the same porcelain sherd, with different micro beam sizes from 100x100µm^2 to 400x400µm^2 (white spectrum 3-30 KeV, 4W1A beam line) and from 2x2 to 2x6 mm^2 (monochromatic spectrum, 10KeV, 4W1B beam line). That is, we first measured five different positions of the glaze of the same porcelain sherd with the micro beam size of 100x100µm^2, and then the other five different positions with 200x200µm^2, and so on. The mean values of the five different position (five times) measurements and their standard deviations of elemental relative contents were determined and shown in Fig. 2a, 2b and 2c, where vertical axis indicates relative elemental contents per unit beam size area.

Because of the angular distribution effect of SR intensity itself in vertical direction [7], i.e. the SR intensity decreased with the increase of the distance from the orbit plane of electron beam in the storage ring of the accelerator [8, 9]. Therefore, the mean values per unit beam size area of the five different position measurements decreased when the beam size changed from 100x100µm^2 to 400x400µm^2 in the Fig. 2a. In our experiment the orbit plane was in the middle of the vertical slit, that is, it was in the middle of the vertical direction of the beam size. The same cases occur in Fig. 2b and 2c, where the horizontal slit was fixed, only the vertical slit changed from 2 to 6mm, i.e. the beam size changed from 2x2 to 2x6 mm^2.

In order to further explain this phenomenon of angular distribution in Fig. 2a, 2b and 2c, especially in vertical direction, a reference experimental result was determined by measuring a homogeneous stainless steel film standard specimen (SS4) [10] with different X-ray spot sizes. The result is shown in Fig. 2d; the vertical axis indicates relative elemental contents per unit beam size area. The curve is the first order exponential decay fitting result. It was shown that the SR intensity decreased with the increase of the distance from the orbit plane of electron beam in the storage ring of the accelerator and also observes the exponential decay. It was shown that the decay of the curves in Fig. 2a, 2b, and 2c correspond with the one in Fig. 2d. It can also be seen that the standard deviations in Fig. 2a, 2b and 2c decreased with the beam size increase. It means that with the same measuring times the larger the beam size is, the better the mean measurement value.

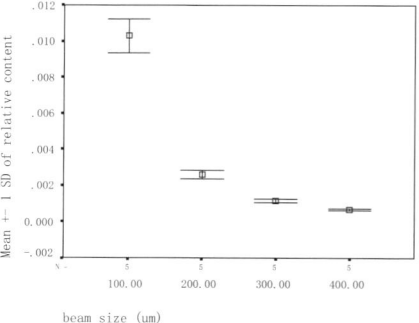

Fig III-3-2a. The mean values (square sign) of five different position measurements (N=5) and their standard deviations (SD) of the relative contents of Fe element are shown. X-axis indicates the measuring times (upper) and beam size (bottom).

Fig.III-3-2b. The mean values (square sign) of five different position measurements (N=5) and their standard deviations (SD) of the relative contents of Ca element are shown. X-axis indicates the measuring times (upper) and beam size (bottom).

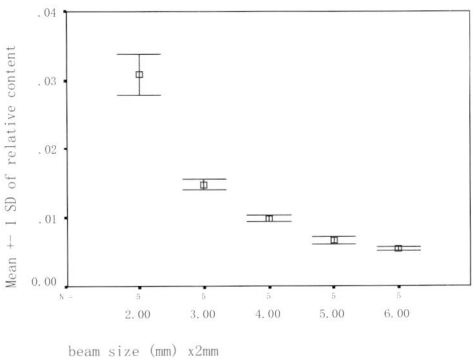

Fig. III-3-2c. The mean values (square sign) of five different position measurements (N=5) and their standard deviations (SD) of the relative contents of Fe element are shown. X-axis indicates the measuring times (upper) and beam size (bottom).

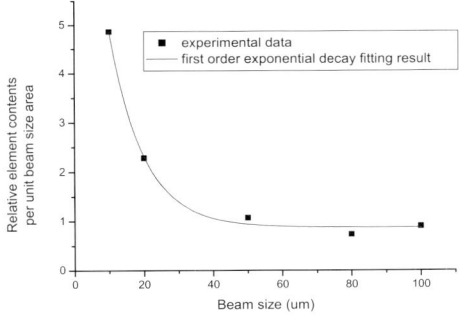

Fig. III-3-2d. The vertical axis indicates relative elemental contents per unit beam size area for elemental iron, horizontal axis is the beam size of the vertical slit, and all the beam spot are the squares.

3.3 RELATION BETWEEN THE SPOT SIZE AND THE NUMBER OF THE TESTING POINT

As the description above in the section 3.1, the scanning area in the experiment is $200 \times 200 \mu m^2$, the micro beam size is $20 \times 20 \mu m^2$. The beam was scanned one by one without overlapping and spacing on the surface of the porcelain. That is, there are 100 different positions (square areas) each one with size $20 \times 20 \mu m^2$. The relative elemental contents in each one position were determined and are shown in Fig. 1.

Therefore, there are also 25 different positions (square areas,) each one with size $40x40\mu m^2$ for the scanning area of $200x200\mu m^2$. Each bigger square with size $40x40\mu m^2$ is composed of four adjacent smaller squares each one with size $20x20\mu m^2$. The relative elemental contents in each one with size $40x40\mu m^2$ can be obtained by adding those of the four adjacent smaller squares each one with size $20x20\mu m^2$. In the same way, there are also 4 different positions (square areas) each one with size $100x100\mu m^2$ for the scanning area of $200x200\mu m^2$. The relative elemental contents in each one with size $100x100\mu m^2$ can be obtained.

The $40x40\mu m^2$ and $100x100\mu m^2$ can be regarded as a bigger beam spot size for the scanning area of $200x200\mu m^2$ respectively. Therefore, the mean values and their standard deviations of the elemental relative contents could be obtained by averaging the different measuring positions (square areas, here 100, 25 and 4 respectively), which are shown in Table 1 and Fig. 3a for calcium as well as Fig. 3b for iron. In this way the problem of the beam size and the angular distribution effect of synchrotron radiation intensity can be avoided. The relation between the spot size and number of the testing point becomes simple.

As shown in the Table 1 and the Fig. 3, the mean values of the elemental relative contents are almost the same for calcium and the same for iron when the beam size increases. Meanwhile the standard deviations become smaller and smaller with increasing beam size. That means that the measuring results of element calcium and iron are already accepted when the beam size $20x20\mu m^2$ is used for the analytical area of $200x200\mu m^2$, but it needs to measure 100 different positions. However, if a larger spot size is used, for example, $100x100\mu m^2$, only 4 different testing points are enough and the measuring precision is higher. Therefore, the beam size of $100x100\mu m^2$ can be regarded as a suitable spot size by which the average elemental distribution of the four measured points can represent the elemental distribution of the materials.

Table III-3-1. The mean relative elemental contents per unit beam size area and their standard deviations.

beam size (um)	20	40	100
number of square area	100	25	4
Ca mean value	2.8352	2.8360	2.8350
Ca standard deviation	0.1727	0.1598	0.1162
Fe mean value	1.3776	1.3776	1.3776
Fe standard deviation	0.0792	0.0746	0.0567

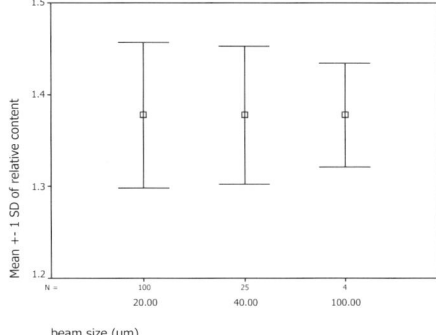

Fig. III-3-3a. The mean values (square sign) and their standard deviations (SD) of the relative contents of element Ca are shown. X-axis indicates the measuring times (upper) and beam size (bottom), Y-axis indicates the relative elemental content per unit beam size area.

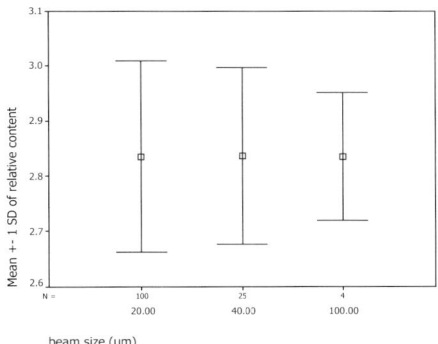

Fig. III-3-3b. The mean values (square sign) and their standard deviations (SD) of the relative contents of element Fe are shown. X-axis indicates the measuring times (upper) and beam size (bottom), Y-axis indicates the relative elemental content per unit beam size area.

Conclusion

Elemental distribution is inhomogeneous for ancient Chinese porcelain, such as the celadon porcelain measured above, especially for element Zr. When measuring the sample, if the beam size is changed, especially in the vertical direction, the angular distribution effect of the synchrotron radiation intensity should be considered. For obtaining the accepted average measuring value of elemental contents, less testing points is needed and higher precision can be obtained when the larger beam size is used. One suitable beam spot size was determined. By using this beam size the average elemental distribution of a few measured points represents the elemental distribution of a whole porcelain ware.

Acknowledgements

The work described in this paper was partially supported by a grant from the City University of Hong Kong (Project No.7001104), and the experiments were carried out at the Beijing Synchrotron Radiation Facility.

References

[1]. He Wenquan, and Xiong Yingfei, (1999) *The Non-Destructive Analysis of Ancient Ceramics*, in Guo Jingkun (eds), The Proceedings of the 1999 International Symposium on Ancient Ceramics-its Scientific and Technological Insights, Shanghai Scientific and Technical Reference Publishers, Shanghai, China, pp563-567.

[2]. M. Mantler and M. Schreiner, (2000) X-ray Fluorescence Spectrometry in Art and Archaeology, *X-ray Spectrometry* **29**, 3-17.

[3]. P. Wobrauschek, G. Halmetshclager, S. Zamini, C. Jokubonis, G. Trnka and M. Karwowski, (2000) Energy-Dispersive X-ray Fluorescence Analysis of Celtic Glasses, *X-ray Spectrometry* **29**, 25-33.

[4]. K. Janssens, G. Vittiglio, I. Deraedt, A. Aerts, B. Vekemans, L. Vincze, F. Wei, I. Deryck, O. Schalm, F. Adams, A. Rindby, A. Knochel, A. Simionovici and A. Snigirev, (2000) Use of Microscopic XRF for Non-destructive Analysis in Art and Archaeometry, *X-ray Spectrometry* **29**, 73-91.

[5]. Y. Y. Huang, K. F. Li, W. He, G. C. Li and K. X. Lin, (2001) Single Fluid Inclusion Study by SRXRF Microprobe, *Nuclear Instruments and Methods in Physics Research* A **467-468** ,1315-1317.

[6]. Y. Y. Huang, J. X. Lu, R. G. He, L. M. Zhao, Z. G. Wang, W. He and Y. X. Zhang, (2001) Study of Human Bone Tumor Slice by SRXRF Microprobe, *Nuclear Instruments and Methods in Physics Research* A **467-468,** 1301-1304.

[7]. A. Iida, (2000) Instrumentation for μ-XRF at Synchrotron Radiation Sources, in Koen H. A. Janssens, Freddy C.V. Adams and Anders Rindby (eds), *Microscopic X-ray Fluorescence Analysis*, John Wiley & Sons Ltd, pp.121-123.

[8]. Tang Esheng, Huang Yuying, Wu Yingrong and Yi Futing (1998) Radiation Character of 3W1 Permanent Magnet Multipole Wiggler, *High Energy Physics and Nuclear Physics*, (Chinese), **22**, 951-954.

[9]. Esheng Tang, Yonglian Yan, Shaojian Xia, Peiwei Wang, Futing Yi, Yuying Huang, Jin Liu and Mingqi Cui (1998) New Wiggler Beam Line at BSRF, *Journal of Synchrotron Radiation*, **5**, 530-532.

[10] H. Shao and Q. Xu (1995) *Nucl. Instr. And Meth. In Phys. Res*.B **104** , 201-203.

Chapter III-4

Study on the Compositional Differences among Different Kilns' Tang Sancai by SRXRF

Y. Lei, S. L. Feng
Laboratory of Nuclear Analytical Techniques, Institute of High Energy Physics, Chinese Academy of Science
P O Box 918, Beijing 100039, China
J. Jiang, Z. X. Zhou
Archaeological Institute of Shaanxi Province
Xi'an, Shaanxi, 710000, China
S. L Zhang, Y. M. Liao
Archaeological Institute of Henan Province
Zhengzhou Henan, 450000, China
Fengsl@ihep.ac.cn

Keywords: Tang Sancai, Tang Dynasty, provenance, SRXRF, statistical method

Tang Sancai is a general name for the color-glazed pottery produced during the Tang Dynasty. It is famous for its distinctive color in China and other Asian countries. Many kinds of Sancai were excavated in the Shaanxi and Henan Provinces. However, the provenance of some Tang Sancai has not been well identified until now. In order to group the Tang Sancai specimens of the unknown provenance to the respective kilns according to their chemical compositions, the major and trace elements of more than 100 Sancai body specimens taken from three important kilns were analyzed by SRXRF. The results indicate compositional differences among the studied Sancai specimens. The characteristics of the trace elements of each kiln's Tang Sancai were identified by statistical methods and the results could be used as database for the identification of unknown Sancai in future.

1. Introduction

Tang Sancai is a general name for the color-glazed pottery produced in Tang Dynasty. One specimen was firstly unearthed at Luoyang in 1899, when the Qing government built the Longhai railway. Since that period many kinds, of Tang Sancai have been unearthed in the Shaanxi and Henan Provinces. Yellow, green and white are the three dominant colors in the Tang Sancai's glaze, but it was not limited to these ones. White or red clays were used as raw materials for producing the body pottery [1]. Later pottery artifacts were likely further processed through two steps of firing for the glazed color on the body.

M. Uda et al. (eds.), X-rays for Archaeology, 217–222.
© 2005 *Springer. Printed in the Netherlands.*

Figure III-4-1. A Tang Sangcai Figure of a Riding Woman
718 AD, Tang Dynasty, 35.2 cm (H). (See also Color plate, p. 304)

The provenance study of Tang Sancai is a critical step to study the history and the development of Tang Sancai. In the Shaanxi and Henan provinces three kiln-ruins making Tang Sancai have already been discovered: Huangye Kiln at Henan [2], and Huangbu Kiln [3] and Xi'an Kiln at Shaanxi (Fig. 2). In order to clarify the Tang Sancai items, whose provenance is unknown, in distinct kilns according to their chemical compositions, we studied the elemental difference between different kiln's Tang Sancai by SRXRF.

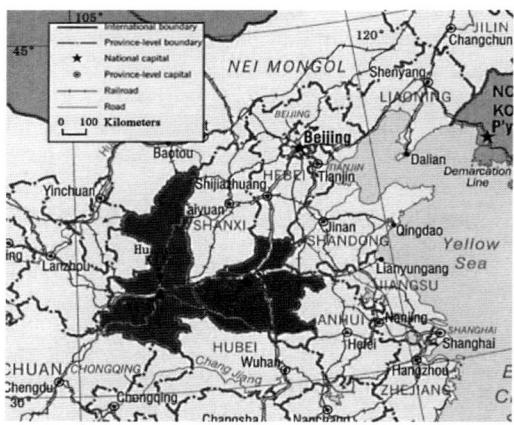

Figure III-4-2. Location of three ruins of Tang Sancai Kilns
✿Huangbu Kiln, ✠Xi'an Kiln, ⋎Huangye Kiln

SRXRF (Synchrotron Radiation X-ray Fluorescence) is an efficient method for element analysis. The main characteristic of this recent analytical technique applied to archaeometry is its high beam intensity, which greatly improves the sensitivity of element analysis and makes it possible to simultaneously determine numerous-elements in one single sample. Elemental analyses allow us to identify the origins of particular kinds of pottery and to compare it with the composition of pottery of the known origin or raw materials.

2. Experimental

2.1 SAMPLE PREPARATION

Thirty-one samples (Nos. H1-H31) of interest were collected from the Huangye kiln, 51 (Nos. T1-T51) from the Huangbu kiln and 79 (Nos. X1-X79) from the Xi'an kiln. Each sample's section was polished. The specimen body from the Xi'an kiln exhibits five kinds of colors, i.e. red, off-white, gray, white and pink. The body from both Huangbu and Huangye has only two kinds of color, white and pink.

2.2 INSTRUMENTATION

The experiment was carried out at the 3W1A beam line of the Beijing Synchrotron Radiation Facility (BSRF)[4]. The experiment arrangements (slits, ion chambers, three-dimensional sample scanning system, Si (Li) detection system, collimation system and microscope) have been previously described [5]. It is routinely used to perform 3-D scan on Sancai' body-sections. Each X-spectrum acquisition took 200 seconds. Approximately area 20×5000 μm^2 was measured for the element determination on each sample. Under these conditions, the concentrations of 15 elements (K, Ca, Ti, Cr, Mn, Fe, Ga, Cu, Zn, Rb, Sr, Y, Zr, Nb and Pb) were analyzed.

2.3 STATISTICAL ANALYSIS

It is known that particular elements are good discriminators between pottery made of different sources while having little variation within a single type [6]. The samples T23 to 42 were all excavated from the Huangbu Kiln. They also have similar glaze and body colors. Therefore, those samples are considered as a single type of Tang Sancai, whose chemical composition should obey the normal distribution. Normal distributions are symmetric with scores more concentrated in the middle than in the tails. Many kinds of behavioral data are approximated well by the normal distribution. However, because of the influence of matrix effect on the analytical data when using SRXRF, the analytical data were checked for normality [7].

In the normality test, the significance was calculated by the Kolmogorov-Smirnov test with Lilliefors significance correction. The Kolmogorov-Smirnov statistic with a Lilliefors significance correction for testing normality is produced with the normal plot and probability plots. This is a test of normality based on the absolute value of the maximum difference between the observed cumulative distribution and that expected based on the assumption of normality. If the significance level is greater than 0.05, then normality is assumed. Table 1 shows that the significance of K, Ti, Cr, Mn, Fe, Zn, Rb, Sr, Y, Zr and Nb is far greater than 0.05. Then it was verified whether the data of those elements could be reasonably approximated by a normal distribution at the 95 % confidence level. The data of other elements including Ca, Cu, Ga and Pb are not in the normal distribution. The significance of Nb is close to 0.05, which implies that the data of Nb are not sufficiently accurate. The same accounts for Ca, Cu, Ga and Pb.

The data of K, Ti, Cr, Mn, Fe, Zn, Rb, Sr, Y and Zr were processed by the factor analysis. Factors 1 and 2 both account for 84.8% of the total variation, indicating that the two factors contain the majority of total variation. In Fig. 3 five colors of samples from the Xi'an kiln are shown. In this figure the data points of the red and grey body samples are almost located in the same field. The off-white body samples are far away from others in Fig. 3, indicating difference in their chemical compositions. Although only a few pink and white body samples are available for analysis, they still exhibited the chemical difference from the other types.

In Fig. 4 both factors 1 and 2 account for 80.4 % of the total variation, indicating the two factors contain the majority of total variation, too. The discrimination between the data points of the Huangbu and other kilns' Tang Sancai is apparently demonstrated, suggesting their chemical difference. However, it should be mentioned that the chemical difference between the samples from the Huangye kiln and the white or pink body from the Xi'an kiln is not evident.

TABLE III-4-1□ Tests of normality for elements contents of Tang Sancai from the Huangye Kiln (T23 ~ 42, 20 samples).

Elements	Sig.
K	0.200
Ca	0.001
Ti	0.200
Cr	0.200
Mn	0.200
Fe	0.200
Cu	0.001
Zn	0.200
Ga	0.028
Rb	0.200
Sr	0.200
Y	0.188
Zr	0.200
Nb	0.090
Pb	0.000

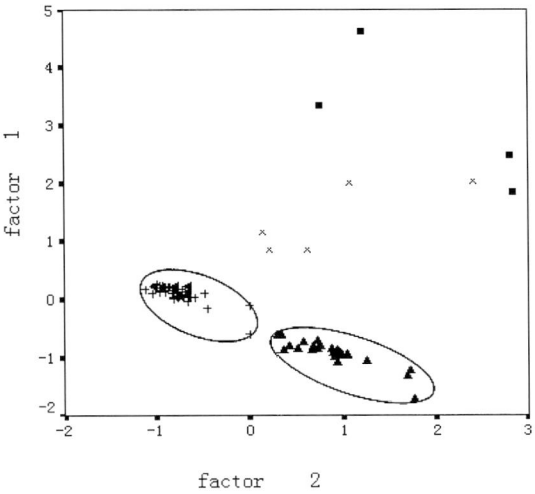

Figure III-4-3. Factor Analysis for Tang Sancai from the Xi'an Kiln.
Color of body: ✈ Red; ✿ White; ✠ Off-white; ☾ Gray; ✿ Pink

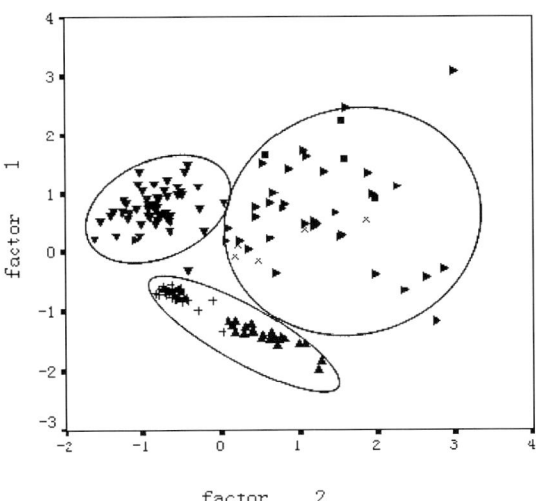

Figure III-4-4. Factor Analysis for Tang Sancai from the Xi'an, Huangbu and Huangye Kilns.
Xi'an Kiln: Color of body: ✈ Red; ✿ Pink; ✠ Off-white; ☾ Gray; ✿ White
✿ Huangbu Kiln; ☾ Huangye Kiln

Conclusion

Most of Tang Sancai from different kilns proved to be chemically separable. The separation is clear with the application of multivariate analysis mainly based on the content of ten elements, i.e. K, Ti, Cr, Mn, Fe, Zn, Rb, Sr, Y and Zr. Archaeologically, this discrimination means that those kilns exhibit differences in their raw materials and manufacturing technology.

The body-colors of red, grey and off-white Tang Sancai from the Xi'an kiln are much different. Tang Sancai with red and grey body from the Xi'an kiln has similar composition. It then suggests same raw materials but different manufacturing technology between the two types of Tang Sancai.

Some specimens from the Xi'an kiln and Huangye kiln have similar compositions. It implies that same raw materials were used for pottery production and the technical exchange with each other occurred during that period.

Acknowledgement

The authors would like to appreciate the great assistance from their colleagues: Huang Yu-Ying and He Wei (Institute of High Energy Physics, Beijing, China). This work is supported by the Natural Science foundation of China (NSFC 10075060&10135050), the Intellectual Creative Engineering of Academy of Sciences (KJCX-N04) and the Foundation of BSRF.

References

1. Yu Fuwei and Zhang Jian.(1994) *Introduction about Tang Sancai*, Cultural Relics of Central China(in Chinese) 1, 61-64.

2. Yang Yubin (1985) Tang Sancai Kilns in Gong county, *Archaeology in Henan* (in Chinese) **10**, 404

3. Shaanxi Archaeological Institute (1992) *Huangbo Kiln of Tang Dynasty* (in Chinese). Beijing, Culture Relics Press, pp. 25

4. Huang Yuying, Zhao Limin and Wang zhouguang(1999) Synchrotron Radiation XRF Microprobe Study of Human Bone Tumor Slice, *International Journal of PIXE*, Vol. **9**, Nos. 3&4,: 175

5. Y. Y. Wu, Z. Y. Chao, Y. A. Xiao, J. X., Pan and E. S. Tang. *Nucl. Instrum. Methods* A. **359**, 1995: 291-294

6. Prudence M. Rice(1987) *Pottery Analysis*, pp. 420

7. Lu Wendai (2000) *SPSS for Windows* (in Chinese). Being: Publishing House of Electronics Industry, pp. 158

Chapter III-5

Study of Chemical Composition in Ancient Celadon of Yue Kiln

Dongyu Fan, Songlin Feng, Qing Xu
Institute of High Energy Physics, Chinese Academy of Sciences,
P.O. Box 918, Beijing, 100039, China
Fengsl@ihep.ac.cn

Keywords: celadon, Tang Dynasty, Southern Song Dynasty, SRXRF, provenance

Ancient celadon samples were produced from Late Tang Dynasty to Southern Song Dynasty (850 ~ 1279 A.D.), which were excavated in the Yue Kiln, located in the Zhejiang province of China. The elemental concentrations in body of celadon samples were determined by synchrotron radiation X-ray fluorescence (SRXRF). The experimental data were analyzed by clustering analysis. The statistic analysis results indicated that the elemental compositions of most celadon samples of Yue Kiln revealed provenance and ages characteristics. Furthermore, the elemental distribution of cross-section between body and glaze in one sample of Five Dynasty was analyzed. The concentrations of Ca and Mn in glaze are higher than those in body, but K, Ti and Fe are lower in glaze. Almost all of elements moved from high concentration region to lower area.

1. Introduction

Chinese celadon originated from the Yue Kiln. The Shang-Lin Lake site of the Zhejiang province is the earliest original and principal one of Yue Kiln and the celadon porcelain. It consisted of more than one hundred kilns. Si-Long-Kou kiln is one of the Shang-Lin Lake kilns sites. The celadon was manufactured and fired in Shang-Lin Lake kiln sites from late Eastern Han Dynasty to Southern Song Dynasty (200~1200A.D.), and was used frequently in the Tang Dynasty and the Five Dynasties (907~960 A.D.). The Institute of Cultural Relic and Archaeology of Zhejiang Province and Department of Archaeology of Beijing University excavated the Si-Long-Kou Yue Kiln site in 1998. A lot of wares and fragments produced from Late Tang Dynasty (850 ~ 907 A.D.) to Fore Southern Song Dynasty (1127 ~ 1279 A.D.) had been excavated. In order to study the provenance and ages character of chemical composition in ancient celadon of Si-Long-Kou Yue Kiln, the samples of this site are collected and analyzed by SRXRF. He-Hua-Xin kiln (about 5 km away from Si-Long-Kou kiln) is another one of the Shang-lin Lake kilns sites. The

M. Uda et al. (eds.), X-rays for Archaeology, 223–227.
© 2005 *Springer. Printed in the Netherlands.*

celadon samples of this site are used to compare the provenance characteristic of elemental composition.

SRXRF can be used for the compositional analysis of ceramics. Synchrotron radiation (SR) was emitted from turning high-energy electrons in Synchrotron Radiation Facility and can be used to excite the specimen and to produce fluorescence. The SR's illuminating power possesses high strength and greatly increases the sensitivity of element analysis. Moreover, this method is suitable to the simultaneous analysis of multiple elements, and is wide in the coverage of concentration analysis. Therefore, it is a very suitable technique for the compositional analysis and study of ancient porcelain, especially for the valuable whole vessel of ancient porcelain.

2. Experimental

Sixteen pieces of Yue Kiln celadon samples (14 pieces collected from Si-Long-Kou kiln and other 2 pieces gathered from He-Hua-Xin kiln) were cut into cubic shape (15mmx30mmx5mm). They were cleaned in deionized water and ultrasonic for 3 times more, and then dried in an oven under 105°C for 5 hours. The national certified reference material (GSR-1) of 500mg was shaped as a disc of 10mm in diameter and thickness of about 1mm. The SRXRF measurements were carried out in a BEPC-3W1A beam line of the Beijing Synchrotron Radiation Facility (BSRF), National Laboratory of China. The X-ray source with 5×10^{13} photons/sec/0.1%B.W was supplied from the Beijing Electron Positron Collider (BEPC). The accelerated electron energy was 2.2 Gev. The beam current was 60mA~107mA. The sample was set at an angle of 45 degrees relative to the incident light beam. A Si(Li) detector was used to get X-ray spectra emitted from the samples. The SR light spot size was 10μm×20μm. All the X-ray spectra were computed by the AXIL software. The concentrations of these elements (K, Ca, Ti, Cr, Mn, Fe, Ni, Cu, Zn, Ga, Rb, Sr, Y, Zr and Pb) in the body of ancient celadon samples were semi-quantitatively analyzed.

3 Results and Discussion

3.1 DISTRIBUTION PATTERN OF THE CONTENT OF ELEMENT

The elemental distribution in transition layer from the glaze to the body was determined in one of celadon samples produced in Five Dynasties and the Si-Long-Kou kiln. The detected distribution curve is shown in Figure 1. It exhibits the elemental concentration variation from the glaze to the body. The concentrations of major elements K, Ca, Ti, Mn and Fe varied in the transition layer with about 50μm. The concentrations of Ca and Mn decreased from the glaze to the body. The decreasing distance is about 150μm. Ca could enter the body deeper than Mn. It indicates that the calcium compounds such as CaO and $CaCO_3$ might become melting state and then penetrated into body during the firing process. The concentrations of K, Fe and Ti decrease in the interlayer from glaze to body. These results indicated that several elements could infiltrate into the lower concentration region from the high concentration area.

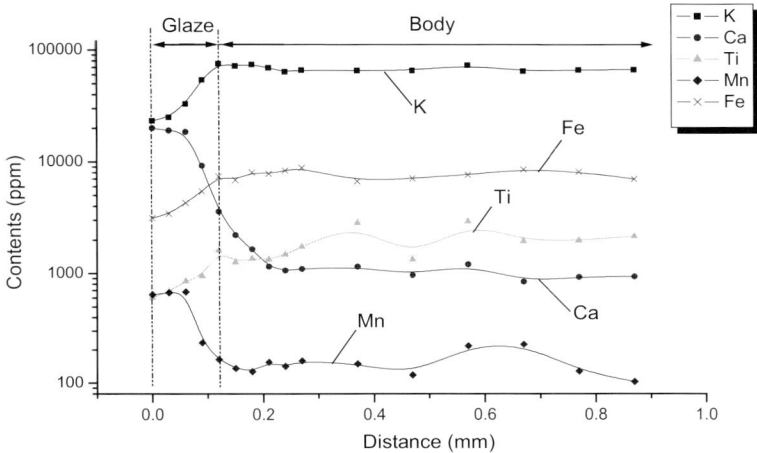

Figure III-5-1. The distribution of element concentration in one celadon sample of Yue Kiln.

3.2 CLUSTERING ANALYSIS

The chemical composition of the celadon consists of major (Si, Al, K, Ca, Fe, Mn, Mg and Na) and trace elements (Cr, Ni, Cu, Zn, Ga, Rb, Sr, Y, Zr, Pb and so on). Fifteen elements in the body of ancient celadon samples were determined by SRXRF. It was difficult to draw some guiding principles from direct analytical results. Therefore, statistic analysis of the experimental data of all the elements was performed, except for Zr for it distributes inhomogeneously in the celadon samples. The clustering analysis method is based on the algorithms that put similar objects into several clusters. It has great advantages that do not need mathematical mode and is suitable to deal with complex and multi-source data, so this method is used widely in archaeological studies. Cluster analysis was performed with the SPSS software package. The cluster method of Between-groups linkage and the Squared Euclidean distance interval was applied in this paper. The clustering chart of elemental concentration in body of celadon is shown in Figure 2. Numbers 1 and 2 denoted the samples of He-Hua-Xin kiln in Late Tang Dynasty. Other samples were collected in Si-Long-Kou kiln.

Sixteen samples are split into 3 big clusters when the interval distance is set at 13. The first cluster is related to both general and refined samples of Fore Northern Song Dynasty (960 ~ 1022 A.D.) The second cluster is composed of the samples of Late Tang Dynasty (in Si-Long-Kou and He-Hua-Xin Site) and Five Dynasties, whose corresponding interval is from 850 A.D. to 960A.D. The third cluster is consisted of the ones of Mid Northern Song (1023 ~ 1077 A.D.) and Southern Song Dynasty (1127~ 1279 A.D.). The samples are distributed into 6 small clusters, as the interval distance is set around 8. The first small cluster is consisted of both refined samples (with mark 8 and 9) of the Fore Northern Song Dynasty. The second is composed of three general samples (with mark 10, 11 and 12) of Fore Northern Song Dynasty. The specimens (with mark 1 and 2) of He-Hua-Xin kiln form the third cluster. The fourth contains the samples of Late Tang Dynasty (with mark 3, 4 and 5) and Five Dynasties (with mark 6 and 7). The fifth and sixth cluster are respectively related to the samples (with mark 13 and 14) of mid Northern Song dynasty and (with mark 15 and 16) Southern Song Dynasty.

Using this clustering analysis, the provenance and age properties of Yue Kiln ancient celadon clearly appear. The samples of different cultural period and provenance were distinguished obviously according to the respective species of the sample's historical period.

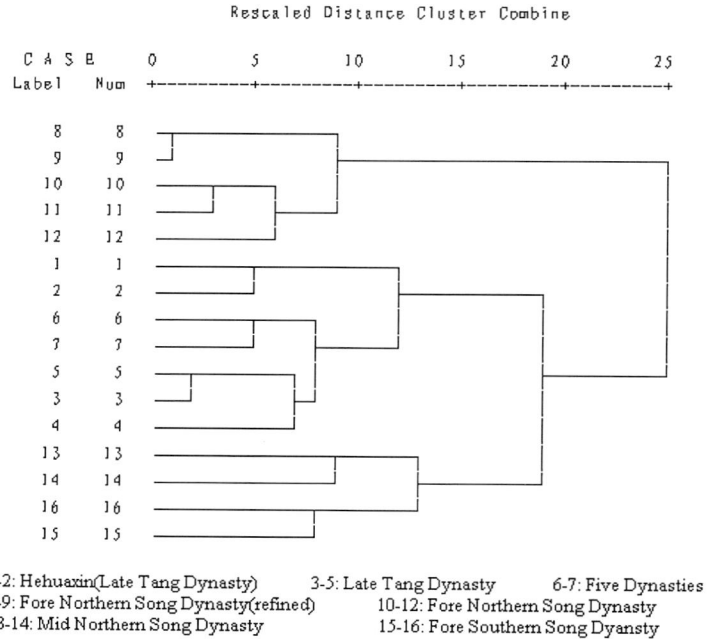

1-2: Hehuaxin(Late Tang Dynasty)   3-5: Late Tang Dynasty   6-7: Five Dynasties
8-9: Fore Northern Song Dynasty(refined)   10-12: Fore Northern Song Dynasty
13-14: Mid Northern Song Dynasty   15-16: Fore Southern Song Dyansty

Figure III-5-2. Dendrogram of cluster analysis of body.

4. Conclusions

The elemental distributions from the glaze to body are obviously different. The main elements such as K, Ca, Ti, Mn and Fe could infiltrate into the lower concentration region from the high concentration area. According to the clustering analysis, the ancient celadon porcelain of Si-Long-Kou and He-Hua-Xin in the Late Tang Dynasty exhibits different provenance characteristics. Most of the samples of the Si-Long-Kou kiln are clustered into a group of their ages. The age characteristics are also represented obviously. Furthermore, it is necessary to analyze more samples to validate these primary results.

Acknowledgments

Financial support was provided by the Chinese Academy of Sciences (KJCX-N04), Natural Science Foundation of China (10075060, 10135050), BSRF and LNAT. The authors thank Prof. Kuishan Quan (Department of Archaeology, Peking University) and Prof. Yueming Shen (Institute of Cultural Relic and Archaeology of Zhejiang Province) who offered the celadon samples, and Prof. Yuying Huang and Dr. Wei He of the Institute of High Energy Physics, CAS for their kind help in experimental work.

References

Feng Xianming, An Zhimin (1997) *Chinese Ceramic History*, Cultural Relics Publishers, Bejing.

J. Victor Owen (1997) Quantification of Early Worcester Porcelain Recipes and the Distinction between Dr Wall- and Flight-period wares, *Journal of Archaeological Science*, **24**, 301-310.

Lu WenDai (2000) *SPSS FOR WINDOWS Statistic Analysis*, Electronics Industry Publishers, Beijing.

Part IV: Radiography

Chapter IV-1

The Use of Medical Computed Tomography (CT) Imaging in the Study of Ceramic and Clay Archaeological Artifacts from the Ancient Near East

N. Applbaum and Y.H. Applbaum

Institute of Archaeology, The Hebrew University of Jerusalem, Jerusalem Israel
Institute of Radiology, Hadassah University Hospital, Jerusalem Israel
appelbau@h2.hum.huji.ac.il

Keywords: computed tomography, CT, radiographic technique, X-rays, archaeology, ceramics, clay artifacts, URIII period, envelopes, curved planes, non-destructive, ancient near east, Pottery Neolithic, Shaar Hagolan Culture, figurines, ceramic technology

Abstract

Computed Tomography Imaging (CT) is highly regarded as an efficient and relatively inexpensive medical diagnostic tool. It has not, however, come into its own in the study of clay and ceramic archaeological artifacts. Our studies demonstrate, however, that Medical CT is, in fact, a singularly powerful and efficient tool for in-depth radiological studies and analysis of a wide, range of archaeological finds. As the images obtained by the CT scans are digital, we have been able to manipulate them in many ways, thereby revealing new dimensions to non-destructive X-radiological studies of archaeological finds. By adapting various image post-processing techniques, developed for the CT as a medical diagnostic tool to our specific needs, we have been able to reduce research and development costs.

1. Introduction

In this paper we present the results of imaging studies we have conducted in Israel of clay and ceramic archaeological artifacts using Medical X-Ray Computed Tomography (CT).

Medical X-Ray Computed Tomography, commonly known in the medical profession as "CT", is a non-destructive radiographic technique. It is widely used by physicians as a diagnostic tool, as it is superior in many aspects to the imaging produced in a conventional radiograph.

M. Uda et al. (eds.), X-rays for Archaeology, 231–245.
© 2005 *Springer. Printed in the Netherlands.*

We will describe the techniques we used and demonstrate how we adapted them for use on clay and ceramic archaeological material. Emphasis will be placed on the advantages that these techniques have over other radiological imaging techniques currently in use by archaeologists.

From a wide variety of studies that we have successfully completed, we chose two examples for this paper:

A. We present the results from our testing of UR III period tablets that had been sealed in clay envelopes. We will demonstrate how we used CT imaging, a totally non-destructive research tool, to view and actually read the inscriptions on the inner tablet without tampering with the outer envelope. Furthermore, we will show that the data we collected from these scans allowed us to understand the actual techniques used by the ancient scribes in forming these envelopes.

B. We also present results from our scanning of a figurine from the Pottery Neolithic period (sixth millennium BC) at the Shaar Hagolan site in the Jordan Valley of Israel. This site has revealed the earliest finds and largest collections of figurines formed from clay material in this region of the world. We will show how these scans revealed to us the ceramic technology the craftsmen used in its production.

2. Background

Computed Tomography (CT) is a popular non-destructive radiological technique. It was developed for, and is primarily used as a diagnostic tool in the field of medicine. CT scanners can be found in every modern medical facility in the world. More recently, however, CT was found to be an indispensable tool in other areas, e.g. industry, where complex pieces of machinery and even pipes are scanned.

Computed Tomography (CT) is proving to be a very practical diagnostic tool for archaeological studies:

1.   The testing process is fast and non-destructive.
2.   The data collected is digital and can be stored for future post-scanning processing.
3.   The stored digital images can be printed, in fine detail, on film or paper.
4.   A wide variety of post-scanning computer applications have been developed for the medical radiological community. We have succeeded in adapting some of these applications for use in our studies of archaeological material. Not having to develop new computer applications has drastically cut research costs.
5.   The availability of CTs in almost every modern community makes it an ideal tool for archaeologists and conservationists. Research projects are invariably delayed by bureaucratic red tape when the shipping of artifacts to distant laboratories for testing requires official permission.

Some archaeologist and museum conservationists have already recognized the potential of CT. Projects have been published demonstrating the use of CT in the study of human and animal bone material (Anderson 1995, Davis 1997), Mummies

(Notman 1986; Pahl 1986) and plastered skulls (Hershkovitz et, al. 1995).

Industrial and high energy CT has been shown to be of great potential in the study of metal artifacts (Mazansky 1993; Bossi 1990).

The successful use of CT as an efficient and very powerful non-destructive analytical tool for the study of clay and ceramic archaeological artifacts is reported in our numerous papers and monographs (Applbaum et al. 1994; 1995; 1998; 1999; 2001; 2003; Jansen et. al. 2001). Aside from our work, only a limited amount of other projects have been conducted and published to date. Furthermore, not always have sufficient results been obtained (Lang and Middleton 1997; Carr 1990; Vandiver et. al. 1991).

3. The Technique

In CT, "*Figure 1*", an X-Ray tube and a series of electronic detectors rotate around an object. The X-rays pass tangentially through the scanned object and are measured by detectors. A digital image is formed from the collected data. Consecutive "slices" through the object form a series of images that provide a full picture of the object in cross section. The advantage of CT is that the overlapping parts of the object do not obscure the image under study, as they do in standard X-rays. Furthermore, because the data is acquired as a volumetric data set, it can be manipulated, not only to form images in different planes but also to generate 3-D images.

The scanning process is performed in two stages; followed by post-scanning processing and interpretation of the data collected.

First, a preliminary scan called a "surview" or "scout scan" is preformed. In this scan a digital radiograph of the object is taken. This image is a 2 dimensional representation of a 3 dimensional object; similar to that which is obtained by conventional radiography. The contrast and clarity of these digital radiographs, however, are of much higher quality than of those obtained with conventional radiography. These preliminary images have, time and again, proven to be invaluable in our study of these artifacts.

The "surview" helps us in planning the CT study. Our first step is to mark, on the surview, the exact area or areas we want to scan. Only then do we decide upon the exact distance between, and the thickness of the slices. Furthermore, based upon our preliminary observations of these 2 dimensional radiographs, we can, if needed, concentrate or focus our "second stage" scanning on specific areas.

The object is then passed through the scanner with "slices" being produced, as predetermined by our study of the surview. In this particular study we scanned thin slices using a high-resolution technique to attain fine detail and good contrast. Our CT scanning process allows us to observe both qualitative and quantitative differences in the densities of the material from which the scanned artifact was made.

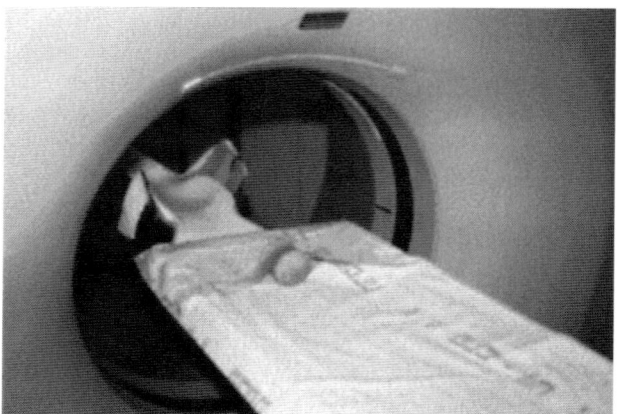

"Figure IV-1-1", Medical Computed Tomography of an archaeological artifact.

To achieve even finer detail of the artifacts under study, we found that, by using an "edge enhancing" algorithm or filter and viewing the artifacts with a wide graphic "window", we were able to achieve superior results.

In the post-scanning stage we viewed the images of the scanned slices on an on a Proview (Algotec, Raanana, Israel) high-resolution screen. We also selected images of scanned slices for printing on film.

Software applications as such allow us to further broaden our understanding of the artifact, in the "post-scanning" stages of our study. They make it possible to examine and analyze the images, obtained during the scanning phase, in depth and with great accuracy. Among the most useful applications we use, are the various filters that facilitate to enhance the scanned images. These filters allow us to "soften" or "harden" the contrast of the images. This often helps us to view voids and inclusions in the mass of the clay or ceramic material that we might have otherwise overlooked.

Graphic tools are used to quantitatively measure different features of the image, such as the length, width or thickness of specific features. We can also measure the comparative density at various points of the material. These comparative studies, helps us determine possible connections between different artifacts as well as to compare variations in quality.

4. Envelopes *"Figures 2, 3"*

At various archaeological excavations, in Mesopotamia and Turkey, 1000's of sealed "Envelopes" have been discovered. These envelopes are clay or ceramic containers within which cuneiform tablets were inserted and sealed for safe keeping.

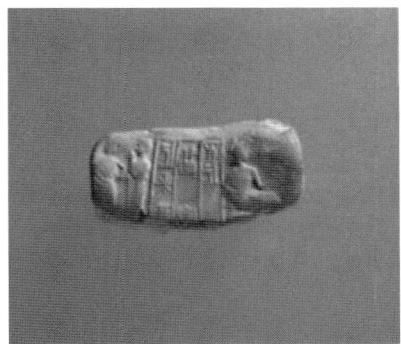

"Figure IV-1-2", Envelope - note cuneiform   *"Figure IV-1-3"*, Envelope – note seal on
inscription and seal.   side of envelope.

The tablets that we have been working on are, strange to say, economic documents in the Sumerian language. To be exact, they are legal receipts from the Ur III period dated to the 21st century BCE. The kings of this dynasty, whose capital is traditionally identified as the Ur of the Chaldees mentioned in the first book of the Bible, Genesis, developed a royal nationwide bureaucracy. Our knowledge of that dynasty is based, in the main, on information gained from the study of these clay tablets are literally. There are thousands, if not more of these Ur III documents in existence. They are found in public and private collections all over the world.

The envelopes we have studied are from the collection of the Institute of Archaeology at the Hebrew University of Jerusalem. A private donor with no information available as to its provenance donated them to the collection.

In antiquity, in the URIII period, it was common practice among scribes to inscribe documents on clay tablets using the cuneiform writing system. Most of these tablets were legal documents. To guarantee the authenticity of the document and discourage possible tampering and/or forgery of the said document, the scribe sealed it in a clay container, thus the name "Envelope" (Chiera 1938:69-74).

Upon sealing the tablet in the envelope, the scribe would write a summary of the text on the envelope (Hallo and Weisberg 1992: 52). Often, the exterior of the envelope bore the seals of witnesses (using cylinder seals) as well as the identity of its owner. Some of these seals contain glyptic designs. The seals themselves are of unique artistic value (Leinwand 1992).

The seals and text on the envelope served as a security system to protect the inner tablet from being tampered with, after it was sealed in its envelope (Collon 1987: 113-114). If the matter was later contested, the envelope was produced in court; the witnesses would be called upon to identify their personal seals, and then the envelope would be broken open. This ancient security system had proven to be fool-proof.

The current state of the art in reading Ur III envelope texts is to copy or photograph the text and seals on the exterior of the envelope. The envelope is then breached so as to "get at" the text inside. Thus, the study of these documents presupposed the physical destruction of the envelope, an important part of the artifact. This destruction is irreversible. The clay envelope, the seals and texts are invariably damaged in the process of opening them *"Figure 4"*.

"Figure IV-1-4",   Envelope after being opened – note irreversible damage.

The aim of our research was to seek a non-destructive technique that would allow us to read the inner tablet without damaging the exterior envelope.

It was obvious from the very beginning that conventional radiology would not be of much use to us. Nevertheless, we tried conventional X-rays. The X-rays revealed, only, that there was a tablet concealed inside the clay envelope *"Figure 5"*.

They did not, and could not reveal that the tablet was inscribed. Our findings reflect the major limitation of conventional X-rays where all structures through which the rays pass are superimposed on the image produced, making it difficult, if not impossible, to distinguish particular features

We also experimented unsuccessfully with conventional tomography. Conventional tomography is a technique by which the shadows of superimposed structures of an X-ray can be "blurred", thereby highlighting the structure to be diagnosed. This allows the diagnostician to see more clearly the structure with which he is presently concerned. The sharpness of the image, however, cannot be enhanced. Conventional tomography, therefore, could not offer us a clear view of the text inscribed on the tablet inside the sealed envelope.

Medical Computed Tomography was found to be the best, and perhaps the only tool available, as of now, that affords us the opportunity of being able to read the text inscribed on the tablet while sealed inside the clay envelope. Whereas, all previous imaging modalities in Medicine, such as Radiography, are simple geometric projections of the object onto detectors (film), computed Tomography produces results vastly superior to that of other systems. To begin with, its X-rays pass through the subject at different angles, eliminating superimposition and enabling individual structures to be viewed in insulation. CT also detects subtle differences in the density of the structures observed and presents them on the computer screen in graduated shades from white, for dense substances such as bone or metal, through a succession of grays to black. Furthermore, CT uses multiple planar projections. These planar projections are mathematically recombined by a digital computer to provide a cross sectional image of the object.

"Figure IV-1-5",   Conventional X-ray radiograph of envelope.

In this study we used a high resolution 3rd generation CT scanner - an Elscint 2400 elite scanner. The envelope was placed with its flat side parallel to the scan plane. The scanner examined the envelope in a series of lateral overlapping "slices".

The object was scanned with an image matrix of 512 x 512 with thin, 1.2 mm slices.

An algorithm, developed for the diagnosis of fine bone detail, was used to reconstruct the raw data. To maximize our imaging parameters, to get the best possible image, we sacrificed contrast, radiation dose and signal to noise ratio. This gave us higher resolution and sharper edges. Because the clay of the tablet is dense and air fills the voids created by the impressions, which are the cuneiform text inscribed on the inner tablet, we were able to achieve the highest possible contrast "Figure 6".

In the scans of the lateral sections, clear high contrast images of most of the inner tablet's face were obtained. It is possible to see and even read most of the cuneiform text. One of the problems we faced while testing the envelope, however, was when we found that the writing surface of the tablet sealed inside the envelope was convex. This resulted in many of the cuneiform signs on the surface of the inner tablet being "cut" by the "slicing" of the scanner. We were only able to view limited sections of the text, inscribed on the surface of the convex inner tablet, in each section obtained in the scanning process.

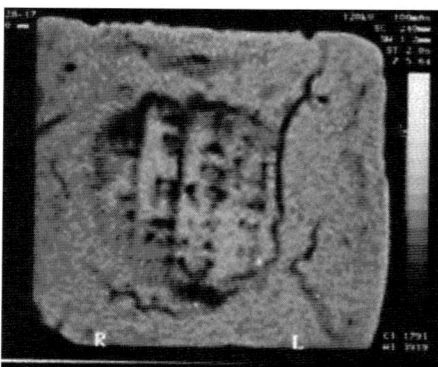

"Figure IV-1-6", CT section of envelope – note lines of cuneiform writing

In order to obtain an imaging of the whole face of the inner tablet with all the Cuneiform signs visible, we utilized post-scanning data processing. Using all the digital data collected and stored during the scanning stage we were able, with the aid of a variety of computer software programs, to reconstruct or reformat the surface of the inner tablet. To be able to image the full text from the reformatted surface, we used an imaging process technique known as "curved planes" multiplanar reformatting (Newton and Potts 1981).

The successful reformatting of curved planes assumes that the lateral scanned slices of the tablet are properly stacked one on top of the other, with minimal movement or change in other imaging parameters. It is then possible to select voxels in planes other than the ones in which they were originally obtained. This technique allows us o obtain images in orthogonal planes or non-orthogonal (oblique) or even curved planes. It is then possible to select voxels in planes other than the ones in which they were originally obtained. This technique makes it possible to obtain images in orthogonal planes or non-orthogonal (oblique) or even curved planes. In general, using curved planes requires a great deal of meticulous planning and great deal of time-consuming effort as the contour of the object must be carefully traced on multiple levels in order to be able to bring out as much of the information as possible. This is especially true with some of the tablets whose surface appears to be uneven.

We are of the opinion that, as far as our scanning procedures are concerned, we have attained more than adequate results. The fact is that we are now able, by reformatting in curved planes, to decipher the inscription on the inner tablet's surface without affecting the outer envelope in any way, whatsoever. "Figure 7".

In addition to the imaging of the inner text, we found that CT also supplies us with valuable information as to the technology used by the ancient scribes in forming these envelopes (Collon 1987:114).

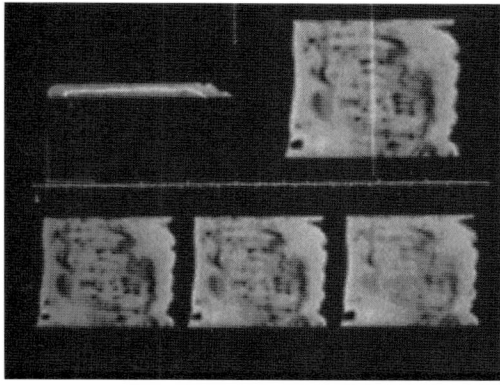

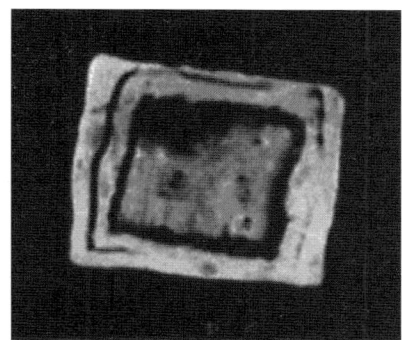

"Figure IV-1-7", Envelope - "curved planes" multiplanar reformatting.

"Figure IV-1-8", CT section of envelope – note the layers formed while sealing the envelope.

We are able to view and measure the thickness or thinness of the envelope in each of the sections. We have determined that in some cases the scribe wrapped the inner tablet in two layers of clay that formed the envelope "Figure 8". The corners were pinched, pressed and squared off. All signs from the forming process of the envelope were then smoothed over and concealed by the scribe. The final product was a square shaped tablet with a smooth exterior, upon which the scribe could write the summary of the internal document and upon which the witnesses could stamp their seals. In the future we plan to publish, with the help of Dr. Wyane Horowitz of the Hebrew University of Jerusalem Israel, the full decipherment of the cuneiform texts from the envelopes we studied.

5. Anthropomorphic Figurine, from the Pottery Neolithic Period Site of Shaar Hagolan *"Figures 9, 10"*.

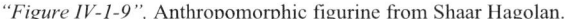

"Figure IV-1-9", Anthropomorphic figurine from Shaar Hagolan.

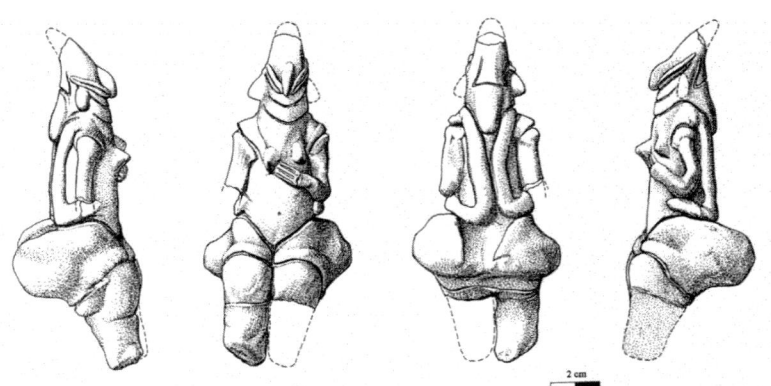

"Figure IV-1-10", Anthropomorphic figurine from Shaar Hagolan – technical drawing.

The site of Shaar Hagolan , in the northern Jordan Valley in Israel, is where the Pottery Neolithic Yarmukian culture had been first identified (Stekelis 1951, 1972; Garfinkel 1993; Kafafi 1993). The largest assemblage of prehistoric art ever excavated in Israel has been uncovered at this site. Previous studies of similar Yarmukian anthropomorphic figurines had concluded that a standard core technique was employed in their forming (Garfinkel 1995:30-31; Mozel 1975; Yeivin and Mozel 1977:196). This conclusion was based solely on external examination of fragments from figurines previously excavated at this site and other Pottery Neolithic Yarmukian culture sites.

The Pottery Neolithic period in this region marks the beginning of all ceramic technology. Their use of only one, standard forming technique in this genesis period is to be expected. After all, as the people of this culture pioneered ceramic technology, we may assume that, at such an early stage of development, they had not yet achieved technological variability. On the other hand, one might expect to find at least some experimentation being conducted by these Shaar Hagolan people, even at their earliest learning stages of the new technology. Such experiments would have led to technological variability with a possible preference of one technique over the other.

In an earlier study we conducted, we used medical computed tomography to help in the conservation and restoration of two parts of a statue, from the site of Shaar Hagolan (Applbaum et al. 1998). There we found clear evidence that the core technique was used "Figure 11". But other figurines from this site that we tested at the time, for conservation and restoration purposes, showed possible signs of other ceramic techniques having been used. Based on this premise we decided to broaden the scope of our post-scanning analyses to include an in-depth study of the ceramic technological abilities of the people of the Shaar Hagolan culture.

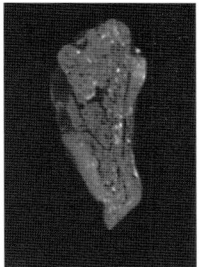

"Figure IV-1-11", CT section along the longitudinal axis – note the use of the core technique.

To further our thesis we present our scanning of a figurine (63X35X21 mm.) that was excavated at the Pottery Neolithic site of Shaar Hagolan during the 1997 season (Garfinkel 1999: 48). It is an almost complete anthropomorphic figurine, having only slight damage to the legs. It is a fine example of the technological and artistic abilities of the craftsman of the Yarmukiam culture.

The study of this object was conducted on a third generation CT scanner. The scanning techniques we used have been explained in our study of the "envelope", above. In this study we first conducted a "scout view" or surview, a two dimensional digital radiograph of the object "Figure 12". As mentioned above, the CT technicians

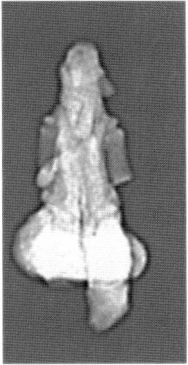

"Figure IV-1-12", Surview (scout view) of figurine

use this image in the planning of the sectional scanning that follows. We may note that in most cases the surview is not very helpful to us because it has limitations similar to those of conventional radiography, namely, two-dimensional projections of three-dimensional objects, causing a superimposition of impenetrable layers. This often leads to a misinterpretation of the surview image. In our study, however, the surview proved very helpful in determining that this figurine has a complex internal structure. The exact nature of this complexity could not be fully determined based on the surview alone. Using both the surview and the second-stage CT scanning, however, we were able to determine the exact technique used in the forming of this particular artifact. We conducted a full set of CT sections, scanning along the longitudinal sagittal plane of the figurine. We could find no sign of the core technique, discussed above, having been used in the modeling of this figurine. In our previous testing of figurines we had found that the core technique could be easily identified in CT scans. Air voids are captured between the core and the material modeled upon the core. This difference in density (air / clay) is clearly viewed in the results of CT scanning. In our current study, however, no core could be identified. Our density studies showed voids of air that were captured in areas where pieces of clay were added to the exterior of the figurine in the final phase of its production.

We also observed an abnormality in a CT section passing through the center of the figurine along its coronal plane "Figure 13". In order to better understand the nature of this abnormality, we decided that an axial sectional view of this area of the figurine was called for.

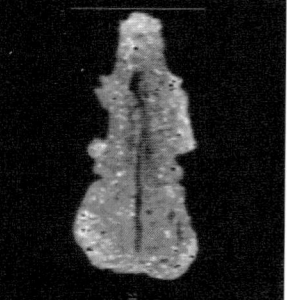

"Figure IV-1-13", CT section along the coronal plane of figurine – note abnormal "fuzziness" along center of image

Having conducted a full scanning series of overlapping thin slices along the longitudinal plane, we already possessed a full data set. We processed this data on an Omnipro Workstation. With using multiplanar reformatting, we were able to reconstruct a view of the axial plane section for any area of the object, without having to rescan the figurine. In the reconstructed axial section we were able to determine that there was a void that caused fuzziness along the coronal plane of the figurine "Figure 14". This void and the low density lines we observed in the axial section were caused by air trapped between the folds of clay at the time when the clay slab was folded and pasted.

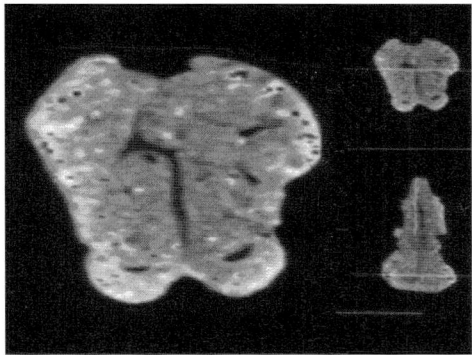

"Figure IV-1-14", Using multiplanar reformatting – reconstruction of the axial plane section.

The exact nature of this void, that caused the fuzziness, could not be understood until it was viewed in the adjoining axial section. In the axial section the void clearly defines the modeling technique used to form this figurine from the Neolithic period. Instead of forming the figurine on a core, the craftsman formed the body of the figurine from a single slab of clay. The slab was folded in half along its longitudinal axis, to form the main body of the figurine. The fold is, therefore, contained inside the mass of the body of the figurine. Buttocks and eyes and other refinements were then pasted on the body. All signs of the voids caused by the folding of the slab of clay that forms the body of the figurine and all refinements that were pasted on the body were then smoothed over and concealed by the craftsman. They cannot be observed in any way by an external viewing of the object. We were only able to identify the techniques used in the forming of this figurine after an exacting study of the voids that were revealed to us by our CT reformatted sectional images.

6. Conclusions

We have demonstrated, as shown in this paper, that standard medical Computed Tomography (CT) can be a very practical and powerful diagnostic tool for the study of ceramic and clay archaeological artifacts. The use of this non-destructive diagnostic tool in research, offers the archaeologist the opportunity to:

1. "See" inside artifacts without damaging them.
2. Determine what techniques were used in the production of ceramic and clay artifacts.
3. Conduct conservation of the most delicate artifacts with great confidence, making use of density scans that determine the parameters of the object.

This, however, tells only half the story. Cost and convenience considerations make the medical CT even more ideal for archaeological research of clay and ceramic artifacts, because of the following:

1. CT's are available, worldwide; they can be found in Medical centers, hospitals and even in private labs. There is no true need to invest in the

purchase of this very expensive equipment. Time can usually be rented on a CT. The cost of testing is, therefore, relatively inexpensive. This drastically reduces research and development expenses.

2.   There is no need to ship artifacts to distant labs for analysis. This eliminates shipping and insurance costs and a great deal of red tape.

3.   Once the artifact is properly scanned and the data stored there is no further need of and convenience of the researcher.

4.   The archaeologist can make use of a wide variety of computer applications that have been developed for the medical profession without incurring the expense of developing new ones.

This paper, limited to just two examples from our extensive work on ceramic and clay artifact, cannot cover the full scope of our research; nor does it explore the full potential of this system, as we now know it

This First International Symposium in Tokyo enabled us to share, with a broader scientific community, our experiences and successes in the testing of clay and ceramic artifacts using Medical CT. It is our hope, that the momentum achieved in this first meeting, will be followed by others, bringing us together time and again in the years to come.

7. References

Anderson, T. (1995) Analysis of Roman Cremation Vessels by Computerized Tomography, *Journal of Archaeological Science* **22**, 609-617.

Applbaum, N., Applbaum, Y.H., and Horowitz, W. (1994) Computed Tomography Imaging of Sealed Clay Cuneiform Tablets, *Imaging The Past - Electronic Imaging and Computer Graphics in Museums and Archaeology*, Abstracts, British Museum, p.18.

Applbaum N., Applbaum, Y.H., and Horowitz, W (1995) Imaging with the Aid of Computed Tomography of Sealed Documents from the Ur III Period, *Archaeology and Science Bulletin* **3**, 8-12. (Hebrew)

Applbaum N., Applbaum, Y., Lavi M., and Garfinkel, Y. (1998) Computed Tomography Imaging as an aid in the Preservation and Conservation of Neolithic Figurines from the Site of Sha'ar Hagolan, *Archaeology and Science Bulletin* **6**, 46-54 (Hebrew).

Applbaum, N. and Applbaum, Y. (1999) The Use of Medical Computed Tomography (CT) in the Study of Ceramic Archaeological Finds, in N. Messika, N. Lalkin and J. Breman (eds.), *Computer Applications in Archaeology*, Conference Proceedings, Tel Aviv, pp. 42-52 (Hebrew)

Applbaum, N., and Applbaum, Y.H. (In preparation) Medical Computed Tomography (CT) of Selected Pottery: Preliminary Results, in A.M. Maeir (ed.), *Bronze and Iron Age Tombs at Tel Gezer, Israel: Finds from the Excavations by Raymond-Charles Weill in 1914 and 1921.*

Bossi, R.H., (1990) Computer Tomography of Ancient Chinese Urns, *Materials Evaluation* **48**, 599-602.

Carr, C., (1990) Advances in Ceramic Radiology and Analysis: Applications and Potential, *Journal of Archaeological Science* **17**,13-34.

Chiera, E., (1938) *They Wrote on Clay*, Chicago ILL, University of Chicago Press.
Collon, D., (1987) *First Impression Cylinder Seal in the Ancient Near East,* London, British Museum Publication.

Davis, R., (1997) Clinical Radiography and Archaeo-human Remains, in J. Lang and A.Middleton (eds.), *Radiography of Cultural Material*, Oxford, Butterworth-Heinemann, pp. 117-135.
Garfinkel, Y., (1993) The Yarmukian Culture in Israel, *Paleorient* **19:1**, 115-134.

Garfinkel, Y., (1995) *Figurines and other Baked Clay Objects from Munhata*, (Les cahiers du centre de recherche francais de Jerusalem 8), Association Paléorient, Paris, Direction General des Relations Cultrelles Scientifique et Techniques du Ministere des Affaires Etranger.

Garfinkel, Y., (1999) *The Yarmukians - Neolithic Art from Sha'ar Hagolan*, Bible Lands Museum, Jerusalem.

Hallo, W.N., and Weisberg, D.B. (1992) A Guided Tour through Babylonian History: Cuneiform Inscriptions in the Cincinnati Museum, *The Journal of the Ancient Near Eastern Society* **21**, 49-90.

Hershkovitz, I., et. al. (1995) Remedy for an 8500 Year-old Plastered Human Skull from Kfar Hahoresh, Israel, *Journal of Archaeological Science* **22**, 799-788.

Jansen, R.J., Koens, H.F.W., Neeft, C.W., and Stoker, J. (2001) Scenes from the Past: CT in the Archaeologic Study of Ancient Greek Ceramics, *RadioGraphics* **21**, 315-321.
Kafafi, Z., (1993) The Yarmoukians in Jordan, *Paleorient* **19**, 101-114.

Lang, J., and Middleton, A. (1997) Radiography - Theory and Practice, in J.Lang and A.Middleton (eds.), *Radiography of Cultural Material*, Oxford, Butterworth-Heinemann, pp. 117-135.

Leinwand, N. (1992) Regional Characteristics in Styles and Iconography of the Seal Impressions of Level II at Kultepe, *The Journal of the Ancient Near Eastern Society* **21**, 141-172.

Mazansky, C. (1993) CT in the Study of Antiquities: Analysis of a basket-hilted Sword Relic from a 400-year-old Shipwreck, *Radiology* **186**, 55A-61A.

Mozel, E. (1975) A Special Human Figurine from the Pottery Neolithic in Israel, *Mitekufat Ha'even* **13**, 70-73 (Hebrew), 74 (English summary).
Newton, T.H., and Potts, D.G. (eds.) (1981) *Radiology of the skull and Brain: Technical aspects of Computed Tomography*, Vol. 5. ,CV Mosby ,St. Louis.

Notman, D.N.H. (1986) "Ancient Scanning: Computer Tomography of Ancient Mummies". in R.A. Davis (ed.), *Science in Egyptology*, Manchester University Press, Manchester, pp. 251-320.
Pahl, W.M. (1986) Possibilities, Limitations and Prospects of Computer Tomography as a Non-invasive Method of Mummy Studies, in R.A. Davis (ed.), *Science in Egyptology*, Manchester University Press, Manchester, pp. 13-24.

Stekelis, M. (1951) A New Neolithic Industry: The Yarmukian of Palestine, *Israel Exploration Journal* **1**, 1-19.

Stekelis, M. (1972) *The Yarmukian Culture of the Neolithic Period*, Magnes Press, Jerusalem.

Vandiver, P. et. al. (1991) New Applications of X-Radiographic Imaging Technologies for Archaeological Ceramics, *Archeomaterials* **5**, 185-207.

Yeivin, E., and Mozel, I. (1977) A Fossil Directeur Figurine of the Pottery Neolithic A, *Tel Aviv* **4**, 194-200

Chapter IV-2

The Radiographic Examinations of the "Guardian Statues" from the Tomb of Tutankhamen

Jiro Kondo

Department of Archaeology, Waseda University, Japan
1-24-1 Toyama, Shinjuku-ku Tokyo, 162-8644, Japan
jkondo@waseda.jp

Keywords: Tutankhamen, "Guardian Statues", Wooden Statues, Valley of the Kings, the Egyptian Museum in Cairo

Introduction

An X-ray examination of the wooden statues, JdE 60708 (Figs. 3 and 4) and JdE 60707 (Figs. 5 and 6), was undertaken by a Waseda University Team consisting of Sakuji Yoshimura, Jiro Kondo, and Nicholas Reeves, on 28-29 April and 2-3 May 1993, with the generous cooperation of Dr Muhammad Ibrahim Bakr, the Chairman of the E. A. O., Dr Muhammad Salah, Director of the Egyptian Museum, Dr Nasry Iskander, General Director of Conservation of E. A. O., Dr Kamal Barakat, Director of Conservation of the Egyptian Museum, Mme Soheir el-Sawi, and other members of the Egyptian Museum.

The Guardian Statues from the Royal Tomb of Tutankhamen

The royal tomb of Tutankhamen was discovered in November 1922, by Howard Carter on behalf of the fifth Earl of Carnarvon. The tomb is now numbered 62 (KV 62), and is located in the central part of the main Valley of the Kings. There were two life-sized wooden statues in the northern part of the antechamber of the tomb. They were standing facing each other and guarding the entrance to the burial chamber. They were carved in wood and 1.9 meters in height. The face, body, arms and legs were painted with black resin. The gilded bronze uraeus was attached to each statue's forehead and the eyes were inlaid with limestone and obsidian in the frames of gilded bronze.

Their headcloths, collars, and kilts were overlaid with gold on the base of linen and gesso. The two statues were not exactly the same. The most notable difference was their headcloths. The one on the west of the doorway (Carter's no. 29: JdE 60708 - Fig.1 and 2) was wearing the *khat* headcloth, while the one on the East

M. Uda et al. (eds.), X-rays for Archaeology, 247–251.

(Carter's no. 22: JdE 60707 - Fig. 3 and 4) was wearing the *nemes* headcloth. On the triangular kilt of each figure, the names and titles of Tutankhamen were inscribed. The figures wearing the *nemes* headcloth represented the royal *ka*, or spiritual double of the king.

The similar life-sized wooden statues were found from several royal tombs of the New Kingdom. In the British Museum, there are three such figures (EA 854, 882, and 883). The statues of EA 854 and EA 883 probably came from the tomb of Ramesses I (KV 16), and the statue of EA 882 seems to have come from the tomb of Ramesses II (KV 7). The kilt of EA 882 has a hollow which is 20 cm in depth. This hollow is considered to have been used for keeping a roll of papyrus. Similarly, hollows seemed to exist in the two guardian statues of Tutankhamen. They seemed to be sealed by pieces of stone plastered in position and gilded over. In a similar way four concealed niches were cut out into the decorated walls of the burial chamber of the tomb of Tutankhamen and sealed by the magic bricks. In order to clarify the role of the guardian statues as the hiding place of religious texts, we decided to perform an X-ray study of the two statues in the Egyptian Museum in Cairo.

The equipment used for the X-ray study

The equipment used for the X-ray study was called "High Technical System X-ray HK-100S", which was produced by Hitex co., Ltd. in Japan. Figure 1 shows the appearance and names of the system. The specifications are as follows:
(1) X-ray
   Input power: AC 240V 2A Single phase, Grounding 100Ω or less
   Output power: 100KV 3mA Intermittent
   X-ray indicator lamp: Provided (A signal-light flickers during X-ray radiation)
(2) X-ray tube bulb and high voltage transformer
   Type of X-ray tube: Fixed positive electrode type
   X-ray tube voltage: 0-100KV Intermittent
   X-ray tube current: 0- 3mA Intermittent
   Focus dimensions: 0.4x0.4mm
   Transformer circuit: All wave rectification type

System layout

We first installed the "cassette stage", "X-ray generator", "hand lift" and "control unit" as illustrated at Fig. 2. The cassette (stage) and the X-ray generator were placed parallel to each other. The control unit was placed at a location in the opposite direction from the direction of emission of X-rays. The cassette (film) and the irradiated object were placed as close to each other as possible. We adjusted the heights of cassette (film) and the X-ray radiation port to be horizontal with the handle of the handlift.

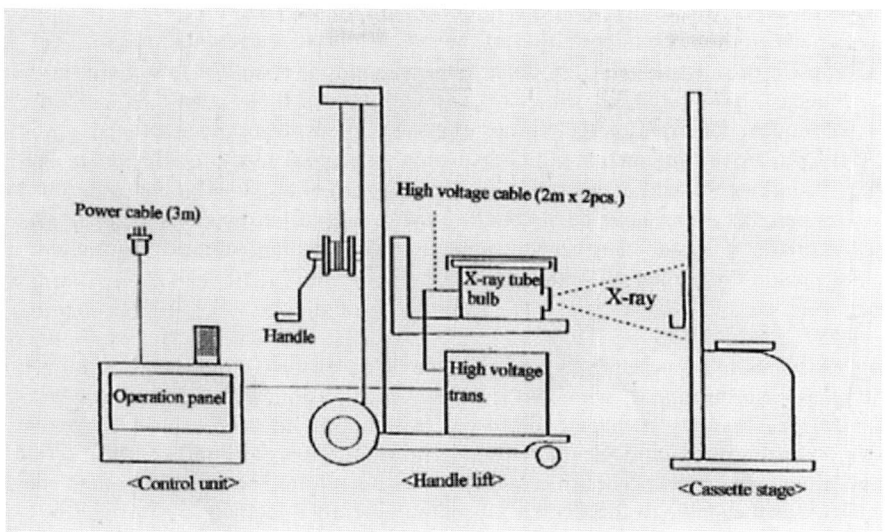

Figure IV-2-1. Appearance and names.

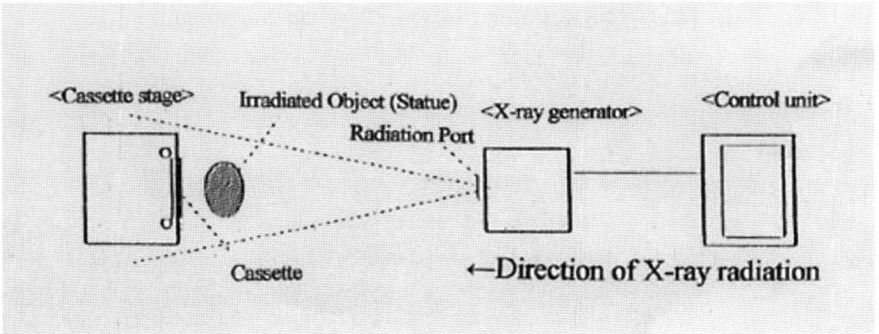

Figure IV-2-2. System Layout.

Background

This work was undertaken for the following three main objectives:

1) To check the present condition of the statues and to learn what kind of preservation work had been done to them, and to prepare for future conservation needs;
2) To reveal construction techniques employed by the ancient Egyptians;
3) To pursue the suggestion which had been put forward by C.N. Reeves in 1985 (Tutankhamen and his papyri, *Göttinger Miszellen* 88, 39-45), that religious text on papyrus might have been concealed within royal funerary on the basis of a similar statue in the British Museum and other smaller statuettes from other tombs in the Valley of the Kings.

A series of non-destructive x-rays was taken of the Tutankhamen statues with the glass cases removed, as shown in Figs. 3-6. It has been cleared that the wooden statues are seriously damaged inside. As was hoped, these revealed the extent of previous restoration, and yielded valuable information regarding the construction of the statues. However, no definite evidence has been found so far that statues contain papyrus, as the statue in the British Museum evidently once had.

Results

The results of this examination have satisfied two of the principal aims of the project. However, the problem of Tutankhamen's papyri had not been solved, and it is proposed that the research be continued along similar lines with the whole series of smaller gilded and resin-coated statuettes from the king's tomb. Such an investigation would provide additional information on the Tutankhamen treasures and on the skills and techniques of the ancient Egyptian craftsman.

References

Reeves, C. N.: Tutankhamen and his papyri, Göttinger Miszellen **88** (1985), 39-45.

Reeves, C. N.: *The Complete Tutankhamun*, London, 1991, 128-9.

Porter, B. and Moss, R. L. B.: *Topographical Bibliography of Ancient Egyptian Hieroglyphic Texts, Reliefs, and Paintings: I. The Theban Necropolis, Part 2. Royal Tombs and smaller Cemeteries*, Oxford, 1964, 507, 535.

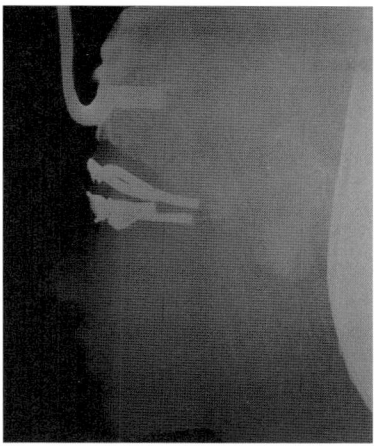

Figure IV-2-3.
Profile of the "Guardian Statue (JdE 60708)".

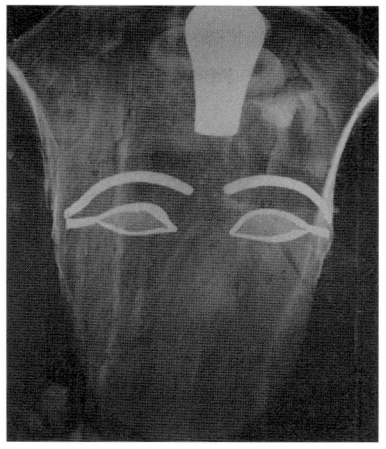

Figure IV-2-4.
Face (front view) of the Statue (JdE 60708).

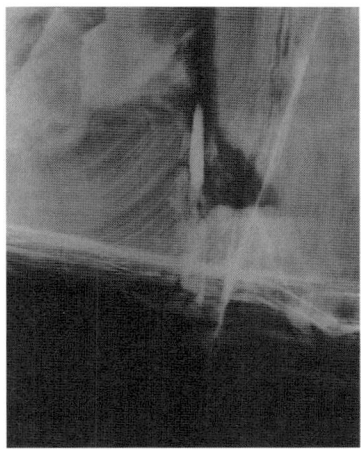

Figure IV-2-5.
Side-view of under-kilt of the Statue (JdE 60707).
*No space for the papyri.

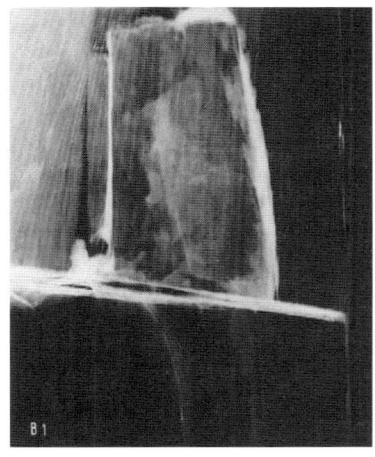

Figure IV-2-6.
Side-view of under-kilt of the Statue (JdE 60708).
No space for the papyri.

Chapter IV-3

Analytical Study of Paintings by X-ray Radiography and Spectroscopy

Kamba Nobuyuki

> *Tokyo National Museum, Conservation, 13-9 Ueno Park, Taitou-ku,*
> *Tokyo, 110-8712, Japan*
> kamba@tnm.jp

Keywords: paintings, x-ray radiography, secondary electron emission radiography (emissiography), paint cross section, scanning electron microscope, energy dispersive X-ray fluorescence analysis, painting techniques and materials, ground

1. Introduction

The various methods of scientific examination of paintings are divided into two classes, namely, surface examinations and point examinations. Surface examinations can in principle be carried out without even touching the paintings. Point examinations may or may not require the removal of a minute paint-sample. Selection of points likely to yield most information should be performed based on surface examinations.

Surface examinations are conducted by such methods as X-ray radiography, secondary electron emission radiography (emissiography), infrared reflectography (black & white, color), ultraviolet fluorescence photography, X-ray excitation, non-destructive X-ray fluorescence analysis etc. Point examinations usually require the removal of a small sample from the paintings, and this should only be accomplished after taking full advantage of surface examinations. The problem of point examinations is that the outcome does not tell anything about the surrounding area.

Therefore, best results are invariably obtained when various methods can be combined, complementing each other and allowing a better interpretation of data.

2. Surface examinations

2.1 X-RAY RADIOGRAPHY

X-ray radiography may be very useful in understanding the technique of the painting, especially in combination with paint cross-sections and infrared reflectograms. In comparing X-ray radiographs with infrared reflectograms and the painting itself, various modifications during the painting process may be elucidated.

M. Uda et al. (eds.), X-rays for Archaeology, 253–258.
© 2005 *Springer. Printed in the Netherlands.*

2.2 SECONDARY ELECTRON EMISSION RADIOGRAPHY (EMISSIOGRAPHY)

Radiography by secondary electron emission is used for the study of pigments. In the secondary electron emission image, pigments are shown in relative lightness or darkness depending on their atomic number. The larger an pigment's atomic number, the brighter the image will appear, the smaller the pigment's atomic number, the darker.

2.3 APPLICATION OF SURFACE EXAMINATION TO THE PAINTINGS OF CHRISTIAN RELICS

When Christianity was first introduced into Japan in the middle of the Sixteenth century by St. Francisco Xavier, it was rapidly propagated and converted an immense number of the people. This new religion soon became one of very important powers in our country, spiritually as well as socially, reflecting its influence in every direction of the national life. But after only half a century's flourishing, Christianity was almost wholly rooted out, in consequence of severe persecutions and the strict prohibition of the authorities, leaving scarcely any remarkable monuments and relics, those of the prosperous days of the past being securely hidden or destroyed.

These of severely damaged paintings were investigated by scientific examinations to recognize how their original figures are, and how their painting techniques and materials are related to or influenced by European paintings. It would be revealed that Seminary's painting instruction and production, which was a school to learn Catholicism through Latin language, music, and manufacturing musical instruments and paintings organized by Jesus missionary. Standing point of view, we have examined severely damaged and discolored copper plate paintings and relatively well preserved paintings on paper such as the Painting of the Madonna with the Infant Jesus and Her Fifteen Mysteries and the Portrait of St. Francisco Xavier (Fig. 1).

X-ray radiography (Fig. 2), emissiography (Fig. 3) and color infrared reflectography of the paintings have played a great role to estimate original figures and a pigment of each color. Infrared reflectography has shown existence of under drawing of the paintings on paper support, which are extremely sophisticated lines drawn by a brush. The evidence will suggest a skillful Japanese painter was engaged in manufacture of the painting. X-ray radiography shows the brushwork for the paint. The evidence implies that the painter was not necessary good at practicing of painting by a European method.

We have not enough evidence to recognize the influence of European painting techniques and materials at present. It will need effort to conduct scientific investigations to the Christian relics toward understanding more details of interaction between European and Japanese culture.

3.   Point examinations

3.1 POINT EXAMINATION IN TERMS OF ANALYSIS OF PAINT CROSS SECTION

The extremely small amount of samples taken form the paintings were first examined through an optical microscope to note the shape of the sample, its size, condition of front and back surfaces and its color. These characteristics were recorded in photographical form. Then the sample was in a sandwich of polyester resin, and after hardening the encased sample at 40 °C under double atmospheric pressure, the sample was split into two samples, using a microtome to cut the sample crosswise. One of these cut samples was then used for the chemical analysis, while the other was kept for reference or for use in the analysis of medium. The cut face of the sample for use in analysis was then successively polished using a polishing powder made up of aluminum oxide with particle diameter from 12 to 0.05 micrometer. This face could then be called a cross section of the paint and the ground, which could allow the painting process to be read through each successive layer.

First, the polished cross section was examined in reflected light through an optical microscope set at around 100x, to determine the color of each layer and the layer construction seen in the cross section. The information was recorded in photographical form. The sample fragment was placed in a low vacuum scanning electron microscope and enlarged until the pigment composition could be examined as a back-scattered electron image. The low vacuum scanning electron microscope precludes the necessity of a conductivity coating such as carbon or platinum. Therefore, repeated observation through electron microscope and optical microscope can be conducted. Further, as it is not necessary to remove the coating, scraping of the sample is minimized. Observation was limited to a back-scattered electron image. However, it was not possible to have a secondary electron image, or magnification to a hundred thousand times original size. In a back-scattered electron image, the elements found in a pigment are shown in relative lightness or darkness depending on their atomic number. The larger an element's atomic number, the brighter the image will appear, the smaller the element's atomic number, the darker it will appear. After preparing the optical microscope image that shows the observed color corresponding to the pigment, and the back-scattered electron image that reveals the relative size of the atomic number of the elements in the pigment, the elements of the pigments were analyzed through energy dispersive X-ray fluorescence analysis equipment (Fig. 7).

3.2 CHARACTERISTICS OF GROUND AND PAINT LAYER OF RUBENS' THE FLIGHT OF LOT AND HIS FAMILY FROM SODOM IN TERMS OF OBSERVATION OF THE OPTICAL AND ELECTRON MICROSCOPE

Rubens' "The Flight of Lot and his Family from Sodom" was investigated to analyze its attribution by comparing two other similar works (Fig. 4).

The ground of the work is a calcium carbonate layer, 150 micrometer thick (Fig. 5 and 6). As a grate number of calcareous nannofossils of around 10 micrometer were observed in the calcium carbonate layer, we can conclude that the calcium carbonate is chalk and that the ground is hence a chalk ground. In observation both through the optical microscope and the electron microscope, we could not confirm a many-layered structure in this chalk ground, and hence it can be considered as single layer.

The ground consists solely of chalk, and as the whole layer has maintained its white color, there is a high probability that an aqueous medium such as animal glue was used as medium. The application of Fuchsine S through the staining method revealed a red reaction, which indicates the presence of protein. An approximately 30 micrometer thick semi-transparent layer of ash brown can be observed in the upper part of the chalk ground. As nothing other than chalk could be detected from this part, we can assume that the semi-transparency is a result of the infiltration of an oil or resinous medium, which reflective index is close to that of chalk, such as drying oil. According to the back-scattered electron image, the reason for the upper layer of the ground to be of a relatively higher density than the lower layer could be because the upper section permeated with oil is harder, and the surface was polished to a smooth finish.

The red color layer from the sample of red robe of the angel is a single layer constructed of relatively finely ground vermilion particles of less than 5 micrometer in diameter, and a transparent red lake made up of indistinctly shaped particles. This layer is about 20 micrometer thick. The red lake is distributed throughout the red layer, although it is relatively denser in the lower part of the layer. There is no glaze layer made up exclusively of lake pigment.

References

1. Kamba, N. and Kojima, M.: Technical and historical examination of the Madonna with the Infant Jesus and Her Fifteen Mysteries, *Bulletin of the National Museum of Japanese History***76**, (1998), 175-247.

2. Kamba, N.: Technical examination of the Portrait of St. Francisco Xavier, *Bulletin of the Kobe City Museum***16** *(2000)*, pl-15.

3. Kamba, N. and Kojima, M.: Technical and historical examination of Higashi-ke Family's the Madonna with the Infant Jesus and Her Fifteen Mysteries, *Buttetin of the National Museum of Japanese History* **93** (2002) , 103-140.

4. Nakamura, T. (ed.): *Rubens and his Workshop – The Flight of Lot and his Family from Sodom*, The National Museum of Western Art, Tokyo, 1994.

Figure IV-3-1.

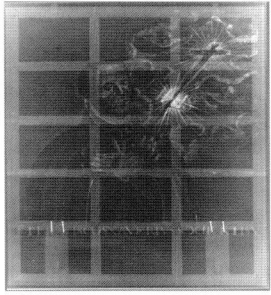

Figure IV-3-2. X-ray radiography of 'Portrait of
St. Francisco Xavier'.

Figure IV-3-3. Emissiography of 'Portrait of
St. Francisco Xavier'.

Figure IV-3-4. P.P.Rubens, 'The flight of Lot
and his family from Sodom', (The John and
Mable Ringling Museum of Art).

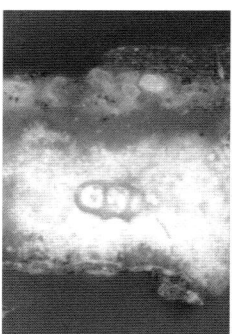

Figure IV-3-5. Paint cross-section of 'The flight of Lot and his family from Sodom', magnification100X.

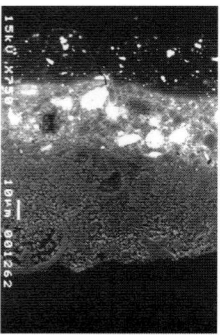

Figure IV-3-6. Back-scattered electron image of the paint cross-section of 'The flight of Lot and his family from Sodom'.

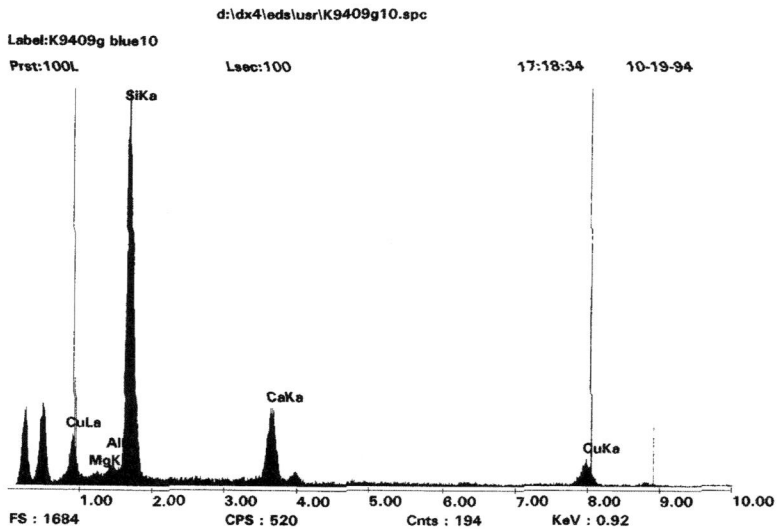

Figure IV-3-7. Energy-dispersive spectrum of the paint cross-section. The spectrum was acquired using a scanning electron microscope.

Chapter IV-4

Radiographic Findings in Ancient Egyptian Mummies

Kazuaki Hirata
Department of Anatomy, St. Marianna University School of Medicine 2-16-1, Sugao, Miyamae-ku, Kawasaki-shi, Kanagawa, 216-8511, Japan
k2hirata@marianna-u.ac.jp

Keywords: ancient Egyptian mummy, radiographic finding, magnified CT scan, embalming technique, sex determination, age determination, palaeopathology

Preservation of human bodies after death is usually designated by two expressions, namely, embalming and mummification. The word "mummification" is derived from the Latin word "mumia" which was mentioned by Dioscorides (first century A.D.) as a black bitumen found oozing from the earth in certain places. This word was applied later to the embalmed bodies in ancient Egypt. Mummification is undoubtedly the most distinctive technique or art, which developed in ancient Egypt.

In 1895, Roentgen, a German physicist, discovered the rays that were to form the basis for the X-ray method. An X-ray examination in 1932 of the mummy of Amenhotep was reported by Derry (1934). The largest series of radiographic examinations of mummies is that of Peter H. K. Gray (1967), who analyzed 133 specimens from museums in Great Britain, France, and Holland. He emphasized the importance of the archaeological aspects of the radiological findings, including the determination of the presence or absence of human bones to detect fraudulent specimens, the determination of sex and age, the correlation of radiological findings with known embalming techniques, and the demonstration of amulets. The palaeopathological aspects of his report included osteoarthritis, growth arrest lines, ante-and postmortem fractures and dislocations, several isolated cases of other bony pathology, dental disease and attrition, vascular calcification, gallstone, and ureteric calculus. Many of the radiological findings in the ancient Egyptian mummies had been described in the past (Harris and Wente, 1980). The author would like to review the several radiological findings in Egyptian mummies.

Radiographic evidence indicates intact nasal septa in Thutmose I, Thutmose III, Thutmose IV, Ramesses VI and Ramesses IX or XI. The septa are missing in Semekhare, Merenptah, Siptha, Ramesses III, and Ramesses V. Visual inspection in Ramesses IX or XI revealed penetration of ethmoid, good evidence for nasal excerebration. Since the septum was intact in this pharaoh, its status cannot be used conclusively to determine a nasal route for brain removal.

M. Uda et al. (eds.), X-rays for Archaeology, 259–261.

X-ray film revealed the scarab and the four sons of Horus in the mummy of Queen Nodjme of the Twenty-first Dynasty. The two-part stone eyes are clearly indicated in the film of the head of Queen Nodjme. Radiograph of the head of Bakenren also shows the artificial eyes.

Frontal radiograph of the dorsal spine of Henttowy shows marked osteophyte formation along the right lateral margin of the mid and lower dorsal spine. Lateral radiograph of the upper cervical spine of Merenptah indicates the hypertrophic lipping particularly marked at the level of C3-4 and C4-5. Antero-posterior radiograph of lumber vertebrae and pelvis of Merenptah reveals moderate rarefaction (atrophic area) of the iliac fossae. The film of thorax and neck shows compressed, demineralized, and osteophytic cervical vertebrae in Merenptah. The antero-posterior film of pelvis of Merenptah shows striated margins of pubic symphysis indicative of maturational stage.

The lateral head film of Ramesses II indicates marked tooth wear and alveolar resorption. Antero-posterior radiograph of lumber vertebrae of the X-ray film of Seti II shows absence of lipping and presence of sharp contours of the centra. The X-ray film of knees shows that the epiphyses at the distal ends of the femurs and at the proximal ends of the tibiae are not united with their respective shafts in the mummy of Thutmose I. The distal epiphyses of the tibiae and fibulae of Thutmose I are also not yet united to their shafts. This radiological finding suggests an age of 18-22 years.

Growth arrest lines (Harris's lines) can be recognized radiographically in the diaphysis of the long bones. It is generally regarded as evidence of previous arrest of growth due to malnutrition and several diseases during childhood. The prevalence of growth arrest lines has been regarded as a valuable indicator available for assessing the state of health and nutrition in the past human populations. Gray (1967) detected a rather high incidence of growth arrest lines in his study of 133 mummies.

Lateral radiograph of skull of Seqenenre Tao reveals multiple skull fractures and multiple fractures of facial bones. Frontal X-ray of the chest of Merenptah shows the absence of the medial head of the right clavicle, comminuted fracture of the right forearm, and an arcuate defect in the border of tenth right rib postero-laterally. Frontal radiograph of the skull of Merenptah shows the cranial defect and a defect in the nasal septum, presumably incident to the transnasal route of evacuation of the cranial contents at the time of embalmment.

The X-ray cephalogram of the head of Ramesses II indicates severe periodontal loss pulp exposure, and periapical abscess. The X-ray cephalogram of Merenptah reveals severe periodontal destruction around the remaining teeth and the apparent cause of the loss of most of his posterior teeth. The X-ray examination of Ramesses V of the Twentieth Dynasty, who died young of small pox (thirty years age), illustrates a healthy dentition with minimal wear and periodontal disease.

Dense fluid level in the posterior skull represents hardened resin, which layered horizontally in the lateral radiograph of the skull of MIA II. Irregular density in the posterior cranial fossa consists of resin-soaked linen. The presence of radiopaque mass in the cranium may indicate a desiccated, shrunken or cranial packing. Lateral radiographs of naturally desiccated skulls of Nubian mummies show the typical radiopaque appearance of the brain in its undisturbed position. Although rapid putrefaction and liquefaction of the brain might be expected, dehydration may produce a consolidation of about one-fourth the brain volume, which adheres to the posterior cranial base. Smith (1902) has described natural preservation of the brain. Density and outline are important in differentiating brain remnants from cranial packing. Desiccated brain tends to be mottled in appearance, with an undulating border. The radiograph cannot indicate cranial packing unless it contains radiopaque material such as resin.

Magnified CT scan of Lady Tashat (25th Dynasty) reveals the funerary mask supported against the inner surface of the cartonnage face by linen packing. The oblong structure lying obliquely immediately anterior to a thoracic vertebral body represents the collapsed heart which dips into a mass of hardened resin in CT scan film at the level of the mid-thorax. Antero-posterior radiograph reveals the second skull placed between Lady Tashat's legs. Lateral radiograph of the second skull shows much of the basiocipital region, maxilla and mandible. Lady Tashat's fingers extended in front of the proximal femurs and flanking the skull. Magnified CT scan of the second skull reveals the eyes are shriveled in their sockets. The hole in the paranasal sinuses, between the orbits, indicates the defect through which the brain was extracted. It can be seen the Many layers of linen wrapping can be seen on this CT scan taken at the level of Lady Tashat's left hip. Denser layers indicate extra coats of resin. Unfused epiphyseal plates are visible Antero-posterior radiograph of Lady Tashat's knees. CT scan of the skull shows a crack in the nasal septum created to remove the brain.

Mummification greatly affected the habits and customs of the ancient Egyptians and, through it, much knowledge was gained in anatomy, medicine, chemistry, and many arts and industries. The radiological interpretation of the X-rays under somewhat restricted circumstances on the mummies in ancient Egypt has indicated findings of both historical and medical interest.

References

Derry, D. H.: An X-ray examination of the mummy of King Amennophis I. Ann. Serv. Antiq. Egypt **34**(1934), 47-48.

Gray, P. H.: Radiography of ancient Egyptian mummies, Med.Radiogr. Photogr. (1967) **4**, 34-44.

Harris, J. E. and Wente, E. F. (ed.) An X-ray atlas of the royal mummies, The Univ. of Chicago Press, Chicago, 1980, 163-345.

Smith, G. E.: On the natural preservation of the brain in the Ancient Egyptians, J. Anat. and Physiol. **55**(1902), 375-380.

Part V: Interdisciplinary Field between Art and Science

Chapter V-1

X-ray Application on Post-Amarna Objects from Dahshur

S. Hasegawa[1], M. Uda[2], S. Yoshimura[3], J. Kondo[4], T. Nakagawa[5] and S.Nishimoto[5]

[1] Institute of Egyptology, Waseda University, 1-104,Totsuka-cho, Shinjyuku-ku, Tokyo, 169-8050, Japan, [2] Department of Materials Science and Engineering, Waseda University, [3] School of Human Science, Waseda University, [4] Department of Archaeology, Waseda University, [5] Department of Architecture Waseda University,
hsgwegypt@waseda.jp

Keywords: Dahshur, Post-Amarna, X-ray, Ramesside sarcophagus, Amarna Blue

1. Historical Background

Here we introduce a study on objects from Dahshur, Egypt. Dahshur is located about 20 km south of Cairo, and it has been well known for the Pyramid area of the Old Kingdom and Middle Kingdom. Waseda University has surveyed this area since 1996 and detected a huge New Kingdom necropolis[1]. One of the largest tombs found here seemed to be constructed by a "Royal Butler, Ipay," and excavations of many shaft tombs around this tomb of Ipay have been carried out. The total assemblage of the objects, such as statue figures, sealing motif, design of the faience ring, and the decoration of glass and pottery wares, showed that it can be dated to the "Post-Amarna" period, at about BC 1300-1000, as is suggested by the following royal names. Tutankhamen's cartouche may imply the period of necropolis formation as being the late 18th Dynasty, and the area may have been reused at least until the reign of Ramesess II, of the early 19th Dynasty, and a Ramesside sarcophagus discovered at the innermost place also supported this dating.

2. Study Subjects

Recent excavations showed that a large tomb had once been constructed at the southernmost hilltop beside the tomb of Ipay, and we will find another tomb with superstructure at the western area by coming excavation. Then we can recall the

M. Uda et al. (eds.), X-rays for Archaeology, 265–269.
© 2005 Springer. Printed in the Netherlands.

necropolis landscape in the New Kingdom period where high officials' tombs were standing like the south area of Unas causeway at Saqqara.

Both the necropolises at Saqqara and at Dahshur were exploited at the desert edge of Memphis. Archaeological evidence and historical documents of the later period also support a making image of Memphite civil life of the New Kingdom, where the royal family, high officials, and priests were carrying out their daily service at the central Ptah Temple and around the northern hilltop. Craftsmen, including masons, potters, faience or glass makers, were centralized around the east-south quarter of the temple precinct, mainly because of wind coming from the northwest, and for its convenience near the Nile port.

Here, limestone, Marl clay at the plateau, and Nile silt clay and vegetal material, such as palm and reed, at the cultivated land, were easily utilized by those craftsmen; and on the other hand, rare materials like precious stones and minerals were transported over the sea, which will reflect the economic network during that period.

3. X-ray Analysis

Therefore, some selected objects were analyzed from Dahshur objects. As a first step of a long-term project, pigments of sarcophagus decoration and funeral equipment were prepared for the examination. The chemical compositions of those objects were investigated at the site using a portable type of X-ray fluorescence spectrometer. The results are as follows.

3.1 OBJECT: SARCOPHAGUS (CF0001) [Fig. 1]

Provenance: Tomb of Ipay, subterranean chamber.
Material: Granite
Analysis: As we mentioned above, a Ramesside high official possessed this sarcophagus. It is composed of a box and a lid, on which underground deities were carved in relief or painted with pigments. Analysis suggests that:
1) Bright red color on the kilt part of Goddess Isis is iron oxides and hematite.
2) Black color on the inscriptions and hair of Isis is iron and manganese oxide.
3) Yellow color of the flesh of gods, which symbolized life and growth in ancient Egyptians' concept[2], is "ochre," which consists mainly of iron oxides and goethite.

Figure V-1-1. Granite Sarcophagus.

Those three colors were mostly obtained from iron oxides, which were the most widespread over the country. In addition, white material attached at the mid of box contained *alum*, which suggests the material is a kind of mortar (consisted of K, Ca, Sr) to be used to fix the rope to lower the sarcophagus.

3.2 OBJECT: POTTERY JUG PAINTED BY RED PIGMENT (EW0326) [Fig. 2]

Provenance: Tomb of Ipay and Shaft 17, subterranean chamber.
Material: Mycenaean whitish clay.
Analysis: This hemispherical shape with a neck, painted decoration, and clay composition that is far from Nile silt and Marl, show that it was transported from the Mycenaean World[3]. Analysis suggests that the red color is iron oxides and hematite, in addition to arsenic (As). Arsenic sulphide (AsS) makes effectively red more brilliant than hematite, and it seems to have been painted after burnishing.

Figure V-1-2. Mycenaean Pottery.

3.3 OBJECT: POTTERY JUG PAINTED BY BLUE PIGMENT (EW0220)

Provenance: Tomb of Ipay and Shaft 17, subterranean chamber.
Material: Nile silt clay.
Analysis: This kind of jug painted in such deep blue or green, with a lid, was frequently used as votive wares for burial in the New Kingdom period[4]. Blue pigment contains much iron, which is compared to the following light blue pigment example.

3.4 OBJECT: POTTERY BASE PAINTED BY BLUE PIGMENT (EW0266) [Fig. 3]

Provenance: Tomb of Ipay and Shaft 17, subterranean chamber.
Material: Nile silt clay.
Analysis: This blue is different from artificially manufactured "Egyptian Blue" that is copper calcium silicate ($CaO \cdot CuO \cdot 4SiO_2$), known for ancient Egyptians from the archaic period. The light blue color appeared in the New Kingdom period and

became popular after the reign of Amenophis III; it is usually called "Amarna Blue." This light blue is quite effective when it is painted on cream slip over clay surface. It is suggested that this blue is obtained by a mixture of calcium sulphate ($CaSO_4$) of white color, and cobalt aluminum oxides (Co (M) Al_2O_4) of deep blue[5]. Here M means Mn, Fe, Ni, and Zn.

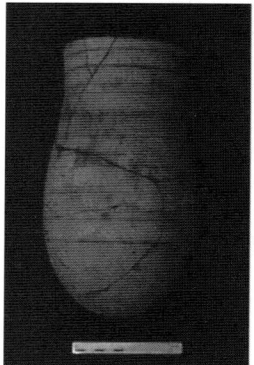

Figure V-1-3. Amarna Blue Pottery.

3.5 OBJECT: INLAY EYEBROW OF COFFIN MASK (EY0044)

Provenance: Shaft 17, subterranean chamber.
Material: Blue glass.
Analysis: This blue color contains the same composition, mentioned above in the example of an "Amarna Blue" pottery vase. Here again, cobalt aluminum oxides (Co (M) Al_2O_4) of deep blue, and the use of white calcium sulphate ($CaSO_4$), would make the deep blue material light. The presence of goethite will result from the remnant of yellow wash on the coffin mask.

3.6 OBJECT: A PART OF GLASS GOBLET (GW0014) [Fig. 4]

Provenance: Tomb of Ipay and Shaft 17, subterranean chamber.
Material: Crystallized blue glass.
Analysis: This blue also contains the same composition as the "Amarna Blue" pottery vase and the former example of glass eyebrow, that is, cobalt aluminum oxides (Co (M) Al_2O_4) and calcium sulphate ($CaSO_4$), where M means Mn, Fe, Ni, and Zn.

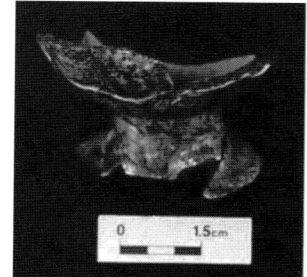

Figure V-1-4. Glass Goblet.

4. Conclusion

Some "Post-Amarna" objects supposed to have been produced in Memphite workshop and found at Dahshur were analyzed. On the transported Mycenaean jug, a use of arsenic for its brilliant red color was found. On the other hand, some pigments might have been acquired from the vicinity of Memphis, in such cases when the sarcophagus was prepared and painted by pigments consisted of iron oxides, hematite, manganese oxide, and goethite.

One of the most conspicuous technologies, which might have been developed at that period, is the use of light blue color, that is, "Amarna Blue," whose characteristic feature is obvious when compared with another kind of traditional "Egyptian blue." "Amarna Blue" was analyzed, and it was suggested that cobalt aluminum oxides (Co (M) Al_2O_4) is used for deep blue, and calcium sulphate ($CaSO_4$) for white could make the deep blue material light. In addition, this device was not restricted to the ware decoration but also seemed to have been applied to ornamental objects, such as a glass of coffin mask and a glass goblet.

5. References

[1] Yoshimura, S. and S. Hasegawa (2000) New Kingdom necropolis at Dahshur - The tomb of Ipay and its vicinity -, Barta, M. and J. Krejci, ed. in *Abusir and Saqqara in the year 2000*, Praha, pp.145-160.

[2] Colinart, S. (2001) Analysis of iconographic yellow color in ancient Egyptian painting Davies,W.V., ed. in *Colour and Painting in Ancient Egypt*, London, p.1.

[3] Davies, W.V. and L. Schofield, ed. (1955) *Egypt, the Aegean and the Levant –interconnection in the Second Millennium BC-*, London.

[4] Nagel,G. (1938) *La céramique du nouvel empire à Deir el Médineh* , Tome1, Le Caire

[5] Uda, M. et.al., (2002) Amarna blue" painted on ancient Egyptian pottery, *NIMB*, **189** pp.382-386.

Chapter V-2

Decorative Program at Malqata Palace, Egypt

Shin-Ichi Nishimot
Department of Architecture, School of Science & Engineering, Waseda University, Japan, 3-4-1 Okubo, Shinjyuku-ku, Tokyo, 169-8555, Japan
nishimot @ waseda.jp

Keywords: Amenhotep III, painting, decoration, bedroom, ceiling, building phases

A severely ruined palace-city founded by Amenhotep III (ca. 1390-1352 B.C.) is preserved at Malqata on the West Bank of the Thebes in Egypt. It consists of various structures in the desert: several residential palaces, the temple of Amen, a festival hall, houses and apartments for attendants, and a desert altar "Kom al-Samak", all of which were constructed from mud bricks and had decoratively painted walls and ceilings. In 1985, this area was placed under the auspices of the Waseda University Mission, and several rooms in the main palace have been excavated. A report on the excavation was published in 1993. Based on this publication, further detailed study of the site has been carried out by the author.

From Room H, the great columned hall located in the middle of the main palace, numerous painted mud fragments have been excavated as reported in the previous publication, and the painting that occupied the entire ceiling has been reconstructed. At the innermost room of the main palace is the king's bedchamber, from which a large numbers of fragments of the ceiling painting have also been recovered. As reported by a previous excavator, the Metropolitan Museum of Art, one of the most remarkable motifs is a series of great vultures representing the Goddess Nekhbet with outstretched wings, under each of which are inscribed the names and titles of Amenhotep III. The series of vultures is surrounded by geometrical patterns, such as rosettes and checkerboard patterns. The first attempt to reconstruct the whole ceiling painting was made in 1988, and a detailed study of images of each fragment and of attempts at re-assembly since 1989 has revealed that there were eight images of Nekhbet, not seven as was supposed at an earlier stage of reconstruction. All nine lines of inscription have also been reconstructed, based on the position in which fragments were found on the floor. The floor of the innermost part of the king's bedchamber, where the king's bed was located, is raised. It has emerged that the ceiling of this upper level was more elaborately painted than that of the lower level. The inscriptions are slightly longer, and the color of the center circle of the rosette pattern is red, rather than green.

M. Uda et al. (eds.), X-rays for Archaeology, 271–274.
© 2005 *Springer. Printed in the Netherlands.*

The fragments of the northernmost ceiling suggest that the ceiling of this part would have been slightly curved down towards the north wall, which is reminiscent of the roof shape of "pr wr", a traditional shrine of Upper Egypt, with a roof sloping down from the front. A paneled pattern and a wavy line are painted on the lower part of the walls, but the rest of the interior decoration of this bedchamber is painted a glossy transparent yellow color, presumably imitating gold, very similar to the shrines of Tutankhamen.

However, there is another large bedroom for the king at the north of the main palace: Room B. From the fact that the inscriptions on the ceiling mention the king's name, it has been determined that this room was also a bedchamber of king Amenhotep III. Why two king's bedrooms are prepared for a sole king? The presence of the two bedrooms for the king in one palace is problematic, but a clue to resolve this question could be found in the traces of the enlargement of the main palace.

Concerning the building phases of the palace complex, R. Johnson already pointed out: "The palace of the king, west villas (the administrative area), and middle palace, oriented to Amenhotep's mortuary temple, were probably part of the original complex, while the north palace, audience pavilion, and Amen temple, all oriented to the enlarged harbor, date to the later jubilees." [2]. Presumably the southern part of the main palace and the west village seem to have been constructed in the 30th regal year of Amenhotep III, and the northern part of the main palace is thought to have been added soon after that. The temple of Amen, the platform and the north palace were erected in the regal year 33-34. The causeway running to the west side of the main palace seems to have been constructed "when it (palace) had been abandoned. [3]

Appendix:

Chemical analysis on the pigments detected at the King's Bedroom and Room B in the palace of Malqata has been carried out by Masayuki Uda, Professor at Waseda University, and the result of the preliminary analysis is as follows.

White:	$3MgCO_3 \cdot CaCO_4$	huntite
Black:		probably carbon
Blue:	$CaO \cdot CuO \cdot 4SiO_2$	Egyptian blue
Green:	(not identified)	
Yellow:	$\alpha FeO \cdot OH$	goethite
Red:	αFe_2O_3	hematite

References

[1] Nakagawa, Takeshi and Nishimoto, Shin-ichi (eds.): *Studies on the Palace of Malqata: Investigations at the Palace of Malqata*, 1985-1988, Chuo Koron Bijutsu Shuppan Ltd, Tokyo, 1993.

[2] Johnson, R.: Monuments and Monumental Art under Amenhotep III: Evolution and Meaning, in O'Connor, D. and Cline, E.(eds.): *Amenhotep III; Perspective on His Reign* , The University of Michigan Press, Ann Arbor, 1998, 76.

[3] Kemp, B. J.: A Building of Amenophis III at Kom el-'Abd, Journal of Egyptian Archaeology **63** (1977), 81.

Figure V-2-1. Room H, reconstruction (by Takaharu Endo).

Figure V-2-2. Ceiling Painting of the king's bedroom (by the author).

Chapter V-3

X-ray Archaeology in China

Changsui Wang
United Key Laboratory of University of Science and Technology of China
USTC, No. 96, Jinzhai road, USTC, Anhui 230026, P. R. China
wangcs@ustc.edu.cn

Keywords: Archaeometry, X-ray analysis, Ancient ceramics, Heiqigu mirror, Bone chemistry

Archaeology is a science of studying the history of human society based on human remains [1]. Its ultimate purpose is to illustrate the rules of historical developments and its close relationships with other subjects. Archaeology is also a developing discipline and a good example is the archaeology in China. The epigraphy of the Northern Song dynasty (960-1127 AD) in an initial stage, experienced a long-term and gradual development, and finally matured in the early twentieth century. With the help of other subjects, especially the application of the knowledge of natural sciences, the objects of research, approaches, theories and purposes of field archaeology have all changed obviously and developed very fast. Even the principal theories for field archaeology, archaeological stratigraphy and archaeological typology, were actually formed based on the fundamental ideas of geological stratigraphy and biological taxology [2,3].

With the development of research, archaeologists discovered that all of the various remains related to the ancestral life, contain abundant information on ancient society. Most of the information, such as its age and provenance, is not visible without the aid of modern instruments and only with the help of modern technological methods they can be detected. This information is known as the potential information of the ancient remains [4].

The materiality of ancient remains determines the significance of the potential information, structure and compositions, which is the basis of the application of X-ray technology in archaeology.

M. Uda et al. (eds.), X-rays for Archaeology, 275–290.
© 2005 *Springer. Printed in the Netherlands.*

1. X-ray Archaeology on Ancient Ceramics [5-7]

Pottery is the first kind of new substances in the human history that does not exist in nature and is produced by chemical reaction. Its creation widened the human food variety, improved the cookery and substantially changed the style of human life. Thus, it was praised as an important milestone of human civilization history. Pottery was closely connected to the habits and customs of ancient people, and contained plenty of information of ancient societies. In the study of Neolithic Age, pottery naturally became the major object of research [8].

There are many aspects of ceramics to study, among which the relatively important and relevant subjects to X-ray archaeology are as follows:

1.1. THE PROVENANCE OF ANCIENT POTTERY

Pottery is usually fired from clay, which is considered, by some researchers, as soil with a thin layer of sand. As we know, clay or soil could be found everywhere, and since the ancient traffic was very inconvenient, it can be assumed that raw materials of pottery came from local areas. Many Chinese scientists, including Professor Zhou Ren, had made chemical analysis of 65 pottery pieces and some other material samples from the Yellow river Valley dating back from the Neolithic Age (6000-2000 BC) to Shang (1600-1046 BC) and Zhou dynasties (1046-221 BC). The results verified that the raw materials of pottery are laterite, black and some other clays, not ordinary loess or soil. These results also showed that the raw materials of pottery are not taken at random but were locally [9]. To tell the difference between various clays and soil, X-ray fluorescence analysis linked up with X-ray diffraction was used.

Clay originated from weathered rocks and rotten organism. Different natural and human conditions produced different types of clay [10]. As for raw materials of pottery, their composition and phase mainly depended on the characteristics of the local mother rocks themselves. Considering the fact that raw materials are chosen locally, the major elements in the pottery are almost the same due to sintering processes, and the pottery phase change can be neglected, it is possible to explore the pottery provenance by analyzing petrography and phase composition of the pottery body. Although petrographic analysis can acquire more information than X-ray diffraction, X-ray diffraction can still be used successfully in some cases.

Ten years ago, by means of the X-ray quantitative analysis, the Chinese pottery provenance in Neolithic age was studied successfully, based on the amount and variety of feldspar in the pottery body [11,12]. For example, the Huating Site at Xinyi County, Jiangsu Province (3300-2500 BC), is a special site where two kinds of pottery were unearthed. One belonged to Dawenkou Culture (4300-2500 BC) and the other was related to Liangzhu Culture (3300-2200 BC). Dawenkou Culture was distributed over the southwest part of Shandong and the northern of Jiangsu, south of the Huating Site. Liangzhu Culture was distributed over the northern of Zhejiang Province and the southern of Jiangsu Province, and its north boundary was 300 km far from Huating Site. So, why pottery distributed in two different areas and with two cultural kinds, were unearthed from the same site? Some archaeologists believed that is a result of war, while other experts believed that this was due only to long-term cultural exchange [13]. The X-ray quantitative analysis was adopted to analyze the content of albite and potash feldspar in the pottery sherds. In combination with INAA and petrographic data measured by Japanese cooperators, (Satoshi Koshimizu, Kazai Manabu and Kushihara Koichi of the Yamanashi Institute of Cultural Properties, Teikyo University), the results confirmed that the feldspar content in pottery of Liangzhu style is much lower than that of Dawenkou style, which indicated that all pottery of Liangzhu style was produced in the Liangzhu area. Obviously, this result is in favor of the "war exploitation" theory.

This method used, is similar to the X-ray phase quantitative analysis method introduced in many reference books, differing in that the incremental step, which was less and hence saved a lot of manpower [14]. The diffraction intensity of albite (002) with a diffraction peak at (d=3.191Å), potash feldspar (002) with a diffraction peak at (3.241Å) and quartz (001) with a diffraction peak at (d=3.344 Å) were recorded and the feldspar content was calculated. However, it should be pointed out that the above-mentioned two feldspar diffraction peak intensities actually concluded that some other feldspars are present. Therefore, strictly speaking, the results obtained indicate that two groups of feldspars with diffraction peaks at d=3.191 Å and d=3.241 Å respectively are present.

1.2 THE PIGMENT ANALYSIS OF THE ANCIENT PAINTED POTTERY

Yangshao culture is worldwide famous for its bright-colored painted pottery with simple, unsophisticated and lifelike patterns. Since the discovery by the Swedish scientist, J.G. Anderson, many papers about this culture have been published. However, papers about the mineral components of pigments and the sintering technique of painted pottery, using modern science and technology, are receiving relatively less attention from modern science and technology.

Through test analysis, it has been known for a long time that the white pigment of the Yangshao culture (5000-3000 BC) painted pottery is sillico sillico-aluminate, and the red and black pigment is hematite or magnetite. In 1991, the original president of the Chinese History Museum, Professor Yu Weichao, led a comprehensive archaeological research on the Bancun site (4500-3000 BC) in Mianchi County, Henan Province, in order to explore and develop Chinese archaeological theories and methods. As an advisor of this team, I studied the painted pottery of the Yangshao culture period. The results proved that the silico-aluminate used for white pigment is a non-crystalline, and its diffused peak is clearly different from that of diatomite. This result, together with compositional analysis, pointed out that the white pigment is a non-crystalline alumyte. On the other hand, the red pigment and black pigment were composed of different materials. Some are hematite and magnetite, while others are uniclinal structured ferric oxide and franklinite or just merely magnetite [15]. These conclusions have aroused the interest of concerned experts. Not only the mineral compositions of these pigments should be further explored in order to get more information about the ancient technique of pottery production, but also the cultural exchange between ancient peoples by means of the provenance of pigment mineral should be explored.

1.3 APPRAISAL OF WHITE AND BLUE PORCELAIN EXCAVATED FROM THE STATE KILN OF YUAN, MING AND QING DYNASTIES [16]

Since the 1990s, several Chinese scholars have realized the important significance of trace elements analysis especially on ancient ceramics. Prof. Cheng Huansheng et al. of the Fudan University studied ancient ceramics, jade, and bronze swords etc. using PIXE, Prof. Li Jiazhi of the Shanghai Institute of Ceramics, Chinese Academy of Science, and Prof. Luo Hongjie et al. of the Shaanxi University of Science and Technology, measured large numbers of ceramics by the use of EDAX. Other scholars including, Prof. Feng Songlin et al. of the Institute of High Energy Physics, Chinese Academy of Science, Prof. Gao Zhengyao et al. of the Zhengzhou University and Prof. Miao Jianmin of the Palace Museum, carried out similar researches. It should be emphasized that Dr. Leung Po Lau of the Hong Kong City University, also studied a number of Chinese ancient ceramics cooperating with most of the above-mentioned investigators and publishing a lot of papers.

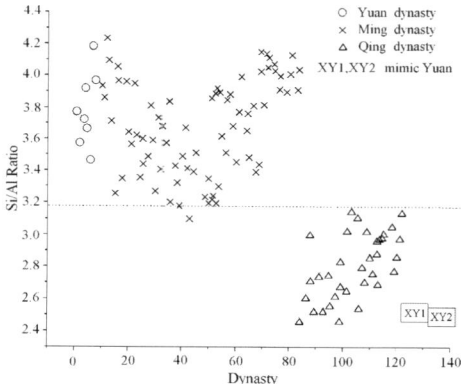

Figure V-3-1   Relationship of the White and Blue porcelain between the composition ratio of Si to Al and the reigns

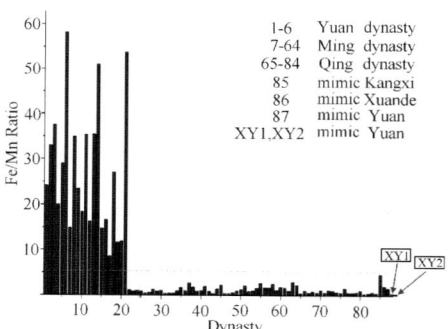

Figure V-3-2   Relationship of the White and Blue porcelain between the composition ratio of Fe to Mn and the reigns.

The emphasis here is only on the White and Blue porcelain produced by the governmental kiln in Jingdezhen city, famous for its exquisite technique and fine art. Recently some perfect copies of this porcelain were made causing confusion in the cultural relics market. It is very important to distinguish true porcelain from imitations, and to determine its age. In 2001, the cooperation agreement for appraisement of the White and Blue porcelain was signed among our University, the Shanghai Institute of Ceramics, the Chinese Academy of Science and the Jingdezhen Institute of Ceramics Archaeology. 168 sherds of these porcelain, whose dates covered all of Ming (1368-1644 AD) Qing dynasties (1644-1911 AD) and some reigns of Yuan dynasty (1271-1368 AD), were provided by the Jingdezhen Institute of Ceramics Archaeology. In the paper of Prof. Li Jiazhi, it was pointed out that all of the sherds were divided into three groups: the first group covered some of the Yuan reigns to the Yongle reign of the Ming dynasty, the second included the Ming dynasty except for the pre-Xuande reign, and the third was related to the whole of the Qing dynasty based on the cluster analyses for the elemental data measured by EDAX. Figure 1 shows the relationship of the White and Blue porcelain, between the body composition ratio of Si to Al and the reigns. Figure 2 reveals the relationship of the same porcelain between the composition ratio of Fe to Mn in the blue colour area and the reigns. It should be pointed out that the samples No. XY1 and XY2 are imitations of the White and Blue porcelain of the Yuan dynasty provided by a donor.

1.4 THE UNIQUE TRANSITIONAL LAYER OF RU WARE [17,18]

There were five famous kinds of porcelains in the Northern Song dynasty (960-1127 AD), in which the Ru ware came in first. Vice-professor Feng Min from our team found several interesting results based on optical microscope, polarizing microscope, SEM, HREM and XRD analyses. For example, the glaze of the Ru ware belonged to the phase separating glaze and crystalline, which is related to its special blue colour. XRD analysis of the glaze surface revealed that there were obvious diffraction peaks of anorthite and diffusion peaks of the amorphous stage. It was also found that there was a transitional layer between the glaze and the body, whose appearance looked like the body in a polarizing microscope (see figure 3), but more compact than that of the body in an optical microscope (see figure 4). This fact implied that some amorphous composition of the glaze has entered into the body and formed a mixed transitional layer.

These results were also proven by SRXRF analysis. Figure 5 depicts the results of linear analyses of Ru ware by Prof. Mao Zhenwei of our team, showing that most of the elemental contents in the transitional layer, were almost found between the body and glaze layer.

Figure V-3-3 Section photo of Ru ware by using a polarizing microscope.

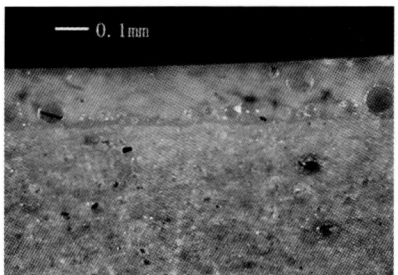

Figure V-3-4 Section photo of Ru ware by using an optical microscope.

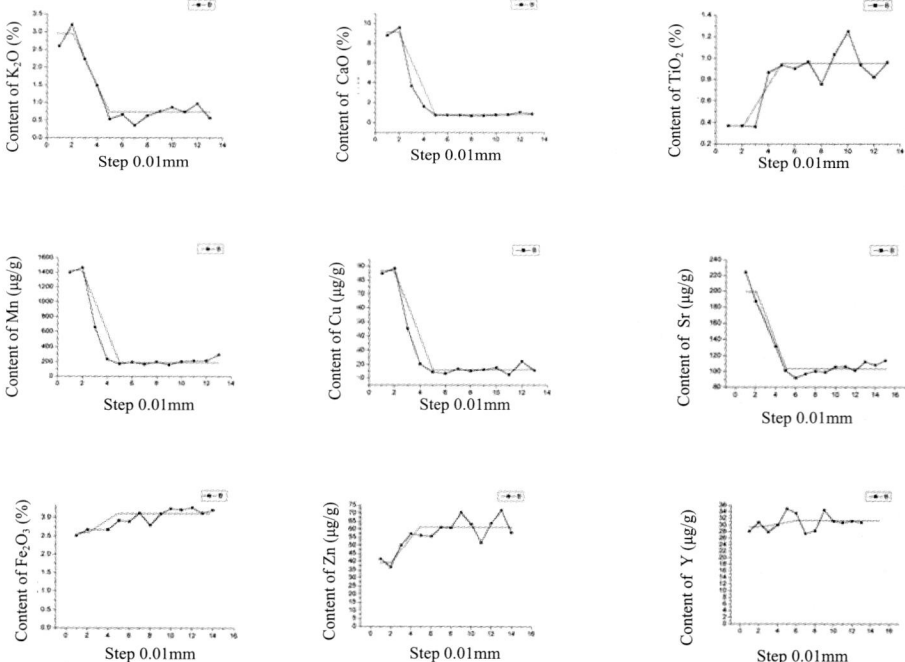

Figure V-3-5 Linear analyses for the section of Ru ware by using SRXRF.

2.   X-ray Diffraction Technology and Ancient Bronze Mirror Study

Metallurgy technology is another milestone of human civilization. Compared with the approximate 3,000,000 years of history of the Stone Age, the mastery and application of smelting and casting technologies of metals is only a flash. However, in the fleeting moment, the world changed with each passing day. This fact clearly shows that metallurgy technology played a significant role in the development of human society. From this point of view, we can see the important academic value of the archaeo-metallurgical study [19,20].

Ancient metallurgy covers a wide range, from bronze, brass to steel, gold and silver, including most kinds of metals and their alloys. This paper only introduces several applications of X-ray technology to the studies of ancient bronze mirrors.

2.1 STUDY OF TRANSMITTING LIGHT MIRRORS IN THE WESTERN HAN DYNASTY [21]

Transmitting light mirrors were invented in the Western Han Dynasty (206 BC - 8 AD) in China. The mirror diameter is about 74 mm. On the back of the mirror, Chinese characters "□□□□□□□□" (The whole world is bright from the sunlight), and patterns like rhombic flower, whorl or the Eight Diagrams were usually casted. When the mirror surface was irradiated by sunlight, the patterns and characters on its back, were reflected on a screen and hence the name "transmitting light mirror".

Can light really travel through the transmitting light mirrors? Many scholars in China and overseas probed into the transmitting light theory and related techniques for centuries. Not until 1975, did the experts in the Shanghai Museum, Fudan University and Shanghai Jiao Tong University solve this sphinx riddle.

These experts in the Shanghai Jiao Tong University discovered that the transmitting light mirror was made of a thin body with thick rims, and the characters and patterns of its back were simple and with reasonable structure. Therefore, during the solidification process, the different residual stress and rigidity of the mirror body was formed due to the difference in thickness of the mirror body. Grinding made the residual stress relaxed. While the mirror body was ground to a certain thickness, the mirror surface had the same curvature difference as the characters and patterns of its back. Consequently, when the mirror surface was irradiated by parallel light, its reflecting shadow had the same patterns and characters as those of the back also known as the transmitting light effect. Putting this theory to the test, the researchers succeeded in making replicas of the transmitting light mirror, which made a great impression on the scientific world.

In the research carried out on transmitting light mirrors, X-ray diffraction played a fairly important role, in particular, in testing the residual stress of these ancient mirrors. This was done first by the study of the phase components with XRD, revealing that the mirror had a casting structure of a bronze alloy with α and δ phases as well as lead particles. Its structure and components were basically similar to other types of ancient bronze mirrors, asserting that the transmitting light effect did not result from the mirror's structure or components. Also as a result of testing the residual stress within the mirror surface on three different locations with different thickness by means of the 0-45° method, and studying the effect of the grinding action on the residual stress, and using the following equation for calculating the main stress parallel to the mirror surface:

$$\delta = \delta_x = \delta_y = (E/[2(1+v)] \times ctg\theta_0 \times \pi /180 \times (2\theta_0 - 2\theta_{45})/[\sin^2(45+\eta) - \sin^2\eta]$$

Where Young's modulus E = 10986 kg/mm^2, Poisson ratio v = 0.33, the diffraction angle 2θ is approximately 161°, η is the angle between the incident ray and normal of the diffraction surface of the crystal [22], The following results, shown in Table 1, were obtained:

The table below demonstrates that before grinding, the stress values are inversely proportional to the thickness of the mirror surface measured. During grinding, the stress in the mirror rim, with the biggest thickness, did not change mostly, but only stress in the thin mirror body showed a great change. Moreover, the thinner the mirror body, the more stress change is observed. This phenomenon reflected that the grinding of the mirror surface decreased the residual stress and increased the difference in structure rigidity.

Table V-3-1 Data of Residual Stress on Mirror Surface.

Serial Number	Mirror thickness	1 mm	Mirror thickness	0.7 mm	Mirror thickness	0.4 mm
	Δ 2θ	δ*	Δ 2θ	δ	Δ 2θ	δ
1	+0.05	+0.95	-0.15	-2.87	-0.30	-5.70
2	-0.10	-1.90	+0.10	+1.90	+0.10	+1.90
3	-0.05	-0.95	-0.15	-2.85	-0.50	-9.50

* The unit of δ is kg/mm^2.

2.2 RESEARCH ON ANCIENT BRONZE MIRROR WITH CORROSION-RESISTING PROPERTY

Apart from the "transmitting light" mirrors, another type of unique ancient bronze mirrors, known for their excellent corrosion-resisting feature, has appealed to the science circle in the world. This type of bronze mirrors, according to structural characteristics and surface color, can be divided into several kinds, such as "Shuiyinqin", "Luqigu" and "Heiqigu".

"Shuiyinqin" bronze mirrors have white bright surfaces, which have resisted corrosion for thousands of years. Mr. Tan Derui and Wu Laiming of the Shanghai Museum and Ms. Shu Wenfen of the Shanghai Material Research Institute cooperated for years and finally detected the corrosion-resisting mechanism and production technology of these mirrors, which was related to the mirror's tin-rich surface with a thickness that varied between tens to several 100 nm [23]. Because this tin-rich layer is too thin, it was difficult to get some information from the layer by using X-ray and hence out of the scope for discussion covered in this topic.

The "Heiqigu" and "Luqigu" bronze mirrors are different from the "Shuiyinqin" bronze mirror. Their surface layers usually are as thick as 100 microns and their corrosion-resisting mechanism and production technology still remains unknown and is a central issue in the field of archaeometry. The application of X-ray technology to the study of these mirrors is described briefly in the next sections.

2.2.1 DISTINCTION BETWEEN "HEIQIGU" MIRROR AND THE GENERAL BLACK MIRROR [24,25]

Black mirrors in ancient China are divided into two kinds: those with a black shiny surface also known as "Heiqigu" mirrors, while those with less shiny surfaces are known as the general black mirrors (GBM). Most "Heiqigu" mirrors were made before the Tang dynasty (618-907 AD), whereas, the GBM were mostly made in the Tang dynasty. The surface layer of both kinds of mirrors has corrosion-resisting property, with "Heiqigu" mirrors superior in this respect, and is historically known to have the best anti-corrosion material on metal surfaces.

Since the exterior appearance, property and production age between "Heiqigu" and GBM are so different, the phase components in their surface layers should also have big differences. 22 black mirror pieces chosen from Hunan, Zhejiang and Henan provinces were analyzed by means of X-ray diffraction technique. It was discovered that the main phase component in the surface layer of both mirrors was neither non-crystalline substance, organic polymer nor local lacquer, but cassiterite SnO_2. It became known that the peak of cassiterite SnO_2 is wider and of smaller d value than the corresponding peak of a cassiterite crystal indicating that the size of SnO_2 crystalline grain is extremely small, and its crystal lattice is seriously distorted. In addition, both surface layers contained δ bronze phase. However, the difference between these two kinds of mirrors was also very evident. The phase components in the surface layer of the "Heiqigu" mirrors were simple and basically contained only two phases that were mentioned above, while those in the GBM usually contained α bronze phase, pure copper crystalline grain and unknown impurities traces, apart from the above-mentioned two phases.

This seemingly simple work actually conveyed abundant information, which chiefly is that the corrosion-resisting property of black mirror is determined by the SnO_2 whose crystalline grain was extremely tiny and its crystal lattice was seriously distorted. Only when both the crystal and the electronic structures of the SnO_2 are known, one could further explain its corrosion-resisting mechanism. The structural difference in the surface layer of the two kinds of mirrors indicated also the difference between their production technologies. There was only δ bronze phase in the surface layer of "Heiqigu" mirrors, which implied that the early processing stage of the "Heiqigu" surface took place in an alkaline condition. In the surface layer of GBM, pure copper crystalline grains always coexisted with the α bronze phase, which indicated that there existed a certain relation between the two phases.

2.2.2 THE ORIGIN OF THE GREEN COLOR OF "LUQIGU" MIRRORS

Except of the difference in colors, "Luqigu" mirrors are basically similar to "Heiqigu" mirrors. The X-ray diffraction pattern of the surface layer of "Luqigu" mirrors showed that there are some pure copper grains and a certain amount of malachite in addition to the main phase of SnO_2 with its tiny grain and distorted lattice. It also implied that the green colour of "Luqigu" mirrors was related to the malachite in mirrors' surface layers. Therefore, the production technology of both "Luqigu" and "Heiqigu" mirrors is closely related, or even identical. Malachite was evenly distributed in the surface layer of "Luqigu" mirror, spreading over the whole surface and even the back. Malachite existed inside the transitional layer of the surface while outside the transitional layer a compact nonmetal layer mainly made up of SnO_2 only existed. Therefore, it is reasonable to assume that malachite or relevant substance was used at the beginning of the production of "Luqigu" mirrors or could have come from the embedding environment.

2.2.3 RESEARCH CARRIED OUT ON SnO2 IN THE SURFACE LAYER OF BLACK MIRRORS [26,27]

In brief, the main phase component in Chinese ancient black mirrors and green mirrors was SnO_2 with tiny crystalline grain and seriously distorted crystal lattice. From the Scherrer equation, the size of the crystalline grain of SnO_2 was estimated to be approximately 3-4 nm. It is well known that research on nanometer crystal and cluster has become the hotspot of physics, chemistry and material fields since the German scientist, Professor H. Gleiter put forward the nanometer crystal concept in the early 1980s. Normally, the surface of nanometer crystal should be more chemically active, so it is quite difficult to make a thick film of nanometer crystal, even by means of advanced equipment. However, the surface layer of black mirror with this SnO_2 is different. It usually is more than 100 microns in thickness and displayed an excellent corrosion-resisting property. This indicated that the study on formation process and the mechanism of black mirror is of definite significance in the fields of physics, chemistry and materials. For this reason, it is necessary to further analyze the size of SnO_2 grain. With the help of Liu Chunlan and Xu Yanping of the Steel Institute, Metallurgy Ministry of China, the surface substance of black mirror was analyzed using X-ray small angle scattering method [28]. A few milligrams of a powdered sample were obtained from the surface of a GBM mirror, and made into a film mixed with celloidin. Assuming the powder shape is spherical, computer-based calculations were obtained using segmenting distribution functions and are shown in figure 6. From this figure, it can be seen that the size of about 65% of powder grain is within the range of 1 to 8 nanometers (1-4nm, 43%; 4-8nm, 22%; 8-16nm, 21%; 16-30nm, 14%), proving that the SnO_2 in the surface layer of black mirrors is a nanometer crystal.

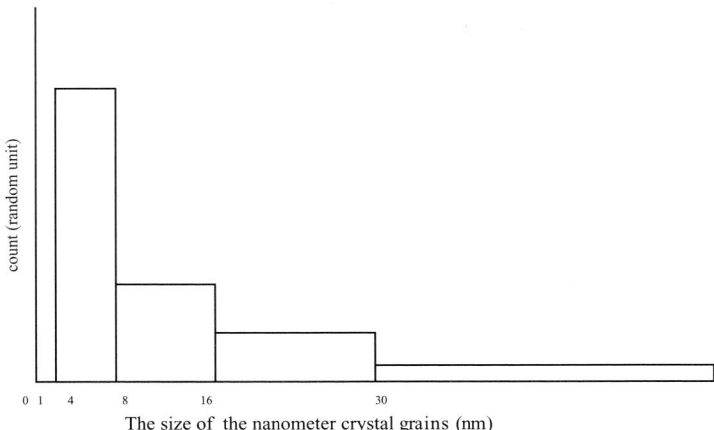

The size of the nanometer crystal grains (nm)

Figure V-3-6: Size distribution of powder grain in the surface layer of black mirrors.

In addition to the XRD analyses and Raman spectrum, TEM and EDAX proved that there was certain amount of doping atoms in SnO_2 crystal grains in the surface layer of black mirrors, and the SnO_2 crystal lattice was seriously distorted. Undoubtedly, the information on the kinds, quantity, doping types and position in the lattice of the SnO_2 crystal are rather important issues in order to probe into the corrosion-resisting mechanism of the surface substance of black mirrors. The work has many obstacles hindering it and results could be obtained only by means of EXAFS.

Professor Lu Kunquan of the Beijing Physics Institute, the Chinese Academy of Science, was very interested in this topic. He tested the K absorption edge of Cu of the surface substance of the black mirrors with EXAFS at Tsukuba Photon factory in Japan. Mr. Hu Xiaojun of the Beijing Metallurgy Institute, the Capital Steel Company, processed the data obtained from Prof. Lu. Unfortunately, due to the complexity of the problem, he could not offer a clear conclusion and/or a reasonable explanation before getting the data from EXAFS of Sn. Nevertheless, this tentative effort is still considered a high level application of X-ray technology in the archaeological field.

3. X-ray Technology on Bone Chemistry [29]

Through the study of the ancient human bone, a lot of important information can be obtained, such as dating to set up a chronological table for cultural production and development, ancient diet in order to reveal the life style, surroundings and migration paths of ancient people, and ancient DNA to provide information about the human origin, genetic relationship, and so on. Problems in these studies arise from bones not preserving their original chemical composition and biological characteristic after long time.

How to identify contamination of Ancient Human Bone? An interesting work of my doctoral student Hu Yaowu, combining XRD and Raman spectra to analyze ancient bones, helped in finding the difference between contaminated and non-contaminated bones.

With large superficial area, the bone is easy to absorb dissolvable and non-dissolvable substances. Some substances can be absorbed on the surface of the hydroxyapatite grain while other substances, such as Fe, Al, Sr, Ba atoms or ions, can replace the Ca atoms or ions in hydroxyapatite lattice and distort its crystal structure. These phenomena have caused bone contamination.

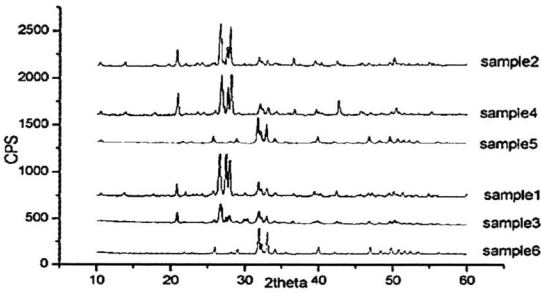

Figure V-3-7: XRD pattern of samples.

Fig. 7 represents the XRD pattern of ancient bone ash. From PDF standard data, some quantity of quartz, albeit and anorthite were found besides hydroxyapatite in samples 1, 2, 3, 4, but there was only hydroxyapatite in samples 5 and 6. The Raman spectra of these samples told the same story. These results made it clear that samples No. 5, 6 were preserved well and had little contamination, while samples No. 1, 2, 3, 4 were highly contaminated. Moreover, other information about contamination can be seen as well from the combination of XRD and Raman spectra. The sharp peaks of hydroxyapatite related to its good crystalline meant big crystal grains and no or few doping atoms or ions in the lattices. It also suggested that samples No. 5, 6 showed light contamination and that sample No. 1 had good crystalline hydroxyapatite even when it was mixed with quartz and feldspar. If the hydroxyapatite can be separated from sample No.1, it could be regarded as an uncontaminated bone enlarging the scope of sample selection that can be obtained after some appropriate pretreatments. Different from the above, samples No. 2, 3, 4 obviously showed broad diffraction peaks of hydroxyapatite with week intensity, indicating that there was heavy distortion in the hydroxyapatite grain, i.e. there were a lot of doping atoms in the hydroxyapatite grain.

4.   X-ray Technology on the Ancient Pigment Research

Besides the studies of ancient ceramics and ancient metallurgy, X-ray technology can be applied to the studies of ancient pigments, jade articles, bijou, stone vessels, glass etc. There have been promising results obtained from these applications. For example, it was proved by X-ray diffraction analysis that the ancient glass made in China was a non-crystalline substance, different from the Potash-Natrium glass. Combined with the XRF analysis, we knew that ancient Chinese glass was Lead-Barium glass. Only the research on Chinese ancient pigment based on X-ray analysis will be presented here.

If we mention ancient pigments in China, the pioneer work of Dr. Rutherford J. Gettens in the Fogg Art Museum of Harvard University, published in 1953 and related to the fresco from the known Dunhuang Grotto, has to be presented first.

Dr. Gettens pointed out that the natural mineral pigments used for Dunhuang frescos were malachite, azurite, carbon black, flake white lead, red lead (Pb_2O_4), kaolinite, gamboges, cinnabar, arsenic sulphide and ochre, and the artificial pigments used were clerodendron cyrtophyllum and indigo. In addition, Dr. Gettens studied the technique of making these pigments and the trade relation between China and the Middle East at that time. A detailed discussion of his work is beyond the scope of this survey, which is focusing on the recent researches carried out in China.

In fact, since 1949, Chinese experts and scholars have carried out immense research on ancient pigments, in which X-ray technology played a key role. The achievements related to XRD technology on the various kinds of ancient pigments are as follows: the fresco found in the Mogaoku Grotto and the West Cave with Thousand Buddhas of Dunhuang, Maijishan mountain of Tianshui city, Yungang Grotto, Laojun Cave of the Binlin Temple Grotto, the Qutan Temple of Qinghai province, the Ancient Temple of Chengde city; the rock painting found in the cave with the Hanging Coffin in Hongxian county, Sicuan province, Huashan mountain of Guangxi province; the white powder from a Tomb of Eastern Han Dynasty, in Shouxian county, Anhui province; the dye of colored challis in the famous Han Tomb No. 1 of Mawangdui, Changsha city and the pigments of painted pottery from the terra cotta museum at Lintong city, Shaanxi Province[30-34].

In these works, Mr. Zhou Guoxin and coworkers of the Paint Industrial institute, National Chemistry Ministry, have performed some profound researches, related to their experience in this field. For instance, according to the analysis of the fresco pigment from the Yungang Grotto, they confirmed that there were many kinds of mineral pigments in the fresco, such as gypsum, kaolinite, ochre, malachite, black carbon, cinnabar, ultramarine blue and yellow iron. They also suggested that the sodium sulphate and magnesium sulphate were taken from the surrounding water. For the first time, they discovered that the usage of cupric acetate arsenate, which is a double salt of cupric acetate and cupric arsenate, in ancient fresco. In his paper on fresco pigments from the Chengde ancient temple, Mr. Zhou explained from various perspectives, the difference and relation of the series of terms such as verdigris, malachite, cupric subcarbonate and cupric subcarbonate hydrate, which is of benefit to the new researchers in this field.

5.   New XRD and XRF Equipments Relevant to Archaeometry

In the last 20 years, X-ray diffraction and fluorescence equipments have been fully and rapidly developed with regard to light source, control system, detector and data processing. All the development has substantially facilitated the research of many subjects and accelerated their development, including archaeometry. Here only the new equipments that are closely relevant to archaeometry will be introduced, rather than all the development of X-ray diffraction and fluorescence equipments.

Apart from broken relics, undamaged relics are usually objects of archaeological research, and it should be strictly forbidden to damage the relics during testing. For this purpose, the Rigaku Company have designed and produced a new type of diffraction instrument, whose light source and detector can be moved in a wide range to test any part of the cultural sample. Lately, the counselor of Rigaku, Mr. Tadahiro Abe designed a more advanced and convenient instrument with a new recording system. At the same time, we made some special sample holders for the non-destructive testing of the bronze mirrors and sword-shaped samples [35]. For XRF, the situation is better than XRD. I remember that when I visited Tokyo National Research Institute of Cultural Properties in 1993, Prof. Hirao Yoshimitsu showed his XRF equipment to me, whose sample room could be refitted to accommodate big vessel measurements. Now several new models of this equipment are sold, and many Universities and Institutes in China have been equipped with such important facilities.

It is a well-known fact in X-ray micro area that diffraction and fluorescence instruments are very useful to archaeometry. There are few X-ray micro area diffraction instrument and few papers related to this instrument in China, despite the fact that the technique is very mature and the availability of many machines such as, DX-MAP2 model of JEOL and PSPC-MDG2000 of Rigaku.

It is obvious that X-ray archaeology can cooperate with other fields, such as X-ray perspective photograph and CT etc., but the scope here is just focusing on XRD and XRF. Due to the limitation of my knowledge in this fast-growing science some of the discussions are incomplete, hence some errors are unavoidable. I hope that experts and readers would generously point out these unintentional errors in order to avoid them in the future.

Acknowledgements

This research was supported by the National Natural Science Foundation of China (No. 10135050, 19975046) and Academia Sinica (No. KJCX-No4). I wish to express my sincere thanks to Mr. Tadahiro Abe, the counselor of the Rigaku company, Mr. Minoru Suzuki of the Yamanashi Institute of Cultural Properties,

Teikyo University, Japan, Professor Pei Guangwen of the Nankai University, China, and Dr. Suzanne Seleem of Loras College, USA, for their kind help.

References

1. Cihai ed. *Cihai* Shangjai Dictionary Publishing House, 1980, 12387;

2. Yu Weichao, Zhang Aibing, Social Science in China, **6**(1992), 147-166.

3. Division of Archaeology ed., *Modern theory and method in Archaeology,* National Museum of Chinese History, Sanqing Press(Xian), 1991;

4. Institute of Archaeology ed., *Collection of China Archaeology,* Academia Social Sinica, Science Press(Beijing), 1993, 495-501

5. Parks, P. A., *Current Scientific Techniques in Archaeology,* Groom Helm, London & Sydney, 1986;

6. Jin Guoqiao, Pan Xianjia, Sun Zhongtian, *Physics for Archaeology,* Shanghai Science and Technology Press, 1989;

7. Tite, M .S., *Archaeometry,* **2**(33) 1991, 139-151;

8. Edited by Zhao Kuanha, *General History of Chemistry,* Higher Education Press(Beijing), 1990, 1-5

9. Zhuo Ren et.al, *Collection of papers about Chinese ancient ceramics* Light Industry Press(Beijing), 1982, 165-187

10. Miziji, *Nature*(Japanese Version), **3**(1987), 143;

11. Wang Changsui, Liu Fangxing, et.al, *Journal of University of Science and Technology of China,* **3**(21), 1991, 108-113.

12. Liu Fangxing, Wang Changsui et.al. *Journal of Archaeology,* **2** (1993), 239-250.

13. Yan Wenming, *Universe of Cultural Relics,* **6** (1990);

14. Xu Shunsheng ed. *Headway in X-Ray Diffraction,* Science Press(Beijing), 1986, 273-274;

15. Wang Changsui, Zuo Jian et al, *Journal of chinese History Museum,* Vol. **74**, **1** (1995), 78-80;

16. Li Jiazhi, Wang Changsui, *Research on Ancient Ceramics,* 2002, in press

17. Feng Min, Wang Changsui et al, *Research on Ancient Ceramics,* 2002, in press

18. Zhu Jian, Mao Zhenwei et al, *Nuclear Techniques,* 2002, in press

19. Hua Jueming et al, ed. *Developing History of World Metallurgy,* Science and Technology Literature Publishing House(Beijing), 1985;

20. Ling Yeqin et al, ed. *Ancient Traditional Casting Technology in China,* Science and Technology Literature Publishing House(Beijing), 1987;

21. Research Team of Western Han Bronze Mirror, Shanghai Jiaotong University. *Acta Metallurgical Sinica,* **3** (1979), 12-22;

22. Zhao Bolin, ed. *Research Method of Metal Physics (Ser.1),* Metallurgy Industry Press(Beijing), 1981, 185-192;

23. Tan Derui, Wu Laiming et al, *The Symposium for Scientific History and Relics Archaeology,* 1986;

24. Wang Changsui, Xu Li, Wang Shengjun, *Journal of University of Science and Technology of China,* **4** (18) 1988, 506-509;

25. Wang Changsui, Xu Li et al, *Archaeology,* **1** (1989), 476;

26. Wang Changsui,Wu Youshi et al, *Chinese Science Bulletin,* **5** (38), 1993, 429-432;

27. Wang Changsui, Lu Bing et al, *Science in China*(Ser. A), **8** (24), 1994, 840-843;

28. Xu Shunsheng, *X-ray for Metallography*, Shanghai Science & Technology Press, 1963,194-207;

29. Hu Yaowu, Wang Changsui et al, *ACTA Biophysics Sinica,* **17** (4), 2001, 1-6

30. Zhou Guoxin, *Archaeology,* **5** (1990), 467;

31. Zhou Guoxin, *Archaeology,* **8** (1991), 755-745;

32. Zhou Guoxin, *Journal of Chinese User of the Rigaku X-ray Diffractometer,* **3** (2), 1990, 103-108

33. Zhou Guoxin, Cheng Huaiwen and Zhang Guowen, *Journal of Chinese User of the Rigaku X-ray Diffractometer,* **6** (1), 1996, 103-108.

34. Wang Jinyu, Li Jun et al, *Conservation & Archaeological Science,* **2** (5), 1993, 23-35;

35. Wang Changsui, Huang Yunlan, *Test of Physics and Chemistry*(Vol. Physics), **5** *(*29), 1993, 36.

Chapter V-4

The Relationship between Arts and Sciences in the Field of Archaeology: from Cooperation to a Truly Equal Partnership

Sakuji Yoshimura
Human Sciences, Waseda University, Japan and Institute of Egyptology, Waseda University, Japan, 1-6-1 Nishiwaseda, shinjyuku-ku, Tokyo 169-005, Japan
maat@tky3.3web.ne.jp

Keywords: arts, science, archaeology, methodology

1. Introduction

I will start by explaining, as someone with an Arts background, my motives for participating in this symposium. The first reason is my close relationship with Professor Masayuki Uda, who organized this international conference. To go into somewhat more detail, the technology that Professor Uda has developed has made a tremendous difference to my own fieldwork in Egypt. The satisfaction that comes from gaining valuable new knowledge using his equipment is one of my chief motives for taking part in this symposium.

Then there is the fact that for the past two decades, since the early 1980s, we have taken advantage of a wide range of advanced technologies on the ground in Egypt. The first piece of equipment we used was an electromagnetic distance measuring device for surveying, which enabled us to do away with plane-table surveying. We stored the data obtained on computer and printed it out back in the living quarters. After a few improvements, we were able not just to compile maps of structural remains but also record objects discovered in the list of finds all on computer. That dramatically enhanced the efficiency of our work as well as boosting objectivity. Next, we developed an automated diagramming system for visually recording dug-up artifacts. The goal of these energy-saving innovations was to reduce the amount of effort that archaeologists have to put into routine tasks and increase the amount of time they have for what they are really supposed to be doing: think.

Consequently, we reached the next stage: harnessing scientific technology in the search for new sites, one of the archaeologist's main tasks. That meant identifying the methods of geophysical prospecting most suited to Egyptian terrain. In 1985, a French team used a micro gravimeter on a pyramid. We chose to use an

M. Uda et al. (eds.), X-rays for Archaeology, 291–293.
© 2005 *Springer. Printed in the Netherlands.*

electromagnetic wave meter instead, which allowed us to peek where we had never been able to peek before. We went on to use a wide range of other technologies as well, all made possible by human ingenuity.

2.   What is Archaeology?

Several years ago, the world of Japanese archaeology was shaken to its core when some major finds of Stone-Age implements were revealed to have been complete fabrications. As a result, the public has come to view the discipline with increasing skepticism. This unfortunate incident demonstrated that archaeology was not being conducted scientifically at all in Japan. Japanese archaeology has been reduced to a flashy stage show with the sole aim of discovering old objects. The only way to set it back on the right course is to rethink what archaeology is all about in the first place.

At Archaeological investigations, how people lived in ancient times based on the evidence of the artifacts and ruins they have left behind. As a discipline, it takes a scientific approach to dating objects and identifying what they were used for. Whether something is old or new is immaterial. However, because the media lionizes any archaeologist who happens to dig up something very ancient, archaeologists find themselves engaging in unscientific deception before they even know it. Nothing testifies to that fact more than the way that Japanese universities count archaeology among the arts. That is an odd place for a discipline like archaeology, which is concerned with the scientific pursuit of truth, since art typically involves fabricating something out of nothing. Evidently, the pioneers of Japanese archaeology did not think that far. Archaeology is supposed to concern itself above all with techniques, so it probably really belongs in the engineering faculty — which is exactly where the closely related discipline of architectural history is found. It needs to maintain that most important of scientific traits, objectivity. The same answer must always come up, no matter who is watching or who is doing the work. In Japanese archaeology today the excavator has exclusive knowledge and the exclusive right to consider the evidence. In some cases, no one but the excavator can even enter the site! That will simply not do. Which is why we want to put archaeology back among the sciences. This symposium is of great significance in achieving that goal.

3.   What Differences do those who Study the Arts Think there are between the Arts and Sciences?

One of the opinions expressed at this symposium is that those in the arts do not study hard enough. To someone in the sciences who has since youth played with numbers and spent much time and energy on uncovering scientific principles, it certainly may seem that we in the arts have an easy time of it. People in the arts are aware of the fact too. They know they should be putting their nose to the grindstone more, and that knowledge gives them an inferiority complex, which in many cases makes them reluctant to work with their scientific colleagues.

That is a tragedy for the progress of learning. In order to discover truth, is it really sufficient just to be good at math, or have an excellent grasp of physics, or be a whiz when it comes to chemical symbols and how chemicals react? I for one do not think so. People are all different, and it is only natural that they should be better at some things that at others. A really good archaeologist, they say, can stand on a site and estimate the size of something without even taking out a tape measure. There are even those whose gut instinct tells them what they are going to find before they even start digging. Such individuals are the exception, but the point is that different people have different roles, and that is fine. What gives scholarship its value, is that it allows to tap different ways of thinking, adopt different perspectives, and harness different methods all in pursuit of a single truth.

The division into arts and sciences did not exist in antiquity. The fragmentation of scholarship into all kinds of specialized fields was perhaps unfortunate. We in the arts and sciences have come together here under a single roof, and this is an opportunity for us to work together in order to discover the truth of what it means to be human — which is after all the ultimate goal of learning. Nevertheless, those in the sciences have an advantage in that they have their own methodology. Therefore, once they have established their objective, they can accomplish everything themselves without having to turn to others for help. They certainly do not need any assistance from the 'methodless' arts. Nonetheless, many in the arts are able to offer a broader outlook on aims and objectives or have knowledge of historical background or context, despite their lack of methodology. When individuals from various disciplines are involved in interpreting whatever numbers come up, there is less room for error. It is time to stop asking which is "better," arts or sciences, or which should help which. Let's all work together. That is the nature of my relationship with Professor Uda.

4.   Archaeology in the 21st Century

The above considerations lead to the hope that archaeology in the 21st century will span both the sciences and the arts. Since the latter half of the 20th century, the two disciplines have been striving to establish a cooperative relationship and have indeed to some extent succeeded in working together systematically. Cooperation inevitably involves one side leading the way and the other lending a hand in a subordinate role. It is understandable that people in the sciences were at first a bit miffed to find themselves treated as mere subcontractors assisting their arts colleagues' research. However, in time they started to develop their own goals and objectives and attain them using their own methods, leaving no room for those in the arts. So now, discontent has spread among the latter.

The only way to eliminate that discontent is for the arts and sciences to share common goals and develop common methods. That, surely, is the path that archaeology should take in the 21st century.

Color Plates

295

(a) The block of Ramesses II holding captives, 19th Dynasty (ca. 1270 B.C.).

(b) The Wooden Coffin and Lid of Neb-Seny, 18th Dynasty (ca. 1400 B.C.).

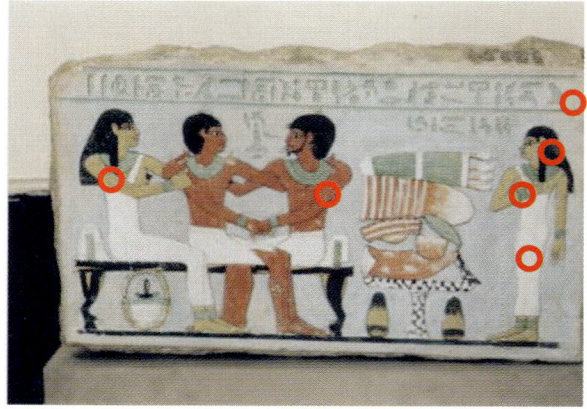

(c) The Funeral Stele of Amenemhat; 11th Dynasty (ca. 2000 B.C.).

Fig. 1-1-1. Objects studied here, which are exhibited at the Egyptian Museum in Cairo. (See also p. 6)

(a) The south wall of Room E.

(b) The north wall of Room I.

(c) The west wall of Room I.

Fig. 1-1-2. The painted tomb walls of Amenhotep III, 18th Dynasty (1404-1364 B.C.) in Luxor, Egypt. (See also p. 7)

Fig. 1-2-9. In-situ EDXRF analysis at the Akrotiri archaeological site in Thera of a yellow (mustard) wall-painting pigment using PXRF-B portable spectrometer. (See also p. 40)

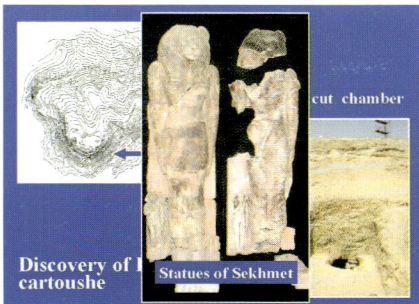

Figure I-3-4(a).
Discovery of Khufu's cartoushe

Figure I-3-4(b).
Dahshur North

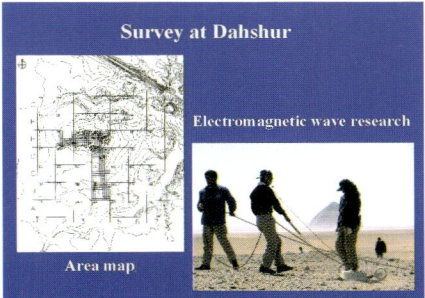

Figure I-3-4(c).
Survey at Dahshur

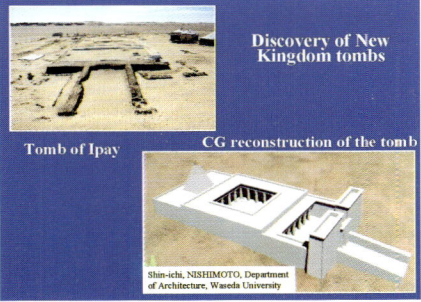

Figure I-3-4(d).
Discovery of New Kingdom tombs

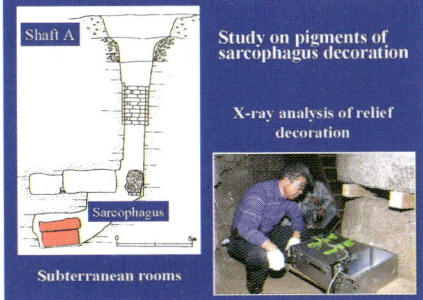

Figure I-3-4(e).
Study on pigments of sarcophagus decoration

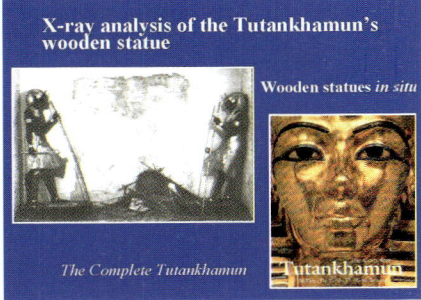

Figure I-3-4(f).
X-ray analysis of the Tutankhamun's wooden statue

(See also p. 62)

Figure II-3-1
Examples of glass à façon de Venise excavated in Ljubljana (photo T. Lauko, National Museum of Slovenia). (See also p. 114)

Figure. II-7-2(a). Cucuteni ceramics cups used for ritual libations.

Figure. II-7-2(b). Cucuteni pottery objects.

(See also p. 166)

Fig. II-8-1   Photograph of the two unknown VASEs. (See also p. 175)

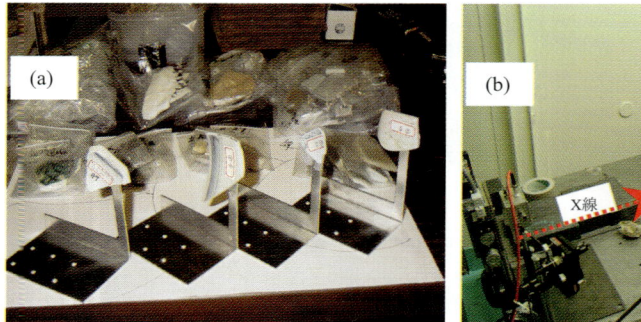

Fig. III-1-2 (a) Samples mounted on a sample holder for XRF analysis. The samples are fragments of excavated porcelain from old Kutani kiln   (b) A view of high-energy XRF analysis of "Old Kutani" China ware. (See also p. 186)

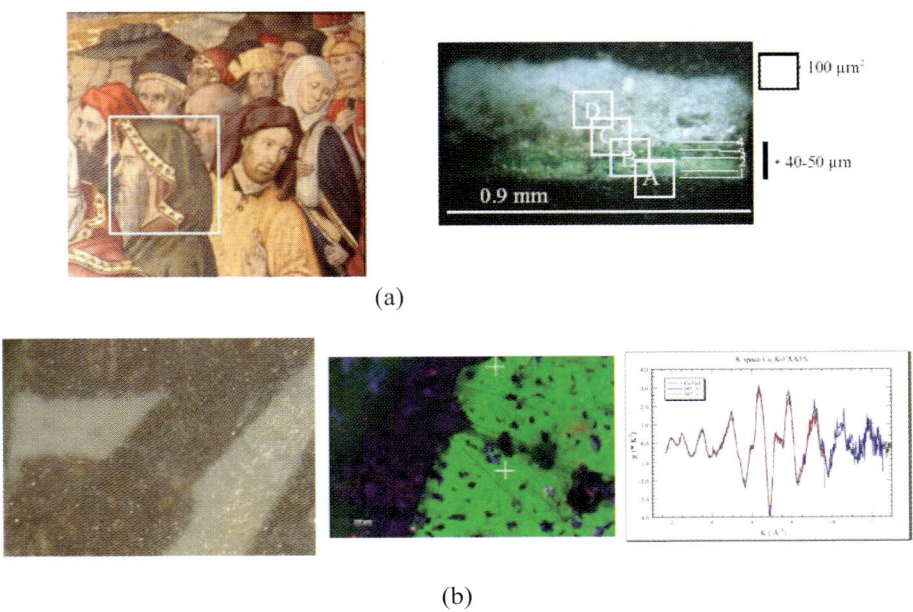

(a)

(b)

Figure.III-2-1.(a) Jaume Huguet's Retaule de Sant Bernadí i l'Angel Custodi, and (b) left: lustre sample, centre: Tricolour elemental map of part of the glazed area where copper is rendered as green, and right: XAS spectra in two positions compared with metallic copper. (See also p. 202)

Figure III-4-1. A Tang Sangcai Figure of a Riding Woman
718 AD, Tang Dynasty, 35.2 cm (H). (See also p. 218)

Index